# A GUIDE TO ENVIRONMENTAL RESEARCH ON ANIMALS



COMMITTEE ON PHYSIOLOGICAL EFFECTS OF ENVIRONMENTAL FACTORS ON ANIMALS

AGRICULTURAL BOARD

NATIONAL RESEARCH COUNCIL

NATIONAL ACADEMY OF SCIENCES

WASHINGTON, D.C. 1971

This study was supported by the United States Department of Agriculture, the United States Department of the Interior, and the United States Public Health Service.

ISBN 0-309-01869-2



*Available from*

Printing and Publishing Office
National Academy of Sciences
2101 Constitution Avenue, N.W.
Washington, D.C. 20418

Library of Congress Catalog Card Number 76-609948

Printed in the United States of America

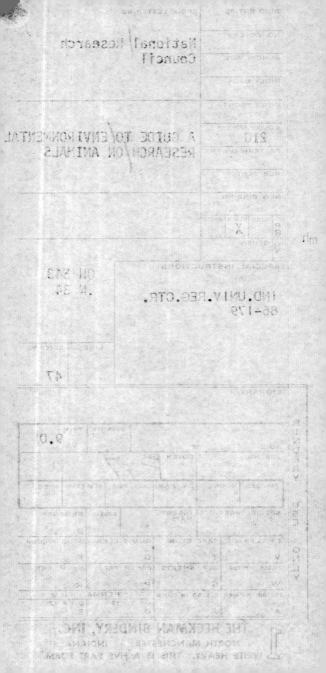

# PREFACE

This guide is intended for investigators, teachers, engineers, and administrators in environmental research on animals, an important aspect of the overall environment picture. We have attempted to show the need for research under controlled conditions and to provide the user with an aid in evaluating environmental factors and related physiological responses and in planning environmental testing facilities. The book is the product of group effort by the Committee on Physiological Effects of Environmental Factors on Animals with the assistance of a considerable number of other persons, whose contributions are gratefully acknowledged.

The sequence of the chapters and organization of their contents is based partly on the steps that an investigator would follow upon initiating a research project in this field. The general observations in the introductory chapter are supplemented by a few comments on terminology. The second and third chapters discuss bioclimatic factors, animal functions, and techniques for their measurement. The fourth and fifth chapters cover animal physiology and disease and the effects of environment. Chapters 6 and 7 are concerned with design and operation of laboratory facilities and the design of experiments. The final chapter emphasizes field research and the interrelationships of field and laboratory investigations.

ROBERT G. YECK, *Chairman*
ROBERT E. McDOWELL, *Vice Chairman*
T. E. BOND
R. W. DOUGHERTY
THAMON E. HAZEN
HAROLD D. JOHNSON
JAMES E. JOHNSTON
C. F. KELLY
NELLO PACE
S. YEARDLEY SMITH
L. C. ULBERG
WILBOR O. WILSON

# CONTENTS

# 1

# INTRODUCTION

The well-being and productivity of animals is affected by and related to their physical environment. In spite of a rapid expansion of research in the field, the factors of the physical environment are still often poorly defined and measured. Combinations of bioclimatic factors are even less well understood. For example, air temperature alone poorly defines the environment unless correlated with air velocity, humidity, and radiation. Furthermore, physiologic adjustment, as by metabolic and physical activity, respiration, or blood circulation is not fully defined and correlated.

Research on the interrelationships of animals and environment serves to answer questions related to human and animal nutrition, human and animal health, animal production and reproduction, animal behavior, preservation of rare wildlife species, management, pollution abatement, and environmental engineering.

Environment is of major concern when attention is directed toward improving animal productivity, preserving rare species, managing wildlife, evaluating the effects of pollution factors on the health and welfare of animals, and utilizing animals for research in basic biological science or in place of man in determining requirements for human health and welfare.

## ENVIRONMENTAL CONTROL

The manipulation and amelioration of environments, long a primary concern in animal production, is an important element in feeding the world's population—particularly in meeting demands for more protein. If animal production is to make its contribution to human food supplies, it must be efficient, which requires a thorough understanding of the influences of various elements of environment on animal functions. The best means of obtaining such knowledge is to assess the components of an animal's environment and their effects on the animal through research under controlled environmental conditions. Advances beyond current knowledge and technology will require increasingly greater understanding and control of the environment. Furthermore, as animal shelters become more sophisticated, livestock facility designers will need answers that require study under carefully controlled conditions and with complete monitoring of climatic variables.

Failure to attain, maintain, or adequately monitor thermal environments may very well distort an experiment or obscure a significant reaction. Some otherwise excellent research of the past has been negated by inadequate consideration of the environmental factors that existed during the experiments. Proper appraisal of environment is essential in all animal experimentation.

Where a controlled environment is needed, it is important that all parties concerned be aware of the difficulties that may arise. Too often, research institutions have failed to obtain the degree of thermal control in research facilities that they had originally sought or their contractor had promised to provide. Precision of control and reliability of operation are costly features of a controlled-environment facility. A central purpose of Chapter 7 is to assist planners in minimizing errors.

## ENVIRONMENT AS A VARIABLE

As the technologies for control of environment and for its modification and measurement advance, new horizons are opened for the separation of environmental variables and a better understanding of the environment–animal interactions. Some environmental factors currently studied in the laboratories are temperature, humidity, thermal radiation, air velocity, barometric pressure, photoperiodism, and chemical substances, including atmospheric pollutants. The interactions of these factors are not well known at this time. Other direct factors that deserve consideration are gravity, vibration, sound, space, and odor.

Indirect effects of the environment on the animal would include the presence of other animals and plants, pathogens, parasites, and the quality and quantity of water and feed. The incidence of pathogens and parasites in relation to various direct climatic factors is one of the many environmental complexes that need study.

Theoretically, any organismic, organ, tissue, cellular or subcellular function that relates to thermal physiology, thermal balance, water distribution, or their interactions is important in the development of the science of environmental physiology and shelter engineering. It is necessary to measure those animal responses that can be critical, exercising judgment as to the animal responses that fulfill the experimental objective. Chapter 2 describes some of the measurements that warrant consideration when an environmental–physiological experiment is being planned. The measurements include body temperature; oxygen consumption; carbon dioxide production; respiratory quotient; radiative, conductive, and convective heat losses; evaporative moisture losses; water and feed intake; urine and fecal moisture loss; blood constituents and organ function; numerous endocrine measures of calorigenic hormones; and hormones concerned with water balance. "Productive" functions may include growth rate, milk-energy production rate, hair and wool growth rate, egg production, and meat production. Physical and biochemical studies of direct cellular response to the environmental temperature, or indirectly via hormones and cell substrates, must receive more attention in the future.

## FACILITIES AND COSTS

Delays or failures to initiate research programs have frequently been due to underestimates of facility costs, unrealistic performance requirements, or insufficient operating funds. In some cases, research facilities have been overdesigned. Simple field experiments might have provided the information needed. The glamour of having a large controlled-environment facility or climatic chamber is obvious. Indeed, it will help attract a high-caliber staff; on the other hand, much research can be conducted through field tests or with relatively inexpensive prefabricated environmental control boxes or chambers. Operating costs are important. There have been instances where expensive research facilities have been constructed, only to remain inoperative or operate much below their full potential because the operational costs were prohibitive.

## TERMINOLOGY

It is increasingly rare for research to be oriented toward the efforts of a single scientist. Generally investigators from several disciplines now work together as a team. Every discipline has developed its own vocabulary, part of which may well be unintelligible or at least unfamiliar to outsiders. The need for team research and the failure of the several jargons to overlap create a serious obstacle to communication. This guide should be helpful in bridging that gap.

The following discussion is intended to identify some terminology differences and thereby to permit understanding despite them.

### Environment

In animal science, *environment* is often used to include all conditions, circumstances, and influences that are not a part of the affected organism. These generally have been considered as being external to the organisms, but there are occasional references to the "internal" environment. When so used, the term generally refers to something apart from the particular function under study or apart from one of the dependent variables.

Animal environmental researchers customarily utilize a narrower meaning of the word. They usually refer to the natural climatic conditions or simulations thereof. Climatic or bioclimatic factors are considered more explicit.

### Bioclimatic Factors

The term *bioclimatic* has generally been favored over the term climatic in trying to differentiate between studies of climate in the meteorological sense and those studies involving the interaction of animals and their environment. It includes more than the common weather phenomena. Artificial light rhythms are examples of bioclimatic factors that may not in fact ever occur naturally. Bioclimatology has been defined as a branch of ecology that studies the interrelations between chemical and physical factors of the atmospheric environment and living organisms. In this definition *atmospheric* is considered to include the conditions within a room.

### Stress

Use of the word *stress* has been and probably will continue to be the subject of philosophical discussion among researchers. Among engineers, stress is defined as the load that is imposed on a subject. The reaction to the load is identified as *strain*. Among some animal scientists, stress is used to describe one or more environmental factors exerting an influence on an animal, and what they measure as the animal response is the resulting strain. Others may use stress interchangeably with strain. The word stress must always be interpreted in context. (See the discussion in Chapter 8.)

### Adaptive Responses

Animals tend to adjust to changes in their environment. This tendency is broadly identified as adaptation to an environment. Adaptation varies as to speed of response and degree of permanence. In extreme cases it will represent an evolutionary change that may be passed to succeeding generations (*acclimatization*). In a lesser case it may represent a very short-term response (*acclimation*). However, the two terms are sometimes used interchangeably. It is a matter of individual opinion as to the point of demarcation between the two. The process is important in environmental research and is frequently discussed in the chapters that follow.

### Heat Transfer

There are wide variations in the usage and interpretations of terms used in reference to the generation, dissipation, and absorption of heat. Engineers consider *heat transfer* as an energy exchange. They subdivide the methods of heat transfer into the classic thermodynamic terms of conduction, convection, and radiation.

Air conditioning and refrigeration engineers usually refer to heat loss (or dissipation) from the animals as part of the heat load. They divide the heat load into sensible and latent components. Latent heat applies to the heat exchanged without change in temperature. In air conditioning, this refers to the latent heat of vaporization (the heat exchanged as a liquid changes to a vapor) or condensation (the converse of vaporization).

Heat dissipation and heat loss are generally used synomously. Literally, the first implies that heat is forced out by a body, and the second that it is pulled away. It is energy transfer between the animal and its surroundings. It is related to the heat released in the life process (heat production). For computing a heat balance it is convenient to identify the various avenues of heat escape as conduction, convection, and radiation. An alternative usage is to separate the latent and sensible components of convection, classify the sensible portion with conductive loss, and call the latent portion evaporative heat loss. In this usage the three methods of heat loss are conduction, evaporation, and radiation. If it is clear to the reader which terminology is intended, no misunderstanding or lack of communication need occur. The subject is reviewed at length in Chapter 2 and briefly in later chapters.

Engineers must exercise care in using animal heat data in their calculations. They must ascertain the exact meaning of their information source and be sure to include heat and moisture losses from feces, urine, bedding, and watering devices. The combination of losses from animals and these other sources is usually identified as the total heat and moisture load.

### Heat Storage

*Heat storage* usually applies to the energy that is gained by a substance and may be readily yielded to its surrounding under one or more of the conditions for heat transfer. A body that is capable of absorbing heat in this manner is called a "heat sink." In calculating heat balances, the storage factor is important. For instance, an increase in animal body temperature accounts for some of the heat that is produced by its life processes. Similarly, an increase in the temperature of the materials within a climatic chamber must be considered in calculating heat exchanges within the chamber.

# 2

# BIOCLIMATIC FACTORS AND THEIR MEASUREMENT

## Atmospheric Factors

### AIR COMPOSITION

Air is a mechanical mixture of many gases. Normal air consists of approximately 21 percent oxygen, 78 percent nitrogen, 1 percent argon, and 0.03 percent carbon dioxide plus trace amounts of inert gases. Unfortunately, air may also contain dust, smoke, fumes, gaseous combustion products, and many other extraneous substances classified as impurities, pollutants, or contaminants. The air inside an animal building or laboratory may be other than normal, either intentionally or accidentally. Odors, obviously, are airborne phenomena.

Little data have been taken on the influence of air composition on animals, but greater emphasis on research along this line can be expected because of increasing public awareness of pollution and its associated problems.

## AIR IONIZATION

"Ionization" is the electrical charge of the particles in the air–dust, microorganisms, gaseous molecules–and may be positive or negative. The term "natural" is used to indicate the balance of positive to negative ions in air that has not been artificially charged by an ion generator. Ions vary also in their "size," and in their "mobility": "Large" ions may be classified as having a mobility between 0.02 and 0.06 cm/sec/V/cm. The "density" of ions is the number of ions per cubic centimeter.

The effect of air with a preponderance of negative ions or positive ions, upon both humans and animals, has been studied by several investigators, with varying results (Kreuger and Smith, 1957; Kornblueh and Griffin, 1955; Dobie *et al.*, 1966; Jacob *et al.*, 1965). The earlier studies by medical researchers on controlling diseases suggested that excesses of certain types of ions might influence growth and health. Studies with swine and quail are inconclusive. In all but one of six trials with quail from 1 to 28 days old, body weights were less for the positive ion group than for the negative or natural ion groups.

In experiments with air ionization, two types of instrumentation are usually necessary: an ion generator and an ion collector with its accompanying electrometer and recorder.

### Instrumentation

#### ION GENERATORS

A simple generator is described by Kornblueh and Griffin (1955). It consists of a 1-mil diameter tungsten wire, 7 in. long, arranged with respect to three brass rods equally spaced on a 5-in. circle coaxial with the tungsten wire. The tungsten wire is charged to 8 or 10 kV. Since there is the possibility of ozone being produced by such a generator, this should be checked.

#### ION COLLECTOR

The collector used by Kornblueh and Griffin (1955) consists of an aluminum duct with a blower at one end to draw air through the duct at 75 cm/sec; two sets of plates are arranged parallel to each other and to the walls of the duct to draw air past them. Five plates, insulated from the duct case, have a potential applied to them. Four collecting plates are mounted equally spaced between the polarizing plates and are connected to the electrometer. The output of the electrometer may be recorded on a suitable potentiometer.

### Interpretation

Ion density measurement is made on the assumption of one charge per ion and derived from the formula

$$N = \frac{I}{q\, va}$$

where

$N$ = number of ions in a cubic cm of air
$I$ = ion current
$q$ = charge per ion = $1.6 \times 10^{-19}$ coulombs
$v$ = velocity = 75 cm/sec
$A$ = area across plates = 49 $cm^2$

Therefore, a reading of $1 \times 10^{-12}$ amp corresponds to an ion density of 1670 ions/ml with a velocity of 75 cm/sec.

The mobility of the ions is derived from the formula

$$M = \frac{d^2\, v}{l\, \mathrm{V}}$$

where

$M$ = mobility collected cm/sec/V/cm
$d$ = 0.782
$v$ = velocity = 75 cm/sec
$l$ = length of plates = 33.1 cm
V = voltage applied to plates

Therefore, with 22.5 V dc applied to the plates, the ions collected have a minimum mobility of 0.062 cm/sec/V/cm (Koller, 1932; Hicks and Beckett, 1957).

## AIR MOISTURE

Moisture in the air influences the rate of evaporative heat loss from animals through both skin and lungs; it also influences the thermal conductance of the animal's coat and, through condensation, affects the properties of the surrounding environment both indoors and outdoors. Measuring the amount or proportion of moisture in the air (psychrometry) is a most difficult problem. The quantity of moisture per unit weight of air changes not only with temperature but also because air is frequently the vehicle for transfer of heat as well as moisture.

### Characteristics

The principles of psychrometry are complex. The purpose of this section is to describe how approximations of these principles can be applied in the field of bioclimatics.

Animals are usually subjected to air temperatures ranging from −18° to 40°C. Mixtures of air and water vapor do not combine as perfect gases, but for many purposes the departure is negligible. These circumstances have led to the informal adoption of a simplified concept of psychrometric relations. For example, different authors of engineering textbooks give differing definitions of relative humidity. For many practical purposes the differences are neglibible and do not contribute to errors in engineering practice. The chief difficulty is that after long usage of approximations one may tend to extend their application to areas in which they may lead to substantial errors. For more rigorous discussions of properties of moist air, see American Society of Heating, Refrigerating, and Air Conditioning Engineers (1967) and Palmatier (1963).

### Measurement

If three independent properties of a sample of moist air can be specified, all the other properties can be inferred. Pressure and temperature are usually easy to measure, thus the practical problem of describing the condition of a sample of moist air becomes one of measuring a third property. Equipment for sensing wet bulb temperature, dew point, relative humidity, or pounds of water per cubic foot are generally used. The choice of a sensor in a particular problem depends on requirements for precision, accessibility, convenience, and cost.

### The Psychrometric Chart

The psychrometric chart is a graphic representation of the temperature and moisture content of an air–vapor mixture at a selected pressure. Its chief use is for anticipating the condition of the air after it has been through various processes in a ventilating system. It may also be used for translating a description of air condition in terms of one set of properties to terms of another set. For example, if air has a relative humidity of 50 percent and a temperature of 60°F, how much water vapor is in the air? The answer is shown directly on the psychrometric chart.

A skeleton psychrometric chart is shown in Figure 2.1. Its main coordinates are dry bulb temperature (the abscissa) and pounds of water vapor per pound of air (the ordinate). This choice of coordinates is

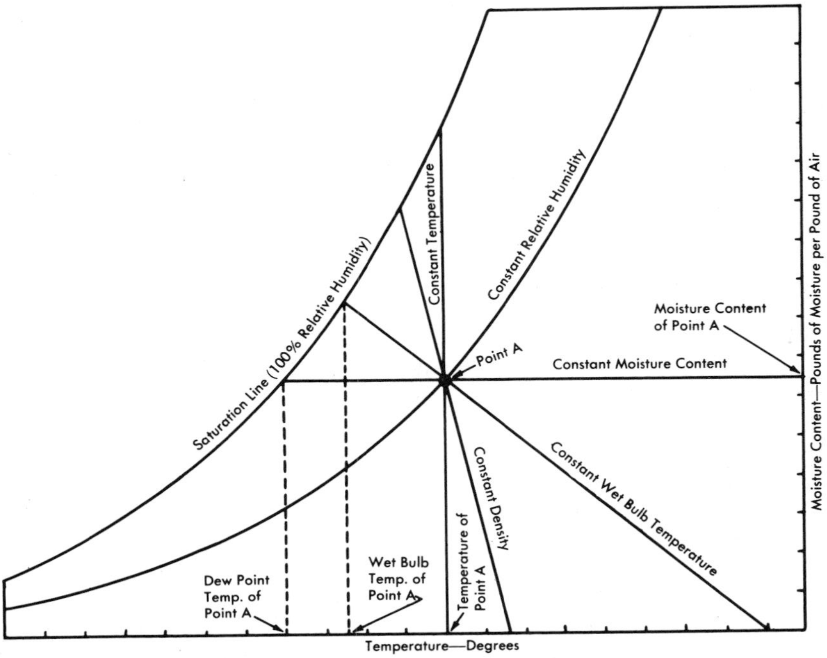


**FIGURE 2.1 Skeleton psychrometric chart.**

purely arbitrary. The chart is bounded on the left by the saturation line. Conditions represented by points to the left of the saturation line do not exist. Other properties of the air–vapor mixture are described by the other lines. The pertinent properties of moist air for our purpose in addition to its temperature, moisture content, and pressure are vapor pressure, relative humidity, wet bulb temperature, wet bulb depression, dew point temperature, and density or specific volume. For other purposes, other properties may be identified, such as enthalpy (in Btu per pound) and density of water vapor (in pounds per cubic foot).

#### TEMPERATURE

The temperature of the air–vapor mixture is sometimes called dry bulb temperature to distinguish it from wet bulb temperature and dew point temperature. The latter two terms are useful descriptions of properties of the air, but they do not describe the air temperature except at the saturation line.

#### WET AND DRY BULB PSYCHROMETER

Two mercury thermometers, one having a saturated wick enclosing the bulb, constitute a psychrometer. To give a satisfactory reading the air must be moving past the bulbs, and unusual sources of radiation on the bulbs must be avoided. One of the problems with a psychrometer is maintenance of saturation in the wick (Henderson, 1952). Another is accumulation of dust, mold, or precipitated salts on the wick.

#### CONTINUOUS RECORDING WET BULB

Various schemes have been used for keeping the wick saturated so that the temperature may be automatically recorded. Saul (1956) passed a sample of the air to be described continuously through a deep bed of saturated vermiculite and recorded the temperature of a thermocouple in the bed.

#### WET BULB TEMPERATURE

If a thermometer bulb covered with a saturated wick is exposed to a stream of unsaturated air, its temperature will drop below the air temperature. The temperature to which it drops, designated as the wet bulb temperature, depends on the air temperature, pressure, and humidity ratio. For any given sample of moist air, there is a unique wet bulb temperature. On the psychrometric chart, all points having the same wet bulb temperature are connected by a line. A common method of determining the state of moist air is to observe its temperature and wet bulb temperature, and the wet bulb lines provide a simple way of translating these readings to other properties such as relative humidity or dew point. When liquid or solid water is evaporated into air without the exchange of external energy, the humidity ratio is increased and the temperature is decreased. The locus of the temperature and humidity change on the chart is almost the same but not identical with a wet bulb line (American Society of Heating Refrigerating, and Air Conditioning Engineers, 1967). The energy content of any two points having the same wet bulb temperature is almost identical. The wet bulb temperature lines are therefore useful in making heat balances in any ventilation system. The lowest temperature that can be achieved by an evaporative cooler is the wet bulb temperature of the entering air.

### WET BULB DEPRESSION

The difference between the dry bulb temperature and the wet bulb temperature is called the wet bulb depression. This is not represented on the chart but may be computed readily. The maximum quantity of water that can be evaporated into a pound of air without the addition of external energy is nearly proportional to the initial wet bulb depression of the air. This relation is useful in computing air requirements for carrying away moisture.

### DEW POINT TEMPERATURE

For any given sample of moist air, there is a lowest temperature to which the sample may be cooled without the condensation of some of the vapor. This is known as the dew point. Lines of constant dew point are not shown separately on the chart since they are identical with the horizontal lines of constant humidity ratio.

### DEW POINT APPARATUS

A glass or metal target is cooled by evaporating alcohol or ether. The surface temperature of the target is measured with a thermocouple; the highest temperature at which a deposit of dew forms on the target is the dew point (Wylie *et al.*, 1965). Brewer (1965) discusses instrumentation and operation for dew point measurements.

### DEW CELL

The moisture content of lithium chloride, like that of other salts, depends on the temperature of the salt and vapor pressure of the atmosphere to which it is exposed. For each salt there is a region on the psychrometric chart in which exposed salt will be dry and act as an extremely poor conductor of electricity. The upper boundary of this region, if plotted on a psychrometric chart, would fall roughly on a line of constant relative humidity. In the dew cell, a thin layer of lithium chloride deposited on the surface of a heating element is either a conductor or a nonconductor, depending on the temperature and dew point. The circuitry is such that if the salt film is a conductor, the element is heated. At some higher temperature it becomes nonconducting and the heating stops, thus maintaining a constant salt temperature. By suitable calibration, readings of the salt film temperature can be translated to dew point temperature of the air (Conover, 1950).

### HAIR HYGROMETER

Many organic materials respond to changes in atmospheric humidity by changes in their physical dimensions. Human hair is used extensively as a sensing element for relative humidity readings or recordings. Synthetic materials are also used (Davey, 1965).

### ELECTRIC CONDUCTIVITY

The electric conductivity of hygroscopic materials changes with moisture content. Almost any hygroscopic material is a candidate for use as a humidity sensor by measurement of its electric conductivity after exposure to moist air. Some are unstable, and some are subject to excessive hysteresis; that is, the magnitude of the reading depends not only on the present ambient humidity but also on the moisture history of the sensor. Perhaps the greatest difficulty in using this method is that passage of the current used for measuring the resistance causes changes in the apparent resistance. This effect, called polarization, is minimized by using alternating current. The most common material used for this purpose is a thin film of lithium chloride deposited on glass or plastic.

The humidity range of an individual sensor is limited, so it is customary to use a composite of several individual elements to cover the necessary range (Matthews, 1965).

### COLOR INDICATORS

Some salts have the property of changing color with changes in their moisture content. For example, cobalt chloride is pink when damp and blue when dry. Each such salt has a specified level of humidity at which the change occurs. For some purposes, the color of exposed salt films is used to indicate air moisture content (Blinn, 1965).

### OTHER METHODS

More sophisticated methods (Richardson, 1965) include *gravimetric* in which the moisture is absorbed from the air and weighed; *spectroscopic* in which selective absorption of light energy is used to measure the density of water vapor; *thermal conductivity* in which heat loss through a sample of air is measured; and *gas chromatography* in which rate of passage of molecules through a filter is observed. A much more complete discussion of methods of measuring humidity and details of some equipment is found in Ruskin (1965).

### MOISTURE CONTENT, HUMIDITY RATIO, SPECIFIC HUMIDITY, MIXING RATIO

All the above terms are used to describe the moisture content of the air in pounds of water vapor per pound of dry air. In Figure 2.1, all points on any horizontal line have the same humidity ratio.

### PRESSURE

The total pressure of the moist air is expressed in pounds per square inch or in inches of mercury. Since most operations in which humidity is important are performed at nearly constant pressure, it is conventional to represent the air properties at a single pressure on a single chart. The most common chart is constructed for standard atmospheric pressure, about 30 in. of mercury. If the pressure departs widely from this, it is necessary to use another chart or one that is constructed for the prevailing pressure. Fortunately, the departures accompanying ordinary changes in barometric pressure or pressure differences from the use of fans are usually negligible.

### VAPOR PRESSURE

If air and water vapor were perfect gases and combined according to Dalton's rule, the total pressure of the moist air would be the sum of the partial pressure of the dry air plus the partial pressure of the water vapor. The actual mixture conforms so nearly to Dalton's rule that the partial pressure of the water vapor, or vapor pressure, is commonly treated as one of the properties of moist air. The vapor pressure is not represented on the psychrometric chart, but to a first approximation it may be considered as proportional to the humidity ratio. The vapor pressure is frequently considered to be the potential that causes vapor to flow to or from the air.

### RELATIVE HUMIDITY

A convenient definition for relative humidity is the ratio, expressed as a percent, of the vapor pressure of a given air sample divided by the vapor pressure that same sample would have if it were saturated at the same temperature and pressure. The relative humidity, more nearly than any other of the simply definable properties, describes the feeling of dryness or dampness associated with air of a given condition. Moreover, hygroscopic materials, which include most organic materials, absorb moisture from and give up moisture to air. The moisture content of such materials tends to come to and remain at a level that is largely, but not entirely, dependent on the relative humidity of the air with which they are in contact. For these reasons, the relative humidity is of much concern in dealing with plant materials or animals. Relative humidity is represented on the psychrometric chart by curved lines such that all points on each line have the same relative humidity.

### SPECIFIC VOLUME

The specific volume is the number of cubic feet occupied by a pound of dry air. Since air moving equipment is usually described in terms of the number of cubic feet per minute that it will handle, lines of constant specific volume are usually included in a psychrometric chart. These lines are also convenient for converting, if necessary, from pounds of vapor per cubic foot to pounds of vapor per pound of air.

## MOISTURE PRECIPITATION AND CONDENSATION

Precipitation in its many forms—rain, snow, hail, sleet—can affect both the thermal and physical environments of animals. Condensation as fog or dew will have similar affects although usually to a lesser degree. Wetting of the livestock will change the thermal characteristics of the coats resulting in change in heat flow to or from the animal. Precipitation accompanied by winds is particularly effective in altering heat flow rates and, depending upon ambient temperature and the temperature of the moisture, can be harmful or helpful. Precipitation in the form of snow or sleet, as well as fog, can affect the vision of the animals and cause actual loss of contact of the animals with their source of feed, under outdoor conditions. Finally, rain can so alter the ground upon which animals are confined that their movement is impaired. Precipitation and condensation, therefore, depending upon amount and accompanying bioclimatic and physical environmental factors, may affect the animal through changing its thermal, visual, and traction conditions. Duration of unusual conditions determines the ultimate total effect upon the animal.

The conditions that cause condensation of water vapor within the atmosphere and subsequent precipitation can be classified under three general headings (Linsley *et al.*, 1958): Cyclonic precipitation is caused by the general convergence and lifting of moist air flowing into a low-pressure area; convective precipitation results from the rising of warm, moisture-laden air due to surface heating or cooling aloft; and orographic precipitation is caused by the mechanical lifting of moist air as it is forced to rise over mountain barriers. The type of precipitation reaching the ground and affecting animal life depends on the atmospheric processes interacting at the time. The discussion following deals with the measurements associated with the precipitation and condensation affecting the local environment, outdoors or in, and leaves the challenging area of causes to the meteorologist, cloud physicist, and air conditioning engineer.

### Precipitation Measurement

There are two basic types of precipitation: liquid—rain and drizzle—and solid—snow, sleet, and hail. Instruments and methods for measuring the occurrence, amount, intensity, direction of fall, particle size, and temperature of both the liquid and solid forms of precipitation and the moisture content (density) of snow are discussed below.

### OCCURRENCES AND DURATION

The onset, continuation, and cessation of precipitation can be detected by a sensor head that is exposed on an outside wall or post. Rain falling on this thermostatically warmed sensor bridges the conductors in two printed-circuit-type grids; the resulting signal can be recorded. When the rain stops, the grids are warmed enough to cause evaporation of the water bridging the grid surfaces, contact is broken, and the output signal ends. Plastic cages to adapt this sensor for snow fit on the sensor head.

### AMOUNT, INTENSITY, AND DURATION

Rain gages are of two classes: nonrecording and recording. An account of the development of rain gages is given by Biswas (1967). Nonrecording rain gages consist of a cylindrical or wedge shaped container with a known area exposed horizontally to catch the falling precipitation. A calibrated stick or window is provided for measuring the amount of water accumulated in the container during a selected time interval. The standard 8-in.-orifice rain gage is in world-wide use today.

Recording rain gages are of two basic types. One type operates on the weight of the accumulated precipitation; the other uses a tipping bucket, which, when filled with a precise amount of water, empties itself. Weighing-type rain gages are available with either analog or digital outputs, or a combination of both, arranged for strip chart or punched paper tape recording. There is also a gage equipped for telemetering digital data. A gage for the continuous measurement of rainfall rate has been developed by Semplak (1966). Captured rain flows through a channel type capacitor that controls the frequency of an electronic oscillator. The output frequency of the oscillator is directly proportional to the rainfall rate.

### DIRECTION OF FALL

Gages called vectopluviometers have been developed to measure the angle of attack or direction from which the precipitation is falling. The elements of such an instrument are five separate storage containers arranged with one in the center and the other four spaced equally around it. The central container has an opening in the top to catch all rainfall, as in a simple rain gage. Each of the other four has an identically sized knife edge opening in the side (one oriented to each of the four cardinal directions) so that rain from one cardinal direction only will be caught. Instruments of this type are not commercially available but have been developed and used in studies in the San Dimas Experimental Forest in Southern California (Hamilton, 1954) and by Agricultural Research Service scientists (Nixon *et al.*, 1965).

### PARTICLE SIZE AND KINETIC ENERGY

Three methods have been used to sample particle size of precipitation. One uses a pan filled with flour or other fine powder exposed briefly to the falling rain (Hudson, 1963b). Balls of the wetted material give an indication of the size of drops in the rain sample. A second more precise method is to expose briefly a traveling aluminum foil strip (Duncan, 1967) or film strip coated with Formvar (Hindman and Rinker, 1967) or other emulsions (Robertson, 1965) to the falling rain or snow. This film is processed, and fixed images of the drops or snow particles are available for further examination. A third method used for estimating hail size and energy of impact consists of exposing a Styrofoam block covered with aluminum foil to the falling hail. Dents in the foil and Styrofoam reveal the size and impact energy of the hail stones (Schleusner and Jennings, 1960). Kinetic energy of liquid precipitation particles is difficult to measure directly. The usual way is to measure the size or weight of the particles and their terminal velocity and to calculate the kinetic energy therefrom (Hudson, 1963a).

### MOISTURE CONTENT (DENSITY)

Several methods are currently in use for measuring the moisture content or density of fallen snow to determine the amount or weight of the water. The classic method is the extraction of a sample, usually a vertical core through the snow pack, and a subsequent weighing of the sample. The volume–weight relationship determines the sample density. "Snow pillows" (Beaumont, 1965; Penton and Robertson, 1967) measure the weight of the snow above them. The snow pack can be surveyed with a passive microwave radiometric technique. Also radioactive sources and radiation sensitive detectors can penetrate the snow pack by a method similar to the techniques for determining soil moisture by neutron scatter (Smith, 1964).

## MEASUREMENT PITFALLS

Horton (1919) pointed out six kinds of errors which can enter into the measurement of precipitation.

1. Observed error caused by misreadings or poor estimates (~0.01 in.)
2. Instrument error related to the catch funnel-measuring tube ratio (~0.01 in.)
3. Evaporation error due to the loss of water by evaporation between the time of precipitation fall and reading the catch amount (~0.01 in./week)
4. Nonhorizontal funnel error caused by a funnel that is not level (5° of ~0.4 percent, 15° of ~3.4 percent)
5. Errors due to location of the gage (variable and can be quite large)
6. Errors due to exposure of the gage (variable and can be quite large)

Two additional sources of error are splash into or out of the rain gage receiver during heavy rainfalls and dents in the receiver or collection containers (Chow, 1964).

Errors due to location and exposure of precipitation measuring devices of all types can be of great significance, and their cause may be hard to determine. Gage catch representative of the area near the gage is desired, but three elements—wind flow, precipitation character, and gage configuration—combine to make truly representative measurements difficult.

Obstructions in the flow of air "carrying" precipitation cause eddies and distortions (turbulence) in the flow around them and down stream from them for some distance. The "carrying" power of this air stream depends on its speed, its turbulence characteristics, and the character, i.e., size and type, of the precipitation particles falling through it. Snow is much more easily carried along than large raindrops, for example. The character and size of the eddies (turbulence) are related roughly to the size of the obstacle producing them and the speed of the wind. Weiss and Wilson (1957) classify these eddies into four general categories according to size, as follows:

1. Climatological (tens of kilometers) related to mountains and large-scale terrain

2. Environment (kilometers to hundreds of meters) related to local topography and groves of trees
3. Site (tens of meters) related to buildings, trees, and bushes
4. Gage (meters) related to the gage configuration

Depending on the size of the area of interest, certain precautions are desirable to minimize the influence of eddies of a certain size. For example, climatological studies are concerned with a relatively large area. For such a use, the gage would be placed so as to eliminate environmental type eddies (Eichmeier *et al.*, 1965; James, 1964) and site eddies (Middleton and Spilhaus, 1953; Weiss, 1963). If this were impossible, which it frequently is, the environment and site surroundings would have to be documented and classified (Brown and Peck, 1962).

In all cases, whether large-scale climatological precipitation studies or local rainfall pattern investigations are of concern, gage eddies must be minimized or eliminated. It has been known for years that gage catch errors are related to wind speed, and many researchers have sought a precise relationship (Lindsley *et al.*, 1958).

Attempts to eliminate gage eddies have been made by burying the gage so that the collector is flush with the ground surface or below it and, therefore, offers no resistance to the wind. Gage surfaces have been oriented into the wind by various techniques (Hamilton, 1954; Story and Wilson, 1944), but some concern as to the meaning of such measurements was raised by Brooks (1945). The most extensive solution to minimizing gage eddies has been the use of precipitation gage shields. Both flexible and rigid shields have been developed and are currently in use (Weiss and Wilson, 1957).

Morgan and Lourenze (1969) investigated gage catch versus ground level precipitation. Two 6-meter-diameter, grass covered lysimeters in the middle of a large grass field served for standard ground-level measurements. Various types of unshielded gages were used in the comparison, with wind speed being the variable.

## Condensation

### FOG

Fog is simply a cloud that occurs at and near the ground. It consists of either liquid droplets (about 20 $\mu$ in diameter) or of ice crystals suspended like a colloid in the air. Estimates of the density of this "colloidal suspension," which can vary greatly depending on the physical processes acting at the time, may be made by eye using the relative visibility of targets at known distances. Early objective measurements of fog density were made with transmission meters or transmissometers (Neuberger, 1951). A recently developed instrument uses a laser beam to determine the drop size distribution (density) of fog (Silverman *et al.*, 1964).

### FROST AND DEW

Frost and dew are phenomena in which condensation or sublimation of atmospheric water vapor occurs directly on exposed surfaces that have cooled, usually through radiation heat loss, below the dew point or frost point temperature of the surrounding air. Visual observation of the whiteness (frost) or wetness (dew) of the surface has been used for years as an indicator of the intensity and amount of solid or liquid water accumulated. There is no standardized method for measuring the absolute amount of dew (Wallin, 1967).

Some methods for semi-objectively measuring dew accumulation follow: (1) The Duvdevani dew gage (Duvdevani, 1947) exposes a specially treated wooden block at night and by visual comparison of the dew accumulated during the exposed period with a set of calibrated photographs gives the equivalent in millimeters of dew. (2) The Hiltner type dew gage exposes a close-meshed sieve to the atmosphere and records the increased weight of this sieve as dew collects on its surface. (3) Dew droplets are absorbed on the surface of filter papers, which are weighed. (4) Polyethylene sheets are suspended over a surface to collect dew and are then weighed (Lourenze, 1964). (5) Weighing or floating lysimeters are used (Pruitt and Angus, 1960). (6) The method described by Hungerford (1967) makes use of buffed acetate sheets exposed for dew collection and weighed to determine the amount collected.

Several instruments have been developed for measuring the duration of dew occurrence (Lomas, 1965). One type of instrument uses a clock-rotated glass plate mounted horizontally. At the commencement of dew formation, a colored pencil is wetted and begins to write. At the cessation of dew, the pencil stops writing. The trace on the glass disk represents the dew duration. A second type of instrument uses gold beater's skin or other humidity-sensitive organic material linked to a recording medium.

### CONDENSATION MEASUREMENT PITFALLS

Measurement pitfalls relating to frost and dew fall into three areas. First, the surfaces of most dew-gathering instruments are artificial and, therefore, may not have the same physical properties of emissivity, thermal conductivity, or heat capacity as the natural surfaces to which their measurements are extrapolated. Second is the problem of proper exposure of these devices so that the measurements made are representative of the area under investigation. Third is the consideration of distillation dew. When dewfall amounts are desired, that is, the amount of water vapor actually condensed from the atmosphere near the ground, one must be careful to consider that part of the amount measured that may have come from the condensation of water vapor from the ground or plant cover (distillation). A lysimeter is probably the best instrument for measuring dew or frost since it eliminates the first and third problems (Lourenze, 1964).

## AIR POLLUTANTS

### Types

Air contaminants may be classified variously as organic or inorganic; visible or invisible; submicroscopic, microscopic, or macroscopic; particulate or gaseous; and toxic or harmless (American Society of Heating, Refrigerating, and Air Conditioning Engineers, 1966). A classification based chiefly on the origin or method of formation of the material consists of the following: (1) dusts, fumes, and smokes, which are solid, particulate matter; (2) mists and fogs, which are liquid particulate matter; and (3) vapors and gases, which are nonparticulate. Air pollution, in a broad sense, includes a wide variety of contaminants, and the detection and measurement of these require a wide range of instrumentation and procedures. Measurement of the concentrations may depend upon particle size, droplet size, chemical content, humidity, and so on.

Very little is known about threshold values for air contaminants with respect to animals. In the United States, threshold-limit values (time-weighted average concentrations) for humans are annually reviewed, revised, and published by the Committee on Threshold Limits of the American Conference of Governmental Industrial Hygienists.

The 1963 values are available in the ASHRAE Guide and Data Book (American Society of Heating, Refrigerating, and Air Conditioning Engineers, 1966). Some values for vapors and gasses that might be found in an animal environment are shown in Table 2.1.

Some methods of detection and analysis of air pollutants are indicated below.

## Measurement

### PARTICLE COUNTERS

Highly sophisticated measuring devices are available to monitor the number of particles in air or liquid. These instruments vary in sensitivity down to 0.3 $\mu$ particle size and will include particles up to 1000 $\mu$ in diameter if desired. The instruments can be set to count the total particles in air or liquid flowing at an adjustable rate for a controlled length of time, or will count particles of selected sizes only. Several models are available.

### DUST MONITOR

A dust monitor is a measuring device that collects dust from an airstream for a prescribed length of time (usually 15 min), records the optical obscuration through a glass collection plate, automatically cleans the collection plate, and repeats the operation continuously. The device is calibrated for known weights and types of dust.

**TABLE 2.1 Threshold Limit Concentrations of Gases and Vapors[a]**

| Substance | Threshold Limit Values (ppm) | Physiological Action |
|---|---|---|
| Ammonia | 100 | Irritant |
| Carbon dioxide | 5000 | Asphyxiant |
| Hydrogen sulfide | 20 | Pungent (poison) |
| Methane | 1000 | Anesthetic |

[a]From Day *et al.* (1965).

### FILTER-TYPE DUST SAMPLERS

The simplest method of sampling dust in air is to filter the particles out of the air with a vacuum cleaner-type mechanism. This type of sampler consists of a 6 or 110 V electric vacuum pump, an air flow control device, a suitable filter holder, and a filter to collect the dust sample. The rate of sampling depends on the size of the vacuum pump and the setting on the air control device. The amount of air to be sampled may be varied by controlling the rate of sampling and the time of operation. Some samplers also include a built-in timer, which may either shut off the pump after a single sample has been obtained or will automatically change the filter and repeat the cycle.

Several types and sizes of filters are available.* A filter holder with a manually replaceable disk-shaped filter (diameter 2.5–25 cm) may be used for single samples. Another type, often used where many samples are to be taken, is the disposable plastic filter holder that comes complete with filter. Tape filters are used in some automatic-sequencing samplers. Others use several disposable filters, sequencing the air stream from one filter to the next. With such sequencing, a very heavy concentration of dust may be disturbed on the samples collected on tape as the tape advances through the machine.

The method of assaying the dust sample will determine the type of sampler and also the filter material used. If the primary interest is to determine the weight of dust in a given amount of air, individual fiber glass filters are convenient, since they can be weighed before and after sampling, and are relatively unaffected by the humidity of the air. However, microscopic examination of the sample is very difficult with fiber glass filters. For this purpose, a membrane-type filter should be used. It is thin and has microscopic holes that occupy up to 90 percent of the filter volume. The hole size may vary from 0.25 to 10 $\mu$, depending on the filter type. The sample penetrates only very little into the filter. The filter may be marked in grids and be colored to aid in microscopic examination. Membrane filters are also useful for collecting samples for weighing.

Weighing of samples requires a very sensitive scale, a removable filter

*Manufacturers of filter-type samplers and filters: Gelman Manufacturing Company, 600 South Wagner Road, Ann Arbor, Michigan 48106; Staplex Company, Air Sampler Division, 777 Fifth Avenue, Brooklyn, New York 10032; Controlled Environment Equipment Corp., 344 South Avenue, Whitman, Massachusetts 02382.

of a type not affected by moisture, and very careful handling. Samples may be evaluated by densitometer or reflectometer comparisons* with clean filters of the same material. The densitometer compares the transmittance of light through the filter and sample with that of a clean filter. One such device is available for use with the tape-type sequencing sampler mentioned above. The correlation is only fair when densitometer readings are compared with weighed samples. This is apparently due in part to the variation in light transmittance through different portions of the filter tape. The reflectometer uses light reflectance from the dust sample in comparison to a standard surface. It may be used for either tape- or sheet-type filters and generally provides more accurate and consistent evaluation than the densitometer. Readings from both devices are not quantitative and are affected by the color, density, and composition of the dust.

Dust samples may be analyzed chemically when collected on certain types of fiber glass and membrane filters. Suppliers provide data on the characteristics of the material used in various filters available for dust sampling and indicate those that may be used for different methods of chemical analysis.

### Gases and Vapors

Many methods of measuring and detecting gases and vapors in an animal environment are given in standard chemistry books (e.g., Scott, 1939). In a study of gases in swine buildings, Day *et al.* (1965) used a cold-trap gas collector to concentrate condensable gases from the atmosphere of the laboratory. Dry ice–acetone was used in a Dewar flask to produce a temperature of –86° C in the cold trap; this temperature is low enough to trap many atmospheric gases (Dal Nogare and Juvet, 1962). A filter of glass fiber paper was used to collect atmospheric solids before the air entered the cold trap. Infrared and ultraviolet spectroscopy were used to analyze the cold trap condensate, and paper chromatography and a pyrolysis technique were used to analyze the filter collect.

Gas chromatography (Littlewood, 1962) is a quick method, once the proper basic unit, detector, column, and carrier gas have been selected; however, highly sensitive units are required for analyzing gases in trace amounts. Day *et al.* (1965) used a thermal conductivity detec-

*Manufacturer of densitometer and reflectometer: Gelman Manufacturing Company, 600 South Wagner Road, Ann Arbor, Michigan 48106.

tor, a silica gel column, and helium as the carrier gas for inorganic gases; they used a flame ionization detector, a silicone oil on a firebrick column, and nitrogen as the carrier for organic gases. McArthur and Miltimore (1961) separated a variety of ruminant gases, including carbon dioxide, hydrogen sulfide, and methane, by using a thermal conductivity detector in conjunction with a molecular sieve and silica gel columns and helium as the carrier gas. Many of the portable gas detectors available can give rapid and accurate measures of small concentrations of gas. (Representative of such detectors are those available from the Union Industrial Equipment Corp., Fall River, Massachusetts.)

## Odors

Odor is that characteristic of a material that excites the sense of smell, the olfactory system. For a substance to be odorous, it must be in a gaseous or vaporous state. Some odors are conveyed on liquid and solid particles having a vapor pressure; temperature and moisture affect most odors. Odors can be characterized by concentration and quality but are difficult to measure directly since the human olfactory system must be relied upon as the final indicator. For this reason, odor-sniffing panels are the main method of measuring odors.

For the measurement of taste and odor in drinking water, members of a panel taste test water diluted with known amounts of pure water. The odorous water is sequentially diluted until the subjects can no longer detect an odor. A similar method can be employed for sniffing odorous air, but in reverse order: The air samples to be sniffed must be odorless at the beginning. Odorous air is then added until the odor is detected by the panel members. The containers can be deodorized by heating to approximately 100°C for about 30 min.

While odor sensing is not fully understood, it is known that sensing involves the dissolving of odor molecules in the fluids bathing a small area high in the nasal cavity. Various chemical analytical techniques have been proposed in an effort to standardize odor measurement methods. Detection and identification of a complex odor mixture at threshold levels is generally beyond the sensitivity of presently available analytical instruments. The odor molecules can be concentrated by adsorption on activated carbon or by absorption in a liquid. The sorbed odors can be desorbed and identified by ultraviolet or infrared spectroscopy. Alternatively, the concentrated odorants in the liquid system may be determined by standard chemical reactions (Turk, 1963).

In a swine environment, Day *et al.* (1965) used a glass fiber filter paper to collect atmospheric solids. The moist solids from the filter were extracted with methanol and became odorless; the extract contained the odor. When the methanol was evaporated, the extract was very viscous and had the strong characteristic odor of the swine building. The authors discuss a method by which this extract could be prepared for infrared spectroscopic examination.

## AIR TEMPERATURE

Air temperature is probably the most important single bioclimatic factor in the physical environment of animals. Since most animals are homeothermic (tending to maintain a constant body temperature), a change in air temperature dictates the need for some adjustment in either respiration, pulse rate, body temperature, surface temperature, or some other physiological function. Air temperature influences the temperature of the surroundings of an animal so that the effective radiating temperature of the surroundings, including the ground or floor on which an animal stands or lies, is affected. Air temperature thus influences three important avenues of heat loss or gain by the animals—convection, radiation, and conduction. These avenues may be needed to maintain a homeothermic condition.

Cold air, in combination with wind or precipitation, will reduce the feed conversion efficiency of most animals. The detrimental effects of high air temperature will usually be reduced if the animal is exposed to wind at the same time. Air with a high humidity has the effect of lowering the level at which air temperature becomes a stress factor for animals. There is still much that is not known about the effect of air temperature in combination with other bioclimatic factors.

This section describes instruments used for measuring air temperature and how, when, and where such measurements should be made.

### Localized Variations

The accuracy needed depends on the purpose of the measurement, but air temperature variation in space and time should be recognized in both formulating the requirements and in interpretating the results. Vertical gradients in air temperature may be as great as 1°–10°C between the ground surface and 2 meters. Temperature may decrease with height (lapse), increase (inversion), or not change (neutral condition). Measurement of the air temperature near an animal will occur somewhere within this 2 meter height, depending on the size of the animal.

Frequently for animals out-of-doors, large temperature gradients will occur in the horizontal because of topography and local obstructions, such as buildings or trees. In addition to vertical and horizontal gradients, the eddy structure of the atmosphere causes large fluctuations of 1°–2°C at a given level in short periods of time. When animals are in a building or laboratory, the air at their back can be 1° or 2°C different than the air at their belly, and the representative air temperature may be different when they are standing than when they are lying. No assurance can be made that any one air temperature is completely representative of the environmental temperature. In work with animals, care should be taken to describe exactly where air temperature measurements were made.

### Temperature Defined

Temperature is a quantitative dimension indicating the average random velocity of molecules. The molecular definition is the basis of the absolute temperature scale formulated by Lord Kelvin, wherein 273.16° is the freezing point of water and 373.16° the boiling point. A degree on the Kelvin scale is identical to a degree on the centrigrade scale devised by Celsius in 1742 and officially known as the International Celsius Scale. Actually the freezing point of water is 0° and the boiling point is 100° on the centigrade (Celsius) scale. In England and the United States a scale devised by Fahrenheit in 1714, with 180° between freezing and boiling point of water, has had general use. A comparison of the various scales is shown in Table 2.2.

### Methods of Measurement

The choice of air temperature sensors depends on the purpose, the accuracy required, and where the measurements are to be made. Care should be exercised in specifying accuracy requirements. Air temperature in a free atmosphere, for example, may fluctuate 1°–2°C over short periods of time at a given height. Air temperature sensors with an accuracy of ±0.1°C would, in most cases, add little more value to the measurements than one with an accuracy of ±1°C. Tanner (1963) suggests that, because of spatial and time variations of air temperature, there is little value in ambient air temperature readings more precise

**TABLE 2.2 Comparison of Temperature Scales[a]**

| Temperature | Kelvin (°K) | Centigrade or Celsius (°C) | Fahrenheit (°F) |
|---|---|---|---|
| Steam point[b] | 373.16 | 100.0 | 212.0 |
| Ice point | 273.16 | 0.0 | 32.0 |
| Absolute zero | 0.00 | –273.16 | –459.4 |

[a]Conversion formulas:
$F^\circ = 9/5\ C^\circ + 32$).
$C^\circ = \frac{5}{9}(F^\circ - 32)$.
[b]At atmospheric pressures.

than 0.5°C. A comprehensive review of air temperature measuring sensors can be found in Tanner (1963).

### LIQUID-IN-GLASS THERMOMETERS

These are basic temperature-measuring elements in use as standards and as laboratory instruments. Mercury-in-glass thermometers are available with varying calibration, range, precision, and accuracy. Specifications in the United States are based on tolerances allowed in certification of thermometers by the National Bureau of Standards (Swindells, 1959) or the American Society of Testing Materials. Quality thermometers have the calibration etched on the glass stem.

Liquid-in-glass thermometers measure rather exactly the integrated temperature of the thermometer itself. Radiant energy will cause errors, particularly in measurements of air temperature. There are numerous methods of shielding against thermal radiation (American Institute of Physics, 1941; Baker *et al.*, 1953; Kelly *et al.*, 1949). The basic principle of shielding is to prevent any thermal radiation from reaching the sensing element. The standard U.S. Weather Bureau Shelter is an example of a simple radiation shield.

A Dewar flask (Thermos), either glass or steel, can be used as a simple method of making periodic comparisons of thermometers, by filling the Thermos with water at successive temperatures. When comparing thermometers in a water bath, the amount of stem insertion must be observed and corrections made. For ambient temperatures of 40°C or less, the correction is usually negligible. This assumption must be verified, however, in individual cases (Baker *et al.*, 1953; Behar, 1954; Swindells, 1959).

### DIAL THERMOMETERS

These consist of a circular scale with a pointer, mechanically coupled to a bimetallic element or to a filled system with a pressure spring (Kebbon, 1961). Bimetallic thermometers with helical or spiral elements are now available at costs nearly competitive with liquid-in-glass thermometers. Similar elements are used in mechanical recording units and indicating devices. Accuracy is within ±0.5 to 1 percent. Dial thermometers are more rugged than glass thermometers. They will withstand moderate temperatures above maximum calibration without damage. Although not entirely waterproof, low-cost dial thermometers can be used outdoors, without protection, for an average useful life of 2 years.

Dial thermometers with remote sensing elements are pressure activated through a capillary tube by the expansion or contraction, with changing temperature, of a liquid or vapor (Behar, 1954; Heinton *et al.*, 1958). Normally the capillary between the instrument and the cylindrically shaped sensing element will be less than 50 feet. Pressure and expansion mechanisms with diverse and complex variations are widely used in monitoring, recording, and controlling. The better instruments are accurate within ±1 percent.

*Vapor-filled units* contain a volatile liquid, part of which is vaporized. The sensing elements are relatively small, and the response is faster than that of liquid or gas units. These instruments can be made for wide range of temperatures with narrow spans (10°C); however, charts and scales are nonlinear. There is difficulty in use where the span crosses ambient temperature, and correction must be made for the difference in level of the instrument and sensing element (Heinton, *et al.*, 1958).

*Gas-filled systems* have wide range, linear response, and are useful for low temperatures. The sensing elements are more than twice the size of liquid–vapor units, and the minimum span is about 80°C (Heinton *et al.*, 1958).

*Liquid-filled systems* also have wide range and linear response. The liquid expansion produces greater magnitude of expansion and pressure than other systems per degree of temperature change (Heinton *et al.*, 1958). Compensation is required for ambient temperature changes at the instrument.

## THERMOCOUPLES

Thermocouples are formed by joining two dissimilar metals. When two such junctions in the same circuit are at different temperatures, an electromotive force is produced in the circuit as a function of the temperature difference of the junctions and types of metals forming the junctions (American Institute of Physics, 1941; Reser, 1940; Tanner, 1963). Common thermocouple wire combinations and their temperature ranges, are shown in Table 2.3.

Thermocouples can be purchased or constructed. Two commercial grades of wire exist, the main difference being in the overall accuracy. Either grade is usually satisfactory for general use. Thermocouple wire and thermocouple extension wire are essentially the same except that extension wire is usually of large gauge with special insulation. Extension wire is used for connecting purchased junctions. Thermocouples are formed by soldering, welding, or brazing (Lorenzen, 1949; Baker, *et al.*, 1953).

If one thermocouple junction is used, then the two dissimilar metal wires are taken directly to the meter so the "reference" junction is the meter terminal. Manual and self-balancing potentiometers with a compensating element can be purchased. This is a temperature-sensitive resistor located in the meter (the reference junction) that biases the potentiometer voltage to correct for the ambient temperature at the reference junction.

With two thermocouple junctions in the circuit, one is used as the measuring junction, and the other serves as a reference junction when kept at a constant or known temperature. A common 0°C reference temperature is a mixture of crushed ice and distilled water in a vacuum

**TABLE 2.3 Types of Thermocouples and Temperature Ranges in Which They Are Used**

| Type | Usual Temperature Range °C | Usual Temperature Range °F | Maximum Temperatures °C | Maximum Temperatures °F |
|---|---|---|---|---|
| Platinum to platinum rhodium | 0 to 1450 | 0 to 2650 | 1700 | 3100 |
| Chromel-P to alumel | -200 to 1200 | -300 to 2200 | 1350 | 2450 |
| Iron to constantan | -200 to 750 | -300 to 1400 | 1000 | 1800 |
| Copper to constantan | -200 to 350 | -300 to 650 | 600 | 1100 |

bottle. Electrically controlled reference temperature sources are available commercially.

Thermocouples for multipoint measurement can be formed with a common return. With this system, however, errors on electrical shorts are difficult to locate. When several thermocouples are connected in parallel, the electromotive force (emf) developed across their common connection will be a weighted mean of the emf's of the individual couples in the circuit.

The contribution of each junction depends on the relative resistance of the circuit to that junction; if the resistances are equal, the emf's will be the arithmetic mean of the individual couples in the circuit (for parallel circuits see Bowen, 1961). This is a convenient arrangement when the average temperature at several locations is desired. The parallel set has the advantage that only one reading is required for the average temperature.

Any number of thermocouples can be arranged in series sets (thermopiles) to increase the total emf proportional to the number of junctions. Many types of thermocouples are described completely by Dike (1954).

### ELECTRICAL RESISTANCE THERMOMETERS

These sensors are based upon the change in electrical resistance of materials with temperature. Metals have a positive coefficient, i.e., the electrical resistance increases with a temperature rise. Precision thermometry has long been based on high quality platinum thermometers (Reser, 1940; American Institute of Physics, 1941). With suitable bridges, laboratory measurements can be made to ±0.015°C. Resistance elements have not been as popular as thermocouples for routine measurements partly because of greater cost of the sensing elements. However, they have several advantages. Several elements can be connected in series and, because of linearity, provide spatial average. Prominent signals are obtained and noise problems are decreased with the increased sensitivity; sensitivity of one to two orders of magnitude greater than that of the thermocouples is easily achieved (Tanner, 1963). A third advantage is that a temperature reference is not needed to record either the absolute temperature or the temperature change, even for unbalanced bridge operation.

The thermistor is a type of electrical resistance thermometer made from semiconducting materials; it has temperature coefficients 10 times as great as metallic elements.

For temperature measurements, the negative-coefficient thermistors are of greater interest. Indicating units are available with spans as narrow as 8°C. Inherent in semiconductors, including thermistors, is the nonuniformity or deviation from a standard calibration. Variations of ±10 percent are normal. Selected units are available within ±0.3°C variation (Baker *et al.*, 1961). Thermistors function equally well with dc or low frequency ac circuits.

A variety of thermal probes constructed from thermistors are available commercially, along with single or multiple-point indicators and recorders. Some probes have a flat (banjo shape) construction useful for measuring animal surface temperatures. Thermistors can be very small and are available imbedded in fine hypodermic needles.

## AIR MOVEMENT

The main effect of wind, or an increase in air movement, is to increase the surface heat loss from an animal by the process of convection. This would generally be beneficial to the animal's productiveness during hot weather, but could contribute to an unwanted loss of body heat during cold weather. Ittner, Kelly, and Bond (1957) showed that wind significantly increased weight gain and improved feed conversion efficiency of beef cattle during the summer in the Imperial Valley of California. Drury (1966) obtained increased growth rates of broilers in a hot environment by increasing the air velocity over them. On the other hand, Bond, Heitman, and Kelly (1965) showed that air movement greater than about 25 cm/sec was detrimental to swine growth at all air temperatures from 10° to 38°C. The same authors also showed that wind increased the heat lost from swine by convection; because high air movement lowered the surface temperature of the animals, heat loss by radiation was reduced. Rate of air movement, therefore, not only increases an animal's heat loss but influences the method by which heat is transferred from the animal to the surroundings.

In animal laboratories, air movement is typically of the order of 15–30 cm/sec—generally referred to as "still" air—unless the air movement is intentionally increased. However, unless the air distribution system of a laboratory is well designed there may be drafts and variations in air movement over the room of 50–100 cm/sec.

Out-of-doors the air movement may range from zero to many kilometers per hour. Wind seldom has a steady velocity and may be highly variable and gusty. Local obstructions can change the velocity of wind movement within a matter of a few feet horizontally or vertically. This great variation in wind in both space and time makes it difficult to measure and specify a particular value for the rate of air movement past animals, particularly over test periods of several hours or longer.

In many animal studies, particularly in the field, a high degree of measurement accuracy of wind velocity, is not warranted. The most useful measure will probably be a value from a continuously recording instrument from which a daily or weekly average value can be obtained.

## Wind Sensors

Basic principles of wind direction and wind speed measurement are discussed by Middleton and Spilhaus (1953). Tanner (1963) also provides good descriptions of wind sensors. The commonly used device for wind direction is the vane. For wind speed, several types of anemometers are available; the type selected depends on the information that is sought. These devices may be classed as (1) pitot tube, (2) pressure plate (velometer), (3) various heated wire methods, (4) windmill type, and (5) the rotating cup type. The first three devices are more or less "special project" types, whereas the latter two have widespread application.

### PITOT TUBE

This is a device for measuring velocity pressure that can be translated directly to velocity. An open-end pipe faces directly up-stream. A manometer connected to the pipe indicates the total pressure against the opening. This pipe is enclosed in a larger pipe with an annular space between. The larger pipe has small holes along the side, the holes facing normal to the direction of air flow. The pressure in the annular space is equal to the static pressure of the air stream and is measured by connecting this space to a manometer. The difference between the two manometer readings is the velocity pressure. Sometimes the two spaces are connected to opposite sides of the same manometer, thus indicating the velocity pressure directly. This method can furnish very exact wind measurement and is therefore used in laboratory-type work and also in wind tunnels. This principle is also used for continuous recording (in England and Germany) since it can provide fine registering of gustiness. A drawback, however, is the need for housing for the bulky recording apparatus, which is usually installed underneath the pitot vane, so that an unrestricted exposure can seldom be achieved.

### PRESSURE PLATE

This method is the simplest and oldest one known, dating back to the Middle Ages. A plate is hinged at the top edge, and the deflected angle is observed when the plate is facing the wind. For higher reliability of measurement, the plate can be mounted on a vane. Smaller models, housed in small boxes with a hole in the wall opposite the plate, are available under the name of velometer. These are portable and require no power source. They are useful for spot checks of air movement, such as at various locations around animals. The accuracy is not very high, especially for the open plates in wind velocities under 8 kilometers/hr, and only instantaneous values can be obtained.

### HOT-WIRE ANEMOMETER

There are numerous forms of "hot-wire" anemometers. The response of these instruments depends on the heat loss from, or cooling of, an electrically heated wire. Hot wire anemometers are particularly suited for measurement of low air velocities 0.5 cm/sec, and portable hot-wire instruments are extremely useful in analyzing the air flow rates in a laboratory when the air velocities are generally low. Some are directional essentially; others are primarily nondirectional—each serves its own purpose. Their value lies especially in quick response, and they are often used for measurements in short time intervals (sec).

### WINDMILL ANEMOMETER

This has been mainly used in indoor engineering projects. The lightweight propeller-type device, consisting of at least eight blades, operates a stopwatch-like counting mechanism, which is attached to it, providing an integrated velocity over a certain time interval, rather than giving an instantaneous value. For outdoor use, mounting a vane is recommended since it may be difficult to hold the instrument perpendicularly to the wind (although it is designed for "hand operation"). This, of course, would falsify the reading, which is unfortunate because this windmill type has some properties superior to most other anemometers: an almost linear relationship with the wind speed, and little inertia that prevents the "overrun" when wind suddenly decreases (which usually troubles the cup anemometers discussed below).

The windmill model has nowadays been developed into a popular recording type, called the aerovane. The transmitter has the shape of a wingless airplane always facing the wind. The "fuselage" contains a Selsyn motor for direction and a generator connected to the propeller for velocity. The wiring to the recorders can be as much as several thousand feet long so that there is the opportunity of installing the transmitter for unobstructed exposure.

### ROTATING CUP ANEMOMETER

The most common velocity transmitter is the rotating cup anemometer. Many models are available commercially varying in ruggedness and reliability. There is also a great variety of arrangements for obtaining the data in the desired form. Instantaneous velocity readings can be obtained from cup anemometers with either a built-in generator or capacitors, which are charged according to the cup rotation speed. Also, built-in Selsyn motors can provide instantaneous information. All three types can be connected to recorders, where their data are especially useful for obtaining velocities of gusts. On the other hand, their inertia is high, so that they are less suitable for very low velocities.

Some anemometer models operate only mechanically, which is often convenient, but their use is limited to obtaining averages for the interval in between two readings. They are either of the odometer type (accumulating "mileage") or they have a stopwatch arrangement. More useful are the models where electrical contacts are closed after a certain number of cup revolutions. This is accomplished by a set of gears the last of which is arranged for "pressing" the contact buttons. Highly sensitive light-weight anemometers have an electric eye, energized through openings in a diaphragm that rotates with the cup rotor, instead of the gear assembly. The electric signals can simply be registered on operation recorders. Many models are arranged for contacts after one-mile wind passage for convenient chart reading, some others for 1/60-mile contacts, for use in the absence of recorders. These contacts may be wired to a flashbulb or buzzer, and the signals indicate one-minute mean wind velocity in miles per hour.

The friction developed in the anemometer bearings and readout mechanism may cause some error. Because of the friction, such anemometers will not rotate unless the wind speed is above a certain value. Cup anemometers tend to slow down more slowly than they speed up in gusty wind (Deacon, 1951). This may cause an overestimation of wind travel. This type of anemometer is good, however, for providing "average" windspeeds over a period of several hours or more, provided it is located so as to give a representative reading of wind speed near animals under test. Cup anemometers often have a dial revolution accumulation that should be read at least once a day (depending on the needs of the test) and preferably twice a day to distinguish night and day periods, for which the wind speed is often quite different.

Many types of anemometers are sold in a package fitted with vane and recorder. Some "package" assemblies include temperature, even humidity, and black globe temperature. These are portable, thus permitting selection of unobstructed exposure.

## Data Availability and Utilization

Despite the large number of anemometers that are installed everywhere, comparatively little wind information is made available, probably for a variety of reasons. Wind data are tedious to work with, especially since wind is a vector. In many instances the anemometers are not connected to recorders; therefore when visual observations are not taken routinely, their value is limited. Recording equipment is highly desirable in obtaining hourly data, but free exposure is often difficult to find when the recording apparatus needs commercial power and, therefore, housing.

However, the large number of field installations at many locations are often near enough to many animal environment test sites to be of value. Evaluation of observations and recordings at these locations may be sparse. However, wind data provided by NOAA, National Oceanic and Atmospheric Administration, are reputed to be free of most possible errors. These are published in the "Local Climatological Data" series and include an hourly statistical evaluation. Information from about 6 to 12 stations from each state are made available in monthly issues. Not all of these stations are located in agricultural areas, because the data of the "first-order stations" are obtained on major airports. The height of the transmitters is known; unfortunately, they are too high for utility in most agricultural problems. If information for lower levels is desired, this can roughly be calculated using the empirical power law:

$$\frac{v_2}{v_1} = \left(\frac{x_2}{x_1}\right)^{1/n}$$

This formula should be used only for averages, not for individual measurements because of the difficulty of assigning a value to $n$. $v_1$ and

$v_2$ are velocities at two levels, and $x_1$ and $x_2$ are vertical distances from the ground at the two levels. $n$ has been found to vary between 2 and 7. Its value increases with height, daytime turbulence, and roughness of the ground. Generally, $n = 4$ would be rather satisfactory in most cases for heights below 15 meters. Over rough ground, e.g., field crops, and during the night, or cool seasons, $n = 3$ may be more representative.

## AIR PRESSURE

"Air pressure," in the context of animal environment, may be considered as "atmospheric pressure." It is the pressure exerted by the atmosphere due to gravitational pull upon the air column above the point of measurement. Sea-level pressure is 760 mm Hg. Air pressure varies with altitude and at any given altitude, with air temperature, humidity, and velocity (see Table 2.4).

### Relation to Elevation

The pressure, as well as the composition, of air is an important factor in the environment of livestock. While air pressure at a given elevation above sea level varies little from day to day or during the day, it does vary with elevation. The most significant factor in change of air pressure with elevation is the decrease of atmospheric $O_2$ partial pressure. This is usually accompanied by changes in air and soil temperature, air moisture content, and radiation, making its effects difficult to evaluate (Monge and Monge, 1966). Several livestock diseases are peculiar to the higher altitudes (see Chapter 5). It has also been shown that high altitudes affect the growth rate of rats (Timiras, 1964) and the fertility of chicken eggs (Smith *et al.*, 1959). The effect of atmospheric pressure fluctuations on gestation length in swine has been investigated (Lewis *et al.*, 1968).

Since much of the world's agriculture is conducted at elevations above sea level and as increasing populations force food production to the foothills and upper plains, air pressure should be considered and measured in livestock environment studies.

### Measurement

Two general types of instruments measure air pressure: (1) the manometer (or barometer), in which the pressure of the atmosphere is bal-

**TABLE 2.4 Altitude–Pressure Relationships**[a]

| Altitude | | Pressure | | | | | |
|---|---|---|---|---|---|---|---|
| Feet | Meters | Atmospheres | in. Hg | mm Hg | psi | Milibars | Density Ratio[b] |
| 0 | 0 | 1.000 | 29.92 | 760.0 | 14.70 | 1013.2 | 1.000 |
| 5000 | 1524 | 0.832 | 24.89 | 632.3 | 12.23 | 842.9 | 0.862 |
| 10,000 | 3048 | 0.688 | 20.59 | 522.9 | 10.11 | 697.1 | 0.738 |
| 15,000 | 4572 | 0.564 | 16.87 | 428.6 | 8.29 | 571.4 | 0.629 |
| 20,000 | 6096 | 0.459 | 13.73 | 348.8 | 6.75 | 465.0 | 0.533 |

[a] Modified from Altman and Dittman (1966).
[b] Ratio of density at given altitude to density at sea level.

anced against a column of liquid, usually water, alcohol, or mercury; (2) the aneroid barometer, in which a partly evacuated chamber expands or contracts with changes in atmospheric or air pressure. A third type, the hypsometer, (rare) utilizes the fact that the boiling point of a liquid decreases with a decreasing pressure.

### BAROMETER

The barometer is the basic instrument for measuring pressure. It is in the shape of a U with one end closed to provide an unbalanced pressure upon the contained fluid. Barometers (or manometers) are of many types and orders of sensitivity, depending upon the density of the fluid, the inclination of the instrument with the horizontal, and the precision of construction (Ower, 1949). Care must be used in mounting and positioning the barometer. The open end must be protected from moving air to prevent pressure variations by induced suctions or pressures. Since temperature changes affect both the bore of the instrument and the volume of the liquid, the barometer must be located in a place of even temperature.

If the scale of the mercury-filled barometer is assumed correct at the "standard conditions" of 0°C and 760 mm Hg, and if pressures are to be read under extreme temperature conditions, a temperature correction should be applied. If $h$ is the height of the mercury column, and $T$°C is the temperature of air around the barometer, then the following corrections should be subtracted from the observed reading.

Brass scale: correction is $0.000163hT$
Glass scale: correction is $0.000174hT$

### ANEROID BAROMETER

This instrument, without liquid, consists of a partly or fully evacuated metal cell. A stiff spring keeps the walls from collapsing under atmospheric pressure. Under varying pressures the walls move in and out, adjusting to a balance between the spring and the air pressure. The walls are connected, through a linkage, to a pen or pointer, which, after calibration against a mercury barometer, gives a visual or recorded indication of the air pressure. The aneroid barometer, while sensitive and adapted to recording, is not as accurate as a well-tended mercury barometer.

# Radiation

## IONIZING RADIATION

### Description

Radiation from systems undergoing nuclear transformation includes emanations such as $\alpha$- and $\beta$-particles. Electromagnetic $\gamma$-radiation is also present. An $\alpha$-particle is the same as the nucleus of a helium atom with two protons and two neutrons. A $\beta$-particle is an electron emitted from a nucleus during radioactive decay. $\gamma$-Radiation consists of discrete bundles of energy, called photons, having no mass or charge. $\gamma$-Photons are emitted when excited nuclei change from a state of higher to a state of lower energy. X-rays emitted when $\beta$-particles or other electrons are subjected to large decelerations are called "bremsstrahlung." They are the same as $\gamma$-rays except for the method of production.

As nuclear radiation is absorbed, it excites or ionizes the atoms of the absorbing material. In biological material, this excitation and ionization renders the atoms and molecules more chemically active than normal, and they may enter into reactions that seriously disturb the equilibrium of the system.

The degree to which radiation penetrates an absorbing material depends upon the type of radiation, the energy of the radiation, and the density of the absorbing material. $\alpha$-Particles are readily absorbed even by air. Consequently they are not a serious health hazard except when emitted within the body. The penetration of $\beta$-radiation is generally confined to the skin, but it can cause serious radiation burning. $\gamma$-Radiation, on the other hand, can penetrate deep into body tissue and affect the workings of internal organs.

### Sources

Small amounts of nuclear radiation occur naturally and are due to disintegration of various radioisotopes and cosmic rays from space; thus, nuclear radiation is a normal part of the environment. Cosmic radiation approaching earth is partially absorbed by interaction with the atmosphere. Some of the particles are charged and are influenced by the earth's magnetic field. Thus the intensity of cosmic radiation varies both with elevation and with latitude. The intensities of radiation in specific localities can differ also because of natural deposits of radioactive materials and fallout from nuclear explosions.

A wide variety of radioisotopes can be obtained for scientific studies. The radiation from each isotope is unique. Cesium–137, for example, emits only $\gamma$-photons with 0.661-MeV energy. The activity of a cesium–137 source decays with a half-life of 30 yr. Iodine–131, on the other hand, emits both $\beta$'s and $\gamma$'s, each having several different energy levels. It decays with a half-life of 8 days. Strontium–90, with a half-life of 28.8 yr, emits only $\beta$-particles; but in so doing, it becomes yttrium–90, which emits additional $\beta$'s and a $\gamma$. In addition the bremsstrahlung from the strontium and yttrium $\beta$'s may result in a wide range of energy levels of X-rays.

The bremsstrahlung is used in other sources to provide X-rays with almost any desired energy spectrum. Various $\beta$-emitters are encapsulated in materials of different mass and different geometrical configurations. Each such combination produces a unique spectrum of X-rays. Carbon–14, hydrogen–3, and phosphorus–32 are used to label compounds that can then be followed through biological systems by the radiation from the labeling atoms.

$\gamma$ and X radiation are useful for determining density of materials since the amount of radiation transmitted by an absorber depends strongly upon its density. The most effective energy level depends upon the composition of the absorber and the geometry involved.

### Measurement

The ionization chamber can be employed to measure all types of radiation that produce either primary or secondary ionization. It operates by collecting ions that are produced within it. Ion chambers are available for highly sophisticated laboratory use and as rugged, portable units for monitoring radiation levels in the field. They can be designed in many sizes and shapes and can be filled with air at atmospheric pressure or nearly any gas at any pressure. The radiation source may be external, internal, or introduced with the chamber-filling gas. Because of the importance of absorption of nuclear radiation in tissues, chambers have been developed that contain, in both the walls and the gas, elements that give the chambers response characteristics equivalent to tissue. Small ionization chambers are employed in pocket dosimeters, which integrate the radiation dosage to which they are subjected. They may be read at convenient intervals and reset with the aid of a charger-charge reader.

The Geiger-Muller (G-M) counter tube is widely used for detection of radiation. Any particle that ionizes in a G-M tube triggers a discharge. All particles produce the same discharge regardless of their energy level. Thus an incident particle that produced only one ion pair would cause a discharge and could be counted. This makes the tube highly sensitive; but in spite of its sensitivity, it is inexpensive and rugged. G-M tubes, also, can be made in a wide assortment of sizes and shapes including a needle counter with the active element in a fine steel tube about 2–3 mm in diameter.

Scintillation detectors are based on the fact that excited atoms of many absorbers, when they return to their normal state, emit light photons that can be detected by sensitive photomultiplier tubes. The variety of scintillators available includes inorganic and organic crystals, solid solutions, organic liquids, and inert gases. Single crystals of sodium iodide doped with thallium are used commonly. These crystals give light output linearly related to the energy level of the incident radiation, and the light pulse is of extremely short duration. These characteristics permit distinguishing between energy levels and separating closely spaced events. Scintillation detectors are generally more expensive and less rugged than ionization chambers and G-M tubes.

Window thickness is important on any detector since radiation absorbed by the window cannot be detected. However, windows of Mylar, mica, or steel are available as thin as 1.4 $mg/cm^2$.

Photographic emulsions are also useful in detecting radiation. Their application to X-ray photography and dosimetry are well known. They are also useful in studying the distribution of radioisotopes within a body and in tracing labeled compounds as they move within a body.

Solid-state electronic detectors are available; they are fast, have excellent energy resolution, and give linear response over a wide range of energies. In order to give optimum performance, solid-state detectors must be operated with sophisticated, low-noise electronic equipment.

## THERMAL RADIATION

An unshaded animal can receive more radiant heat from the sun and from his surroundings during the day than his metabolic production

(Stewart, 1953; Riemerschmid and Elder, 1945); this indicates the importance of radiation in an animal's thermal environment.

An unshaded animal out-of-doors will normally be exposed to thermal radiation from one or more of several sources. The radiant fluxes from these sources are broadly classified as: (1) *direct beam* solar energy from the sun; (2) *diffuse sky radiation* that has been scattered, reflected, and diffused out of the original solar beam; (3) *atmospheric radiation* emitted by particles or gases in the atmosphere; and (4) *terrestrial radiation* emitted by, and reflected from, surrounding terrestrial objects.

A shaded animal, though not exposed to the direct solar beam, may receive solar radiation indirectly, as diffuse sky radiation or as reflected energy from the ground, the shade, and surrounding objects. This "indirect" solar energy may amount to as much as 25 percent of the total thermal load on a shaded animal (Bond *et al.*, 1967).

## RADIATION CHARACTERISTICS OF SURFACES

Anything that is warmer than absolute zero (–273.2°C) radiates energy at a rate proportional to the fourth power of the absolute temperature and to the surface emissivity.* The emissivity of a surface is the fraction of energy emitted by the surface compared with that of a "black body." Absorptivity, transmissivity, and reflectivity are also material properties and are the fraction of incident radiation that is absorbed, transmitted, or reflected. Absorptivity, reflectivity, and emissivity can all be defined for single-wavelength (monochromatic) radiation, for specified spectrum bands, or for all-wave radiation. Normally, in animal work, we are interested primarily in the emissivity of surfaces at earthly temperatures (near air temperature), whereas we might be interested in the reflectivity and absorptivity not only at these temperatures but also at high (shortwave) temperatures represented by the solar energy.

It is relatively easy to measure the radiation to or from an animal surface, but if one is interested in knowing what percent of the radiation falling on an animal is absorbed by it, then the absorptivity must be known. Likewise, values for reflectivity and emissivity should be known. These data are available for many materials, but only limited data are available for animals.

*The ending "ivity" is usually reserved for properties that characterize a material. The ending "ance" refers to a property of a specimen that would change with its size or shape.

Some values for solar absorption of cattle hair coats are provided by Stewart (1953) and Riemerschmid and Elder (1945). Kelly *et al.* (1954) show some surface radiation characteristics for swine and indicate how these may be applied in determining the radiant energy reflected or absorbed by animals.

### SHAPE FACTOR

In a radiant energy problem concerning animals, one is usually concerned with the energy exchange between an animals' surface and the surfaces of objects surrounding the animal. Radiant energy is generally emitted in all directions from a flat radiating surface; also energy reflected from this surface is usually reflected in all directions, depending on the characteristics of a surface. As a result, an animal would receive only part of the energy leaving a particular surface, depending on the shape factor of the animal with respect to the surface. The shape factor of one surface with respect to another represents the fraction of the total energy leaving one surface that is intercepted by the second. Kelly *et al.* (1950) provide a good treatise on the calculation and use of animal shape factors. For greater detail on shape factor theory see Eckert and Drake (1959).

## Sources

A researcher concerned with thermal radiation as a factor of the animal environment will primarily consider only radiation of wavelengths from 0.1 to 100 $\mu$ (1 $\mu = 10^{-4}$ cm) within the electromagnetic spectrum. Figure 2.2 shows this portion of the spectrum with the common designations used in discussing certain limited ranges of wavelengths or wave numbers* of thermal radiation. X-rays, $\gamma$-rays, and cosmic rays are in the shortwave region to the left of the ultraviolet; at the other end of the spectrum beyond the infrared region are the micro and radio wavelengths.

*Figure 2.2 shows both wavelength and wave number. The latter designation is preferred by some. Wavelength and frequency are related by $\lambda\nu = c$, where $c$ is the velocity of light ($3 \times 10^{10}$ cm/sec), $\nu$ is the frequency in cycles per second [1 cps = 1 Hertz (Hz)], and $\lambda$ is the wavelength in centimeters. A frequency scale is often useful, but a scale of frequencies ($c/\lambda$) involves inconveniently large numbers. Since $c$ is a constant, a scale of "wave number" is often used, and is defined as $1/\lambda$ in $cm^{-1}$. If the wavelength is in microns, rather than centimeters, the wave number is determined by $10{,}000/\lambda$.

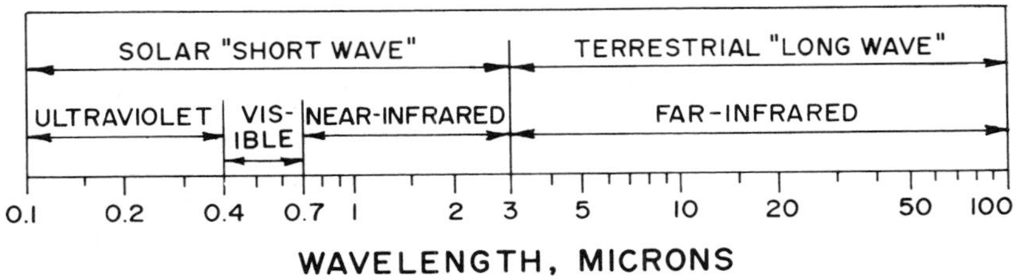


**FIGURE 2.2 Wavelengths and wave numbers associated with common subdivisions of energy within the electromagnetic spectrum.**

## SOLAR RADIATION

The radiant flux density of solar radiation outside the earth's atmosphere has the generally accepted value of 2.0 calories per square centimeter per minute ($cal/cm^2/min$). This flux is modified in passing through the earth's atmosphere; some is scattered by air molecules and particles; some is absorbed by atmospheric ozone, carbon dioxide, water vapor, and cloud particles; and some is reflected back into space. According to Fritz, about 35 percent of the solar energy intercepted by the earth is immediately reflected back into space. Another 19 percent is absorbed by the clear and cloudy atmosphere, so that about 46 percent of the extraterrestrial solar energy reaches the earth's surface.

Calculations from the data of Moon (1940) for an air mass of 2 (sun about 48° from vertical) will show that of the solar radiation at the ground, about 2 percent will be ultraviolet (0.1–0.4 $\mu$), 45 percent will be visible (0.4–0.7 $\mu$), and 53 percent will be near infrared (0.7–3.0 $\mu$). These figures change, of course, with composition of the atmosphere and the altitude of the sun.

## TERRESTRIAL RADIATION

This radiation originates as energy emitted by any object and is due to its temperature. According to the Stefan-Boltzmann law of radiation (Eckert and Drake, 1959), any body at a temperature other than absolute zero will emit radiation at a rate proportional to the fourth power of its absolute temperature. Since terrestrial objects, the ground, buildings, man, clouds, water vapor, and so on are normally at low temperatures (compared to the sun), radiation from these is in the range of 3–100 $\mu$. Figure 2.3 shows a plot of solar radiant flux at the earth and

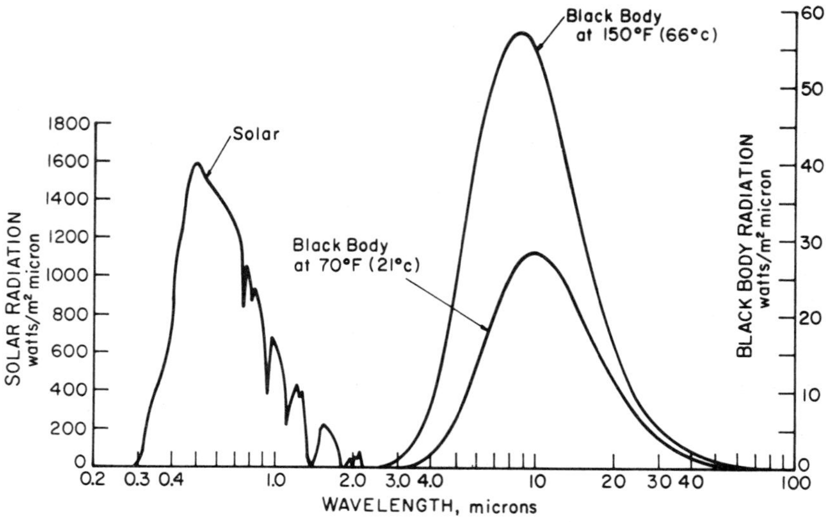


**FIGURE 2.3 Spectral distribution of solar radiation at the ground for a zenith sun (Moon, 1940) and radiation from a blackbody at 21° or 66°C.**

the radiant flux from objects at two temperatures that might be typical of what we think of as "terrestrial" temperatures.

It is obvious from Figure 2.3 that the solar and terrestrial energy curves would not overlap unless the terrestrial object was at an uncommonly high temperature. This provides a convenient means of separating thermal radiation into two convenient categories, which are often considered as "shortwave" and "longwave" energy. This same classification is carried over into most radiation detection devices and to most radiation properties of objects (emittance, reflectance, and so on).

### Detection and Measurement

Researchers interested in thermal radiation as an animal environment factor are faced with the problem of deciding what kind of radiation to measure or account for. This may be ultraviolet, visible, near-infrared or the far-infrared; or the solar or "shortwave" may be required as a unit; or the terrestrial or "longwave" as a unit; or perhaps, the total radiation may be required. Measurement and detection of each of these requires different instrumentation.

One may want to know the radiation from a certain direction, or hemispherically, such as the manner by which a flat surface at the ground is irradiated by the atmosphere above, or spherically as an animal would be irradiated by all of its surroundings. Hence, the directional characteristics of radiation detectors must be considered.

The term "radiometer" implies any instrument for measuring radiation, regardless of wavelength. As used here, however, the term applies to any instrument measuring wavelengths that are important in transferring energy from the sun and earth (0.1–100 $\mu$). Names for radiometers with limited spectral ranges have been somewhat standardized. Pyrheliometers measure direct-beam solar radiation normal to the direct rays of the sun. Pyranometers measure the total solar radiation (direct-beam plus diffuse) incident on a horizontal surface. These are sometimes called solarimeters. A photometer measures visible light (0.4–0.7 $\mu$) and is sometimes called an illuminometer. A spectrophotometer measures light as a function of wavelength. A sunshine recorder measures duration of sunlight. An infrared radiometer measures terrestrial radiation. A pyrgeometer measures the balance between outgoing longwave radiation from the ground and the downcoming longwave radiation from the sky.

Many excellent references provide detailed descriptions of commercially available instruments of all the types mentioned above. Only a general description of some of the more commonly used commercial instruments is given here. For further details on radiation detection devices see American Institute of Physics (1949), Gates (1962), Platt and Griffiths (1964), Reifsynder and Lull (1965), and Tanner (1963).

## SOLAR RADIATION (0.1–3.0 $\mu$)

A simple plate-glass window will not transmit radiation of wavelengths much longer than 3 $\mu$ but will pass wavelengths less than 3 $\mu$. Figure 2.3 shows there is little solar radiation longer than 3 $\mu$. Most solar, or shortwave detectors consist of a thermopile, thermistor, or bolometer behind a high-quality glass or quartz window, which allows only shortwave radiation to be transmitted to the detector. These may be designed to accept radiation directionally from a limited field of view, hemispherically on a flat surface, or spherically on a spherical surface.

*The Eppley Normal Incidence Pyrheliometer* is a true pyrheliometer designed with a narrow view angle (6°) to view only the solar disk. The sensing element is a temperature-compensated, 15-junction thermopile mounted in a brass tube sealed with a quartz window. The window eliminates radiant flux longer than about 3 $\mu$. There are provisions for mounting three additional filters to further restrict the wavelength range. A millivolt potentiometer (range 0–6 mV) is required to record the response. It must be kept pointing directly at the sun either by hand or by an automatic equatorial mount. It has a response of about 3 mV/cal/cm$^2$/min. This is not generally a very useful type of instrument in studies of animal environment. Other instruments of this type are the Abbot silver-disk pryheliometer and the Link-Feussner actinometer (see Reifsnyder and Lull, 1965).

The *Eppley Pyranometer (180° Pyrheliometer)* is the standard U.S. Weather Bureau detector for measuring solar direct-beam and diffuse sky, hemispherical shortwave radiation incident on a horizontal surface. The detector is a circular thermopile (either 16 or 50 junctions) protected by a hermetically sealed spherical glass envelope that limits the response of this instrument to wavelengths less than about 3.5 $\mu$. The 16-junction model has a response of about 2.5 mV/cal/cm$^2$/min; the 50-junction model has a response of about 8 mV/cal/cm$^2$/min. A poteniometer with a range of 0–5 mV is required for the former; a range of 0 to ~ 15 mV is required for the latter model.

This type of instrument is helpful in animal environment studies because the data from it provides a reference point that can be compared with that taken other times of the day, on other days of the year, or at other locations. It provides one means by which the shortwave radiation environments for two different tests can be compared. Reference can also be made to U.S. Weather Bureau data from selected stations around the country.

With the thermopile facing the sky, the instrument measures the shortwave solar and diffuse radiation incident on a flat surface from the upper hemisphere. If a small shade is provided to block out the direct solar beam, only the diffuse sky radiation will be measured. With the thermopile facing the ground, this instrument will measure the shortwave energy reflected by the lower hemisphere (see *Annals of the International Geophysical Year*, 1958).

In use, the glass envelope must be cleaned daily, the instrument must be carefully leveled, and the site must be selected so there will be a minimum of obstructions in its field of view. It is quite fragile and must be handled with care.

Other instruments of this type are the Moll-Gorczynski solarimeter and the Robitzsch bimetallic actionograph (see *Annals of the International Geophysical Year*, 1958, for greater details).

The *Bellani Spherical Pyranometer* measures solar radiation spherically, whereas instruments described above measure the radiant flux density across a flat surface. Pyranometers of the Bellani type have a spherical receptor and may be useful in many biological problems where one may be interested in knowing the total shortwave solar radiant load incident on a spherical-type object. For an animal this would include the direct and diffuse energy from the upper hemisphere as well as reflected shortwave energy from the ground and surrounding objects of the lower hemisphere.

This instrument was described by Bellani in 1836, but many modifications have been introduced. One such instrument commercially available is described by Courvoisier and Wierzejewski (1954). This consists of two concentric glass spheres. The inner sphere contains pure ethyl alcohol, which slowly vaporizes due to absorption of the radiation. The vapor distills through a thin tube and condenses in a sealed, graduated tube. The quantity distilled is a linear function of the shortwave energy falling on the sphere. Much of its utility is due to its simplicity, and to the fact it requires no power or recorder; on the other hand, the level of condensed alcohol must be manually observed and recorded at the beginning and end of a known period of time—this is usually one day. The alcohol must be returned to the center sphere at the beginning of each day by tilting the instrument. An additional drawback in areas of large total daily sunshine, such as in the interior valleys of California, is that the present models do not contain enough alcohol to record throughout the whole day. This is being remedied in a new design, however.

### TOTAL RADIATION (0.1–100 $\mu$)

Total radiation instruments include hemispherical and spherical detectors, directional (narrow-view) radiometers, and net flux detectors. Since solar radiation detectors are usually protected by a glass-type filter, wind and convection currents are seldom a problem. Total radiometers, however, should generally not be filtered since they should measure all wavelengths; wind effects on unprotected detectors can cause serious errors. To reduce this type of error total hemispherical and net flux detectors are of two general types. In one type, the receiver is force ventilated to override the effects of natural air flow; in the other, the receiver is shielded from the natural air movement, generally by some form of thin polyethylene that has a reasonably constant transmission (about 75 percent) over much of the thermal radiation spectrum (Figure 2.4).

Total radiometers must measure all wavelengths of energy important

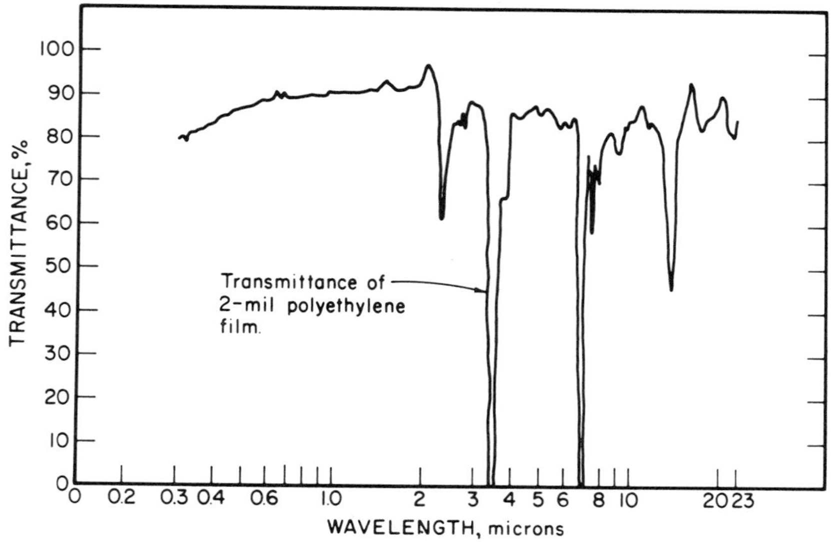


**FIGURE 2.4 Transmittance of 2-mil polyethylene film (Bond *et al.*, 1967).**

in thermal radiation (0.1–100 μ). Lampblack and many black lacquers are satisfactory for thermopile coatings in the range (0.3–3 μ) transmitted by glass (for shortwave measurements) but are mostly unsuitable for longwave measurements. Several types of coatings are suitable, but one that is readily available, and recommended by the Radiation Commission of IAM (*Annals of the International Geophysical Year*, 1958), is Parson's Optical Black Lacquer, which has a constant absorptivity of 0.98 for all wavelengths of thermal radiation.

TOTAL HEMISPHERICAL RADIOMETERS The *Beckman and Whitley* (total hemispherical ventilated) radiometer is used to measure the total thermal radiant energy (irradiance) incident on a flat surface from the entire hemisphere above it (or below it if the sensor is inverted). The uncovered and unfiltered sensor is a 300-junction thermopile imbedded in the middle plate of a 3-plate Bakelite sandwich. The upper surface is coated with Parson's Black Lacquer, and the underside highly polished and protected by an aluminum shield to minimize exposure of the underside to radiation. Convection effects and deposition are minimized by aspiration of 50 cfm over both the top and bottom plate.

The response is in the range of 10 mV/cal/cm²/min, so a millivolt recorder is required with a range from –5 to 20 mV. A temperature-compensated model is available with a response of about 2.5 mV/cal/$cm^2$/min.

This is a useful instrument for animal environment studies because it evaluates the total incoming radiant flux on a surface from the surrounding hemisphere. When this is used in conjunction with a shaded and unshaded Eppley 180° pyranometer, one can get, by subtraction, the hemispherical longwave irradiance and the direct-beam hemispherical shortwave irradiance. By inverting both instruments, these same subdivisions can be made of the energy from the ground and lower hemisphere.

The *Schulze Radiation Balance Meter* consists of two total hemispherical radiometers that are placed back to back so that the downcoming radiation from the upper hemisphere and the outgoing radiation flux from the lower hemisphere can be measured separately. It can also be used to measure net radiation flux by connecting the two thermopiles in opposition. The receiver is coated with Parson's Optical Black Lacquer and is covered with a polyethylene hemisphere about 2 in. in diameter. Since no ventilation is required, no external power source is needed.

The *Georgi Universal Radiometer* is based on the same principle as the Schulze radiometer, but only one detector is used. The detector is a horizontal thermopile and is protected by either a hemispherical glass filter or a hemispherical polyethylene cover. With a glass cover it detects only shortwave flux; with the polyethylene cover it serves as a total hemispherical radiometer.

TOTAL SPHERICAL RADIOMETERS An animal is exposed to radiation from every direction, that is, from the entire spherical envelope of his surroundings. For this reason it is often of interest to know the spherical irradiation on the animal. There are no instruments that are completely satisfactory, but there are some that are quite useful if their limitations are recognized.

The *Panradiometer*, described by Richards, Stoll, and Hardy (1951), was developed to measure the total spherical radiant heat load on an object. It consists of three silver spheres 6.5 mm in diameter—one painted white, one black, and the third polished. The cool globes (the silver one and the white one) are heated to the temperature of the black globe by internal coils. With all globes at equal temperature, convection and conduction losses are the same for each globe. The remaining heat source, radiation, will affect each globe differently because of differences in surface emissivities. Equations are provided, using the currents of the heating elements, to permit calculation of total radiant heat load (0.4–100 $\mu$) and the solar heat load (0.4–3 $\mu$). The longwave or terrestrial radiation can be determined by subtraction. In still air it takes about 4 min for the system to come to equilibrium. With air currents of 1–2 mph, equilibrium is attained in 2–3 min.

A black globe thermometer measures total radiation spherically. Vernon (1930) used globes covered with various materials to measure radiant heat; Bedford and Warner (1934) determined equations for converting globe temperatures to mean radiant temperature; and Bond and Kelly (1955) showed examples of the use of the globe thermometer in agricultural research.

As commonly used, the black globe thermometer is a copper sphere coated with black paint. Its center temperature is measured by means of an inserted thermometer or thermocouple. The common size is 6 in. in diameter, but other sizes can be used (Bond and Kelly, 1955). Pereira, Bond, and Morrison (1967) used a table tennis ball as a black globe of small diameter and showed its thermal characteristics.

The black globe thermometer can be used to determine the spherical radiant heat load at a point, but it does have some disadvantages. Air velocity at the globe, as well as globe and air temperatures, must be measured. Also, its equilibrium response time is rather slow. This may be as much as 17 min at an air velocity of 30 fpm (depending on diameter of globe), and 9 min when exposed to an air velocity of 400 fpm. This slow response time is a disadvantage under many circumstances; in a relatively constant environment, however, this may not be a problem. In a rapidly fluctuating out-of-doors environment the slow response time will provide an averaging effect that may often give satisfactory results. As long as its limitations and thermal characteristics are known (Bond and Kelly, 1955), the black globe thermometer can be a useful tool.

TOTAL DIRECTIONAL RADIOMETERS Directional radiometers can be used to determine spherical radiation if sufficient measurements from the surroundings can be made to ensure a good sampling of the incoming radiation from all parts of the spherical envelope. This method has been used by Stoll and Hardy (1952), Kelly, Bond, and Ittner (1950), and Bond, Morrison, and Kelly (1967a). Details of this method are given in the last paper and will not be included here. This system of measuring spherical radiation is useful in that it indicates high and low intensity sources of radiant energy from various parts of the surround. Its biggest disadvantage is that it is time consuming.

Directional radiometers are designed to measure radiation intensity from a limited field of view, usually with the intent of measuring radiation from a particular source.

*The Gier-Dunkle Directional Radiometer* has as its sensing element a silver-constantan thermopile located at the bottom of a cone-shaped housing. The radiometer response is about 0.452 $cal/cm^2/min/mV$. Fundamentally, such radiometers indicate the net heat transfer rate by radiation between the receiver element and the viewed object. If a black reference shield with known temperature is viewed, the net radiant transfer between it and the receiver will also be known. The thermopile irradiation can then be found by adding the black-body radiation of the shield to the difference between the two net radiant transfers (object and shield). Bond *et al.* (1967a) give details of the use of such radiometers as well as applicable formulas and methods of calibration. These authors also show how two such radiometers used simultaneously will, if one is covered with a quartz filter, divide the incoming radiant flux into solar ($< 5\ \mu$) and terrestrial ($> 5\ \mu$) energy.

*The Stoll-Hardy Directional Radiometer* is commercially available and is similar in principle to the Gier-Dunkle radiometer. It utilizes paired thermistors as the radiant energy transducers. The thermistors are arranged at the base of an aluminum cone that restricts the view of the radiometer to 1.1 steradians. (The steradian is the unit solid angle; the total solid angle about a point is $4\pi$ steradians.) A filter is available to eliminate wavelengths shorter than 3 $\mu$ from reaching the sensor. This instrument is calibrated to give a temperature reading (source temperature) rather than energy units. The temperature reading is based on a surface with an emissivity of one. It is useful for making temperature measurements of animal surfaces if one can assume an animal surface emissivity of one.

The *Barnes "Opti-Therm" Infrared Thermometer* sensing head collects and focuses infrared radiation through an optical system to a thermistor bolometer. By means of a rotating chopper, the sensor alternately compares the infrared energy from the target object with the infrared energy from an internal temperature-controlled reference. Direct temperature readout is provided through an amplifier and readout circuitry. Cone angles of 30°, 3°, and 0.7° are available.

TOTAL NET RADIOMETERS Instruments for measuring the total net radiation usually consist of a flat thermopile sensor that is blackened on both sides to absorb radiation of all wavelengths. Both sides are exposed to radiation so that the top and bottom are heated, but at a different rate. This causes a thermal gradient across the plate, which the thermopile measures. Convective, or wind, heat transfer effects are reduced by blowing equal airstreams across the two surfaces, or by shielding both surfaces with a plastic (usually a form of polyethylene). See Tanner (1963) for an excellent discussion of net radiometers.

The *Fritschen Miniature Net Radiometer* has a sensitive element that is about 0.5 in. in diameter (Fritschen, 1963), and is suspended between 2-mil polyethylene hemispheres that act as wind and rain shields. No ventilation is required. It has a fast response time, about 2 sec, and the commercial version has a battery-operated recorder.

The *Suomi-Kuhn Net Radiometer* is sometimes called an "economical" radiometer because it is relatively easy and inexpensive to construct (Suomi and Kuhn, 1958; Swan *et al.*, 1961). It is unventilated and has thermal insulation 2–4 in. thick separating two blackened plates. Two polyethylene sheets 1 mil thick cover the top and bottom. The temperature gradient between the two plates can be measured with a mercury or dial thermometer, or the gradient can be sensed and recorded with thermistors or thermocouples. It can be used to measure incoming and outgoing fluxes separately. Because of the flat polyethylene shields, the cosine response is poor at incidence angles greater than about 65°. Response time is long—about 14 min.

The CISRO *Radiometer* (*Funk Net Radiometer*) consists of a sheet of plastic approximately 1 × 1.5 in. separating 250 pairs of junctions in a thermopile; all of this is bonded between two sheets of aluminum. Exterior surfaces are painted with Parsons' Optical Black Lacquer and enclosed within two hemispheres of thin polyethylene. The polyethylene is kept spherical with compressed nitrogen. Dew depositions on the hemispheres makes them opaque to longwave radiation, so an external heating ring about 6 in. in diameter is located in the plane of the radiometer. It has a response of about 4 mV/cal/cm$^2$/min.

Several other net radiometers are available, e.g., Gier and Dunkle, and Suomi. See Tanner (1963) and Reifsnyder and Lull (1965) for descriptions, calibration, and errors.

## INSTRUMENT USE IN ANIMAL STUDIES

The literature contains numerous examples showing the use of various thermal radiation detectors in animal environment studies. Kelly *et al.* (1948) used a directional radiometer to measure the surface temperature of swine. Knowing the radiant flux rate from the surface of swine, they assumed a surface emissivity of 0.95 and were able to calculate the surface temperature of the animals by the Stefan-Boltzmann law (Eckert and Drake, 1959). This is a different temperature, usually a little lower, than one would get by measuring the skin temperature of an animal with thermocouples. Thompson *et al.* (1951) determined the "hair" temperature of dairy cattle by the same means. This was compared with "skin" temperature measurements with thermocouples.

Kelly and Ittner (1948) used a flat-plate radiometer to evaluate the benefits animals would receive when thermal radiation was reduced by shades. Bond and Kelly (1955) show several uses of the black globe thermometer in agricultural research. Hahn *et al.* (1961) used black globe thermometers and directional radiometers in studying the effect of building side walls on the radiant heat load of animals. Garrett *et al.* (1967) used black globes, directional radiometers, pyrheliometers, and pyranometers to study the influence of shade height on animal responses. They measured thermal radiation from the surroundings and from the animals. With filtered radiometers they measured both solar and atmospheric radiation. Bond *et al.* (1967b) used a combination of several radiation detectors to measure the solar, atmospheric, and terrestrial radiation received by shaded and unshaded animals. Lee (1953) prepared a manual that covers the application of many of the instruments mentioned here to animals and animal environment.

### REMOTE SENSING BY MULTISPECTRAL RADIATION

Thermal infrared data obtained through remote sensing has been used for inventorying agricultural and forestry resources, including animals, from the air.

The various spectral bands in which aerial reconnaissance of the earth's surface is performed correspond closely to the spectral "windows" wherein radiant energy is freely transmitted by the atmosphere, e.g., the 8–14 micron band. It is commonly referred to as the "thermal infrared" band, since thermal emission (from objects at temperatures approximating that of the earth's surface) is very high here and reflectance (considering the sun to be the illuminant) is very low.

In contrast, the 0.7–1.0 micron band is commonly referred to as a "photographic infrared" band.* Here reflectance is very high and thermal emission is very low–just the opposite of the situation in the 8–14 micron band.

*So-called because properly sensitized halides, when incorporated in photographic emulsions, are directly activated by these infrared wavelengths.

Intermediate between these two bands are others (e.g., the 3–5 micron band) in which both reflectance and emittance play important parts.

SENSING BY PHOTOGRAPHY On thermal infrared imagery the brightness or tone with which an object is registered is the main clue to its identity. This image characteristic depends primarily on the amount of energy being emitted by the object at the time of reconnaissance. This, in turn, is directly proportional to the emissivity of the object and (assuming broad-band reconnaissance) to the fourth power of its temperature expressed in degrees Kelvin.

The object's emissivity is governed partly by its molecular composition and partly by its gross structure; its temperature at any given time is, of course, governed not only by its emissivity but also by its ability to absorb, transmit, and reflect energy and by the thermal forces that have been acting on it in the recent past.

One infrared camera, the Barnes model, uses high-resolution scanning and advanced radiometric techniques to make thermograms showing temperature distributions over the surface of the target. It is made up of an optical head and an electronics unit. The optical head is essentially a scanning system attached to a standard 8-in.-diameter radiometer head. A Polaroid camera back is attached to the Barnes instrument and furnishes a finished print or transparency 10 sec after a scanning cycle is completed. The resulting thermogram appears as a raster of horizontal lines similar to that appearing on a television screen and is the register of 60,000 individual temperature measurements.

The scene on the film is a thermal image of the target. The "grayness" of each image point is a measure of the temperature of the corresponding point on the surface of the target. A reference gray scale appears on each thermogram as well as the black-to-white temperature span.

In essence, the scanning system gathers incoming radiation from the target and point-by-point sends this information to the radiometer head, which compares the received radiation intensity with that of a standard black body. The absolute value of the radiation is thus derived and is used to appropriately modulate a glow tube. In turn, the glow tube scans light-sensitive paper, thereby producing a thermogram.

Specifically, the scanning system contains a large plane mirror that scans its target in horizontal lines. Cams are used to move the mirror, which sweeps a field 20° wide and either 5° or 10° high, depending on the switch the user throws. The Barnes camera uses a fast horizontal return in its sweep cycle.

The radiometer head generates an electrical signal proportional to the thermal radiation in the instantaneous field of view. A filter chops off wavelengths shorter than 1.8 $\mu$, thereby cutting out the effects of light and near-infrared radiation possibly emitted from nearby objects and reflected to the scanning system by the target.

The radiometer head itself is made up of elements that collect and focus the radiations coming from the scanning mirror, an infrared detector, a temperature-controlled black-body radiation standard, and a preamplifier. The detector compares the incoming radiation with that of the radiation standard, and converts this absolute radiation measurement into electrical signals that are preamplified and fed to the electronics unit.

## LIGHT (Visible Radiation)

The portion of energy in the spectrum of electromagnetic waves between wavelengths of 0.45 and 0.70 $\mu$ is considered to be "light" radiation (see Figure 1.2). This is the range of wavelengths in which radiant energy is visible to the human eye. It is presumed that the eyes of animals are generally sensitive to this same range of wavelengths. The common expression for luminous intensity* is the "lux," equal to 1 lumen per square meter (foot-candle = 1 lumen per square foot). A "lux" represents the amount of light received at a distance of one meter from a standard candle. Full sunlight with zenith sun produces an illuminance of 100 to 150 kilolux (9300–14,000 ft-c) on a horizontal surface. For a full moon the value is about 21.5 lux (0.02 ft-c).

It is known that light is an important microclimatic factor affecting reproduction efficiency of most farm livestock (Dutt, 1960). Lights have long been used to bring pullets into lay sooner as well as increase egg production (Clayton and Robertson, 1960; Morris and Fox, 1960).

*The basic light units is the candela. A black body at the temperature of freezing platinum is said to have a luminous intensity of 60 candelas per square centimeter. A source of unit luminous intensity radiating equally in all directions is said to emit one lumen per unit solid angle (steradian), and the normal unit of illuninance on a plane surface is the lux (lumens per square meter) or kilolux. Measurements are sometimes quoted in foot-candles or lumens per square foot. One foot-candle equals 10.76 lux.

English tests (Brande *et al.*, 1958) found no effect of hours of light on production of pigs. Actually, there is much that is not known of the effects of light quantity, quality, and periodicity on livestock; in short, it is a field that is open to investigation.

Light can be measured by any of the different forms into which the incident radiant energy is converted. These are luminosity, heating effect, photoelectric effect, and photochemical action. Luminosity is a highly subjective method relying on the ability of a viewer to see numbers or other items and will not be discussed. Photochemical methods measure summations of light energy over long periods with solutions that slowly decompose in the presence of light—this is not generally practical for animal studies.

### Heating-Effect Sensors

Most of the radiometers discussed under thermal radiation (above) might be called heating-effect types of light sensors. If radiometers are filtered so as to be sensitive only to energy in wavelengths of 0.45–0.7 $\mu$, then they could be used to measure light intensity (providing they had a sufficiently high sensitivity).

The more common types of sensors that utilize the heating effect (also sometimes the photochemical effect) are those used to measure duration of sunlight. Perhaps the most notable of these is the Campbell-Stokes recorder. It has no moving parts and requires no power. It consists of a glass sphere mounted concentrically in a section of a spherical bowl; the diameter of the bowl is such that the sun's rays are focused on a card held in the bowl. The spherical lens focuses the solar rays and burns a trace on the card.

### Photoelectric-Effect Sensors

Instruments of this type are classified as photometers, illuminometers, or simply light meters. The most satisfactory type of sensor (Reifsnyder and Lull, 1965) is the selenium barrier-layer photo-cell; it requires no external source of power. Its widespread use stems largely from the low cost of the basic photo-cell plus the fact it is sensitive to the same range of wavelengths as the human eye. A commercially available model used by the U.S. Weather Bureau is the Illuminometer, a recording photometer. It has a translucent target window for totaling daylight over the entire sky.

Most common exposure or light meters are of the photo cell type. These are often in arbitrary units but they can be calibrated to read in lux or foot-candles. These range in price from a few dollars to several hundred. An inexpensive meter of this type is exemplified by the Gossen "Lunasix" commonly used as a light meter for photographic work. It is, however, calibrated to read the light intensity in units of lux.

For greater details on light, its measurement and controls, the reader is referred to Reifsnyder and Lull (1965), Ditchburn (1963), Platt and Griffiths (1964), Daubenmire (1964), and Bainbridge *et al.* (1966).

## NOISE (Sound Radiation)

Noise is often defined very simply as any unwanted or undesirable sound. The effects of noise upon humans may be classified into four general categories–annoyance, disruption of activity, loss of hearing, and physical or mental deterioration. The literature on the effect of loud sounds upon both humans and animals was reviewed by Bond and Winchester (1963). Their report also includes the results of their studies of the effect upon swine of aircraft noise, and other loud sounds at intensities as high as 135 decibels (db), and at various frequencies. "No evidence was found that the rate of growth, feed intake, or efficiency of feed utilization, or of reproduction was influenced by loud sounds or that ear, adrenal, or thyroid injury detectable by gross examination or microscopic methods took place." On the other hand, reports of adverse effects upon human hearing (Glorig, 1958) and physical and mental health (Parr, 1966) are numerous. Noise for control of rats has been investigated (Sprock *et al.*, 1967). It is evident that with the development of airplanes and machines, interest in the effects of loud noises will increase.

Measurement of loudness of sound is a difficult task, since the loudness must be related to the subjective sensation of loudness by the human ear, a sensitive instrument with incredibly wide ranges of perception of both frequency and pressure. For example, the faintest sound to the ear involves a radiating power of about $10^{-13}$ watts or a sound pressure of about 0.0002 dynes/cm$^2$ at a frequency of 1 kHz. The loudest sound the ear can perceive without pain is more than a billion times this. In order to handle this wide range, a logarithmic scale for sound pressure is used. This scale is commonly referred to as the decibel scale. The decibel, a dimensionless term, is used to express the logarithm of the ratio between two intensities or pressures. The sound intensity level (SIL) in decibels is defined as follows:

$$\text{SIL} = 10 \log_{10} I/I_0$$

where the term $I$ is the measured sound intensity (watts/cm$^2$), and $I_0$ is an arbitrarily chosen reference intensity. Since pressure is proportional to the square root of intensity, the sound pressure level (SPL) in decibels is expressed as follows:

$$\text{SPL} = 10 \log_{10} P^2/P_0{}^2 = 20 \log_{10} P/P_0$$

where $P$ is the measured sound pressure (dynes/cm$^2$) and $P_0$ is the reference pressure. The usual reference pressure, $P_0$, is 0.0002 dynes/cm$^2$ ($\mu$ bars) since this corresponds to the lowest sound pressure an average person can hear at 1 kHz.

The relation between sound level and frequency, band width, proximity, and other variables introduce further problems in sound or noise measurement and should be studied by the researcher.

The basic instrument for measuring sound is the sound-level meter, consisting of a nondirectional microphone, a calibrated attenuator, an amplifier, an indicating meter, and weighting networks. There are many variations in these components as to range sensitivity and quality, and as to their use in the field (Peterson and Beranek, 1963; Snow, 1959; Beranek, 1949).

For a discussion of the calculation methods of detailed analyses of noise evaluation as accepted by Recommendation 532 of the International Organization for Standardization, see Ohme (1967).

# Heat Exchange

## AIR HEAT EXCHANGE

The animal physiologist is primarily concerned with heat production inside the animal body, while the engineer is concerned mostly with the heat loss from the outer body surface of the animal or that gained by it. In hot weather climates the aim is to reduce heat production or increase heat loss; in winter the opposite is needed to increase heat production and reduce heat loss.

Bioclimatic factors of the environment most directly effect the transfer of heat from the exterior surface of the animal (heat production of the animal is discussed in Chapter 4). Some of the physical laws of heat transfer discussed below do, however, apply to transfers of heat and moisture within the animal body.

The loss or gain of heat from the body of an animal can usually be reduced or idealized as a three-component system concerning the transfer of heat between an animal, its surroundings, and the climatic environment. From this system, one can analyze the influence of the surroundings and climatic factors on the external heat loss of the animal. The laws governing the four modes of heat transfer—radiation, convection, evaporation, and conduction—indicate quite explicitly the factors that can be altered to change the rate of heat transfer in the idealized system. One must remember, however, that biological systems are seldom ideal because of the influence of the somewhat intangible physiological functions. The dynamic nature of the animal physiological reactions to its environment (changing surface temperatures, blood flow rates, and so on) complicates the heat transfer relations that, in any but steady-state conditions, become very complex. The assumption of steady-state conditions is made for the relations presented here. These can be very useful and meaningful as long as one recognizes and accounts for the dynamic nature of the system.

### Radiation

The steady-state equation for radiant flux, $q_r$ (kcal/hr $m^2$) is usually shown as

$$q_r = AF_a F_e \sigma(T_{as}^4 - T_s^4)$$

where $T_s$ = absolute temperature ($T$ + 273) of surroundings, °K
$T_{as}$ = absolute temperature ($T$ + 273) of animal surface, °K
$\sigma$ = radiation constant = $4.93 \times 10^{-8}$ kcal/hr m$^2$ °K$^4$
$A$ = area of radiating object, m$^2$
$F_a$ = shape factor
$F_e$ = emissivity factor

This equation shows that the amount of radiant heat transferred can be changed by a change in any of the parameters. The problem of calculating radiation loss or gain is somewhat more complex than the mere substitution of numbers into the formula.

### AREA (*A*)

Formulas are available for estimating the surface areas of most livestock (Brody, 1945) and poultry (Leighton, Siegel, and Siegel, 1966). The exposed body area, however, may be quite different from the total area. Kelly *et al.* (1948) determined that for radiation calculations, the exposed area for swine was 75 percent of the total area. Mount (1964) determined that the effective radiating area of a small pig (2 kg) was 75 percent of its actual surface area at an air temperature of 30°C and 67 percent at 20°C. Some judgment must be exercised in determining how much area is exposed under the circumstances that are being considered.

### SHAPE FACTOR ($F_a$)

In problems of radiant heat exchange, the shape and relative orientation of the two or more radiating surfaces exchanging heat modifies the equation by the factor $F_a$. This really refers to the percent of the area of one surface that is within the view of the other. If an animal is alone in a chamber with all the walls and floors at constant temperature, then $F_a = 1$. For any other condition the shape factor is less than 1 and, depending on the surface configurations, may be very difficult to evaluate. See Eckert and Drake (1959) for the definition and technical application of $F_a$; see Kelly, Bond, and Ittner (1950) for methods of approximating $F_a$.

### EMISSIVITY FACTOR ($F_e$)

The emissivity factor (Eckert and Drake, 1959) depends on the emissivity of all surfaces involved and their relative sizes and shapes. In many cases, particularly within an experimental laboratory, $F_e$ can be approximated as equal to the total emissivity of an animal's surface. Usual values assumed for this range from 0.93 to 0.95 (Bond *et al.*, 1952; Kelly *et al.*, 1948; Mount, 1964; Smith, 1964).

### ANIMAL SURFACE TEMPERATURE ($T_{as}$)

According to Kelly *et al.* (1948), the true radiative temperature of a pig's surface is somewhere between the temperature of the outer ends of the bristles and that of the animal's skin. For the mean skin temperature of small pigs (2 kg), Mount (1964) measured the surface temperatures of the ear and the back and weighted them as 16 percent ear and 84 percent back at an air temperature of 20°C; at 30°C he gave a weighting of 20 percent ear and 84 percent back. The most reliable temperature is probably one obtained with a radiometer. This is always lower than the skin surface as measured by a thermocouple. Thompson *et al.* (1951) shows the same to be true for dairy cattle. This brings up the point that if one must use a radiometer to get a true radiative surface temperature, he might as well use the value measured by the radiometer and not calculate the radiation.

### TEMPERATURE OF SURROUNDINGS ($T_s$)

This, of course, can be measured with a thermocouple, thermistor, radiometer, or by some other method. However, if the various parts of an animal's surroundings differ in temperature, one can determine the mean, or average, radiative temperature of the surroundings by the black globe thermometer (Bond and Kelly, 1955) or the panradiometer (Stoll and Hardy, 1952), for example.

It is generally difficult to calculate the radiant heat transfer between an animal and its surroundings in any but the simplest of conditions. An easier approach is that of Kelly and Ittner (1948) or Bond *et al.* (1952) who used radiometers to measure the radiant heat loss directly from animals. An excellent laboratory approach is that utilizing the gradient-layer calorimeter (Benzinger *et al.*, 1958).

## Convection

Heat exchange between a wall and a fluid when the flow is forced along the wall by external means (wind blowing over on animal surface) is called heat transfer by forced convection. The heat exchange between the wall and a fluid when the fluid is set in motion by temperature differences between the wall surface and the surrounding fluid is called heat transfer by free convection. The main resistance to heat transfer is usually concentrated in a thin film immediately adjacent to the wall surface (often called a boundary layer). Heat is really conducted through a stagnant layer of this film next to the wall and then transferred to the moving particles and carried away by convection into the main portion of the fluid stream. Thus, heat is transferred from the blood stream through the vessel walls to the body by an interplay of both conduction and convection. The same is true for heat lost from the animal surface to the passing air stream.

Only the relatively simple steady-state form of convection heat loss is discussed here. The boundary-layer transport of heat is accounted for in the coefficient *C*.

The rate of steady-state convection heat exchange can be described by the equation

$$q_c = CA_c v^n \ (T_{as} - T_{air})$$

Convection heat transfer (kcal/hr) is a function of the temperature difference between an animal's body surface and air ($T_{as} - T_{air}$), air velocity ($v$), the effective convection surface ($A_c$) and a convection coefficient ($C$).

The convection heat loss from swine was measured by Kelly *et al.* (1948) with small heat flow meters pressed firmly against the hair. This gave the total rate of heat loss by both radiation and convection. The radiant loss was determined by measurements with directional radiometers, and the convective loss was determined as the difference between the heat flow meter values and the directional radiometer values. Mount (1964) developed a laboratory method for determining the convective and radiative heat losses from animals.

Methods for calculating the convective heat losses from animals have been presented by Kelly *et al.* (1948), Smith (1964), Mount (1964), and Wiersma and Nelson (1967). For purposes of calculation, the convection equation shown above is usually written

$$q_c = h_c A_c \Delta t$$

where $h_c$ is the surface thermal conductance, or the convection coefficient of heat transfer (cal/hr m$^2$/°C temperature difference between skin and air). Often values for $h_c$ are determined by assuming an animal has a cylindrical shape; then $h_c$ can be computed from classic heat transfer equations such as those in Eckert and Drake (1959). Smith (1964) followed this procedure. Wiersma and Nelson (1967) developed an experimental relation for cattle of the following form:

$$h_c = Nk/D \text{ cal/sec cm}^2/°\text{C}$$

where $D$ = characteristic diameter (cm)
$k$ = thermal conductivity of fluid (cal/sec cm$^2$/°C)
and $N$ = is the Nusselt number = $0.65\, Re^{0.53}$. The Nusselt number = $h_c D/k$. $Re$ = the Reynolds number = $Dv\rho/\mu$, where
$v$ = air velocity (cm/sec), $\rho$ = air density (g/cm$^3$), and
$\mu$ = air viscosity (g/sec cm).

The animal surface area exposed to convective losses may be somewhat less than the total area because of the shape of the animal, or because it may be lying rather than standing. Mount (1964) suggested a value for $A_c$ equal to about 75 percent of the total area for small pigs. Kelly *et al.* (1948) determined that about 80 percent of the surface of a pig lying down was exposed to air and heat losses by convection.

## Evaporation

The transfer of heat by evaporation can be described by the equation:

$$q_e = KA_e v^n (P_s - P_a)$$

Evaporation heat loss (kcal/hr) can be increased by increasing the air velocity ($v$), by increaing the vapor pressure of water ($P_s$) at the animal's skin surface, by reducing the vapor pressure of water in the air ($P_a$), or by increasing the wet area ($A_e$) of the animal.

The classic method of measuring the moisture given off by an animal is that described by Sanctorius in the year 1614. The "insensible" weight loss of an animal is measured over a period of time on a scale.

The body gives off water vapor, carbon dioxide, and, especially in ruminants, methane. Oxygen is absorbed by the body with the result that the insensible loss = $H_2O + CO_2 - O_2$. When carbon dioxide, methane, and oxygen are determined from respiratory exchange equipment, and insensible loss is measured by changes in weight, the animal's evaporative loss can be determined from the above equation. This method was used and described for dairy cattle by Thompson *et al.* (1949).

Yeck and Kibler (1956) used a "hygrometric" tent to measure the total amount of water vaporized by dairy cattle (surface and respiratory moisture). This was accomplished by careful measurement of the water content of air entering and leaving a moisture-tight enclosure with an animal inside. Mount (1962) used a similar system with swine.

Morrison *et al.* (1967) used a moisture-proof enclosure like that of Yeck and Kibler (1956) for swine but fitted a mask to their pigs so that both the total moisture and skin moisture could be measured. The difference between the two was moisture vaporized from the animal's respiratory tract.

Capsules have been used to measure evaporation from isolated portions of animal surfaces (Bianca, 1965).

## Conduction

The relationship representing steady-state conduction heat transfer is:

$$q_k = UA(T_{ab} - T_s)$$

The rate of heat transferred by conduction (kcal/hr) depends primarily on a temperature difference between two boundaries of a system. In considering conduction loss from animals, it is convenient to consider the body as a complete system so that the boundary temperatures are the internal body temperature ($T_{ab}$) and the surface temperature ($T_s$), although convection will also play a part in the transfer of heat to the surface. $U$ is the over-all coefficient of the animal body (kcal/hr $m^2$/°C), and $A$ is the body area. The coefficient $U$ accounts for conduction across the film layer (as discussed above) on either side of the boundary.

External conduction heat loss or gain by an animal body is usually thought of as occurring when the animal surface is in contact with a surface (a floor, for example). The temperature difference is that between the animal and object surfaces. Conduction may play a part in the transfer of heat from an animal to air surrounding it, but the steady-state transfer is most easily calculated as a convection loss.

The overall net transfer of heat and mass from an animal is usually measured by means of a calorimeter, either directly (Kelly *et al.*, 1963) or indirectly (Brody, 1945; Kleiber, 1961).

Conduction is perhaps the least important mode of heat loss for most animals although it may be important for an animal like a pig that is lying on some material 80–90 percent of the time. Kelly *et al.* (1948, 1964) describe methods using heat-flow meters to determine the conduction loss from swine to concrete floors.

## SOIL HEAT EXCHANGE

The exchange of heat at the ground surface exposed to sunshine establishes the basic environmental condition for growth of all plants (Brooks, 1959); it also establishes the conditions of an important part of an animal's environment. This exchange is a very complicated system involving multiple factors of each of four modes of heat transfer to and from air (see earlier sections of this chapter). The ground absorbs shortwave (solar) radiation directly from the sun, diffusely from the sky, and from reflection by surrounding surfaces; it absorbs longwave thermal radiation from surrounding surfaces and from the air. The ground gives off (emits) longwave energy at a rate dependent on its temperature and emissivity characteristics. Energy is conducted into or out of the ground according to the temperature gradient between soil and air and thermal conductivity of the soil. Heat is convected from the surface into the atmosphere by eddy diffusion and vertical thermals (see Brooks, 1959), and evaporation of soil surface moisture causes energy to be transported from the soil into the atmosphere.

It is true that the temperature of air to which an animal is exposed may have been increased or reduced, by convective or evaporative means, in its movement across a soil surface. The main concern for those interested in animal environment will be with radiation exchange, since it is mostly by this method that an animal will exchange heat with the ground (conduction would only be important when an animal is lying down, in contact with the soil surface).

Radiation exchange between any two surfaces is related by the forth power of the absolute temperatures of the exchanging surfaces. Because of this, a resarcher concerned with an animal's environment will be mostly interested in the surface temperature and condition of the soil, rather than in temperatures below the ground surface. The discussion in this section covers only surface temperatures. Brooks (1959) and Tanner (1963) provide good discussions of temperature measurement within the soil.

### Role of Soil Surface Temperature

The effect of solar radiation on heating a soil surface is indicated in Table 2.5.

The ground surface will radiate energy according to

$$E = \epsilon \sigma T^4$$

where $E$ = emissive power of the soil, kcal/hr m$^2$
$\epsilon$ = emittance of soil (dimensionless)
$\sigma$ = $4.93 \times 10^{-8}$ kcal/hr m$^2$ °K$^4$
$T$ = absolute temperature, °K = °C + 273

The part of the animal exposed to the soil surface (about half of a standing animal) would receive the emitted radiation, plus any solar and sky radiation reflected from the ground to the animal.

Some values for solar absorptance (1 minus the value for albedo) and longwave emittance of a few surfaces are shown in Table 2.6.

If some of the typical surface temperatures in Table 2.5 are com-

**TABLE 2.5 Temperature of Surface of Bare Ground before and after Shading at Various Times during the Day[a]**

| Shaded Time (min) | Temperature of Ground Surface (°C) 11 a.m. | 12 noon | 2 p.m. | 4 p.m. |
|---|---|---|---|---|
| In sun | 51.6 | 62.4 | 66.6 | 67.2 |
| 5 | 40.0 | 42.0 | 44.2 | 45.4 |
| 10 | 38.0 | 40.0 | 43.6 | 44.8 |
| 15 | 36.7 | 39.5 | 43.0 | 43.0 |
| 30 | 36.7 | 38.5 | 40.0 | 41.0 |
| Air temp | 33.3 | 35.0 | 36.7 | 40.0 |

[a]From Kelly *et al*. (1950).

**TABLE 2.6 Solar Absorptance (0.3–2.5 $\mu$) and Longwave Emittance (2.5 $\mu$ and up) of Various Surfaces[a]**

| Surface | Solar Absorptance | Longwave Emittance |
|---|---|---|
| Water | 0.94 | 0.95–0.96 |
| Desert surface | 0.75 | 0.90 |
| Sand, dry | 0.82 | 0.90 |
| Sand, wet | 0.91 | 0.95 |
| Moist ground (70–95% bare) | 0.88–0.91 | 0.95 |
| Ground, dry plowed | 0.75–0.80 | 0.90 |
| Grass, high, dry | 0.67–0.69 | 0.90 |
| Vegetable fields and shrubs | 0.72–0.76 | 0.90 |
| Alfalfa, dark green | 0.97 | 0.95 |
| Ice | 0.31 | 0.96–0.97 |
| Snow (ice granules) | 0.33 | 0.89 |
| Snow (fine particles) | 0.13 | 0.82 |

[a]From Brooks (1959).

bined with the emittance of a typical soil (0.90), one can see the effect of soil surface temperature on the heat load to which an animal is subjected. For example, if the soil surface is 62°C, it would emit radiation at a rate of about 560 kcal/hr $m^2$; a shaded surface at 38°C would emit about 417 kcal/hr $m^2$. About half of a standing animal would be exposed to the surface and receive about 280 and 208 kcal/hr $m^2$ from the high and low temperature soils, respectively. This, of course, would be in addition to any reflected solar radiation. Bond *et al.* (1967b) measured the radiation from a plowed ground surface in the Imperial Valley of California and found average values during the day (9 a.m. to 2 p.m.) of 161 and 548 kcal/hr $m^2$, respectively, for reflected solar energy and emitted longwave energy.

Ground surfaces vary in temperature by reason of differences in thermal conductivity, density, solar absorptance, and other characteristics. Ittner *et al.* (1958) showed a variance in ground surface temperature between areas only a few feet apart:

| | |
|---|---|
| Hard ground, tramped by cattle | 51.0°C |
| Hard ground, in road | 53.5°C |

| | |
|---|---|
| Soft ground, not tramped by cattle | 55.2°C |
| Dry rotten manure in feed lot | 64.8°C |

## Measurement of Soil Temperature

Soil temperature can be measured with thermocouples, resistance thermometers, thermistors, mercury-in-glass thermometers, mercury-in-steel thermometers, bimetal thermometers, and radiometers. For a detailed discussion of most of these instruments and their use see Tanner (1963). The main problem in surface temperature measurements with these instruments is that if they are placed against the surface, the sensing element can be influenced by radiation from the sun and surroundings and by convective loss or gain of heat to or from air passing over the sensor. Most researchers advise that the sensing element be placed "just under" the surface so it will actually be covered by a thin film of dirt. This indicates that the sensing element should be as small as possible; therefore, thermocouples and thermistors are among the best of the sensors.

### THERMOCOUPLES

Thermocouples have the advantage of being small, so their radiation and convection losses are relatively smaller than for most other sensors. They also have the advantage of having a signal that can be continuously and remotely recorded. Thermocouples, such as 24-gage copper-constantan, are small enough that they can usually be inserted just under the surface. If the soil is loose, they can be placed on the surface and lightly covered with a thin layer of soil.

Because of their fast response, thermocouples are good for spot checks of surface temperature. Their response can usually be read or recorded before they are greatly affected by radiation or convection. Thermocouple probes similar to those suggested by Kelly *et al.* (1949) or Thompson *et al.* (1951) are useful for spot checks of soil surface temperatures. Guaranteed limits of error for common thermocouple wire is usually about ±0.8°C, but premium grades can be purchased with limits of error of about ±0.4°C. Usually, the time accuracy of thermocouples is much greater than indicated by the manufacturers' limits of error. Larson *et al.* (1959) determined that the standard deviation of 20 copper-constantan thermocouples used in soil temperature work was about 0.1°–0.15°C.

### THERMISTORS

Thermistors are a form of resistance thermometer whose electrical resistance varies with temperature. They have the same advantage of thermocouples in respect to their small size. They can be obtained in forms of beads, rods, disks, and probes. They have greater sensitivity than thermocouples, and there is no need for a junction maintained at a reference temperature.

The radiation area of thermistors is small by virtue of their size. A number of commercial sources are available that supply thermistors in a variety of shapes and sizes, along with recorders. To minimize radiation errors in soil temperature measurements with thermistors, Platt and Griffiths (1964) suggest dipping the element in lead carbonate solution (90 percent reflection), then into white paint, and finally waterproofing with a benzine solution of white filtered paraffin. They indicate that contact temperature measurements within 0.01°C have been possible with thermistors.

There is a question of stability with some thermistors, therefore they should be calibrated frequently.

### THERMOMETERS

Mercury-in-glass thermometers are available that can be used for soil surface temperature measurements, but the size of bulb and general accuracy limits their usefulness for this purpose. There are models available with long, thin flat bulbs that can be placed in close contact with soil. According to Platt and Griffiths (1964) there are two schools of thought concerning exposure of the bulb: One holds it best to leave the glass completely showing, and the other advises covering it with a very fine layer of soil, a difficult practice in loam or clay soils. A second model is available that is shaped much like a golf club, with the bulb at 135° to the stem. Larson *et al.* (1959) checked 20 good-quality mercury-in-glass thermometers and obtained a standard deviation among thermometer readings of about 0.15°C.

Mercury-in-steel thermometers are similar to the mercury-in-glass ones, but the bulb is connected by means of a fine flexible metal pipe to a Bourdon tube. As the mercury expands, the Bourdon tube changes shape and moves a pen on a dial. The lag of the instrument may be quite large (up to 6 or 7 min). Larson *et al.* (1959) compared 20 such thermometers and obtained a standard deviation of about 0.45°C among readings.

Geiger (1959) suggested that the performance of the mercury-type thermometers could be improved for soil temperature studies if the bulb of the thermometer is imbedded in a small copper plate, which in turn is placed on the soil surface.

### RADIOMETERS

Some directional radiometers (see *Radiation* in this chapter) offer a good method of obtaining soil surface temperatures. The soil emissivity must be known or assumed in order to obtain temperatures from radiometer energy measurements; also, some means must be available to exclude solar, or shortwave, energy from the radiation measurements. The radiometer must not measure the solar energy reflected by the soil. The Stoll-Hardy radiometer can be obtained with a germanium filter that excludes this reflected solar energy. One model of the Barnes radiometer measures energy only in the wavelength range of 8–13 $\mu$. Atmospheric water vapor absorbs very little energy in this range of wavelengths, so the radiometer "looks through" any atmospheric water vapor and the readings are not affected by water vapor.

Both the Stoll-Hardy and the Barnes radiometers are calibrated to read directly the temperature of the surfaces they view. This calibration assumes a surface emissivity of unity. For many soils this assumption would be close. Where the emissivity is known to be less than unity, corrections must be made in order to obtain true temperature measurements. If temperature differences (with time or distance) are obtained for the same soil, the exact value of the emissivity is unimportant.

## INTEGRATION OF MULTIPLE VARIABLES

To determine the total effect of the bioclimatic factors on an animal a summation of the heat transfers must usually be made by calculating the transfers for each mode—radiation, convection, and so on—from the data obtained by the instruments used to measure the individual factors. Two approaches to simplifying these calculations are instrumentation that directly integrates the variables measured into an abstract figure that is related to the individual bioclimatic factors and presumably also to the effects these individual factors would have on the animals, and abstract single environments calculated from responses of instruments used to measure the individual bioclimatic elements.

Each of these instruments or methods will be of interest to the investigator of the effects of bioclimatic factors on animals.

## Instruments

None of the instruments described herein takes into account all of the bioclimatic factors. However, for quickly characterizing an environment with regard to the major factors affecting heat transfer, some of these instruments are of interest.

### GLOBE THERMOMETER

The globe thermometer is perhaps the simplest instrument for integrating the influence of air temperature, air movement, and radiation—three of the most important variables that influence the responses of animals and plants. The "globe thermometer temperature" is the temperature indicated by a thermocouple at the center of a black globe usually about 6-in. in diameter. A discussion of the theory and heat transfers involved is given by Bedford and Warner (1934). Bond and Kelly (1955) report studies of the effect of diameter and color of globe upon globe temperature and response time. Pereira *et al.* (1967) discuss the use of a blackened table-tennis ball as a globe thermometer.

### KATA THERMOMETER

The Kata thermometer was designed for measuring the rate of heat loss from a surface at approximately body temperature. It is essentially an alcohol thermometer, with an enlarged brightly silvered bulb and specially engraved stem. In use, the thermometer is heated above a certain temperature by immersion in warm water, the bulb is wiped dry of the liquid, and the length of time for the instrument to cool 2.5°C (from 37.5° to 25°C) is measured (McConnell and Yagloglou, 1924). The rate of heat loss due to convection is then calculated (Ower, 1949). The bulb may also be fitted with a sock and the rate of cooling of a wetted surface estimated (wet bulb thermometer principle).

### EUPATHEOSCOPE

The eupatheoscope is also an instrument developed to simulate heat loss from the human body. It is a cylinder wrapped with electrical resistance tape through which a measured current flows. The circuit controls the current flow so that the surface temperature is at 35°C. Painted black, and being somewhat the shape of a human torso, air velocity, air temperature, and mean radiant temperature influence the heat loss; the current necessary to maintain the surface temperature is a measure of this loss (Willard *et al.*, 1933).

#### THERMO-INTEGRATOR

This is an instrument for measuring the influence of air temperature, air movement, and radiation (Winslow and Greenburg, 1935).

#### SOL-AIR THERMOMETER

This is cork in the shape of a building, whose surface temperatures are influenced by air temperature, air velocity, and radiation. The cork, of both low heat capacity and high heat resistance, developes surface temperatures representing those of a rectangular structure in the same environment.

### "Equivalent Environments"

Thermal environments represented by a single number, although calculated from bioclimatic factors measured separately, are often the most comprehensible. (Such a number has sometimes been used to establish zones or criteria for heating or cooling for humans.) Several of these composites will be mentioned.

#### EFFECTIVE TEMPERATURE

Effective temperature is an emirically determined index of the degree of warmth sensed on exposure to different combinations of temperature, humidity, and air movement. It cannot be measured directly but is determined from dry and wet bulb temperature and air motion observations and reference to an effective temperature chart.

#### MEAN RADIANT TEMPERATURE

This is a method of averaging the radiant temperatures in an environment with several thermal radiation levels.

### OPERATIVE TEMPERATURE

This is defined as the sum of the radiation constant multiplied by the mean wall temperature, and the convection constant multiplied by the mean air temperature, divided by the sum of the two constants (Stewart, 1953).

### TEMPERATURE-HUMIDITY INDEX (THI)

This is an empirically determined index weighting dry bulb and wet bulb or dew point temperatures for comparison with animal performance. There are various weightings, i.e., THI = 0.4 (*db* + *wb*) = 15; THI = 0.55 *db* + 0.2 *dp* + 17.5; or THI = *db* – (0.55 – 0.55 *rh*) (*db* – 58), where *db* = dry bulb temperature of air; *wb* = wet bulb temperature of air; *dp* = dew point temperature of air; *rh* = relative humidity of air.

For a complete discussion of the use and meaning of the temperature–humidity index, see Chapter 8.

# Multipurpose Data Gathering Methods

## RADIO TELEMETRY

Telemetry means "measurement from a distance"; it includes transmission of information by wires, tubes, or other connections as well as by wireless (radio) waves. Radio transmission of information approaches the ideal in animal physiological measurements because the animal does not have to be tethered or his movements unduly restricted. Biological telemetry (radio telemetry) provides a method for transmitting data by radio and presents a dc voltage output to a recorder or other readout equipment. It can be most advantageously employed where wired connections are impossible or where it is important that the subject, such as an animal, not be disturbed.

Radio telemetry (Slater, 1963; Stattlemen and Buck, 1965; Caceras, 1965; Harris and Siegel, 1967; Mackey, 1968), on the other hand, does not usually simplify the acquisition of data. Rather, because of its inherent complexity, it can make data acquisition considerably more difficult. Furthermore, because the use of radio telemetry in animal research is relatively new, it is still almost as much an art as a science. There is a tendency for each researcher to impose a different set of requirements on the telemetry system, calling for a special design, or at least a modification of existing equipment. Although radio telemetry promises obvious advantages, one should resist the temptation to use it when a simpler technique is adequate.

All telemetry systems (Figure 2.5) have a transmitter and receiver. Modern, sophisticated systems can also transmit more than one channel of data simultaneously (multiplexing). Some will monitor nearly all physiological functions (blood pressure, blood flow, oximetry, respiratory rate, pulse, EKG, EEG, hear sounds, temperature, and so on).

The receiving and recording equipment is usually standard, although special decoding of the received signal is often required. Most of the research and design efforts are, therefore, concentrated on the transmitter, the associated sensors, and input signal conditioners. A transmitter may be attached to the outside surface of an animal, swallowed, inserted into a natural opening, or surgically implanted. Although placing the transmitter inside the animal can prevent destruction or damage, and eliminate having leads through the skin, it imposes stringent requirements on the telemeter itself. For long-term implantation, the most difficult problem is keeping moisture from damaging the electronic components. An entirely satisfactory sealing technique has not been developed, but the electronics of the more reliable transmitters are hermetically sealed in metal cans. Other problems include adverse tissue reaction to the implant, short battery life, and short transmitting distance. Coatings are available that apparently do not cause adverse tissue reaction, but these are not usually adequate moisture barriers.

Battery life can be extended by pulsed operation, by using the transmitter only when data are required, and by designing a low-energy

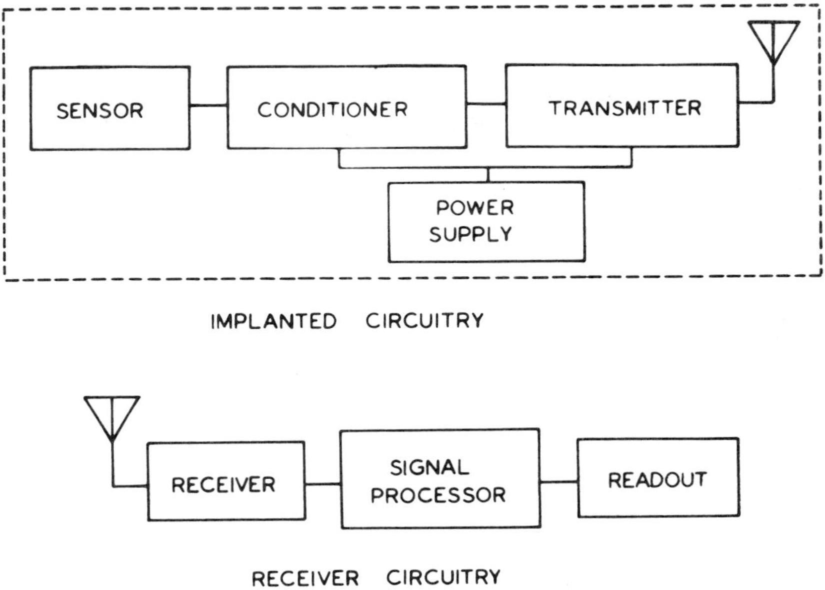


**FIGURE 2.5 Block diagram of a biotelemetry system.**

circuit, usually at the expense of reduced transmitting distance. Transmitting distance is also a function of antenna size; and this probably will be the limiting factor in overall telemeter size as the size of other electronic components is reduced by manufacturing techniques.

Nearly all research and design efforts have been concentrated on the transmitting units, but the generalized criteria of an implant transmitter can be summarized as (1) small size and weight (less than a few percent of the subject's weight); (2) minimum body reaction (packaged with nontoxic materials and with proper shape to reduce tissue reaction to the presence of the telemetry unit); (3) high sensitivity and wide dynamic range (to handle signals ranging from microvolts to millivolts); (4) good fidelity (signal frequency response from direct current to several kilohertz or higher); (5) low power consumption and long life; (6) reliability, rigidity, and ease of handling; (7) optimum transmission range to enable free movement of the subject and the use of units with the same frequency at nearby locations without interference; and (8) compliance with regulations for radio transmission established by the Federal Communications Commission.

The cost increases drastically as the size and weight of the transmitter are reduced. Usually the battery life and transmitting distance also decrease with transmitter size. Because of limitations in size and weight, most of the currently available transmitting units use one or two transistors to provide transducer, conditioner, and transmitter functions. The transmitted radio-frequency carrier may be frequency-modulated continuous waves (CW) or modulated pulses. Amplitude modulation is not used because of the introduction of serious errors during relative motion between transmitter and receiver.

Most transmitter circuits belong to one of four groups (Figure 2.6). The circuit in Figure 2.6a uses a piezoelectric crystal as the feedback element in an oscillator to achieve good carrier-frequency stability; it is widely used for animal tracking and narrowband transmission over a range from 30 meters to several kilometers; both CW and pulsed operation can be obtained by varying the value of resistor $R_1$. The circuit in

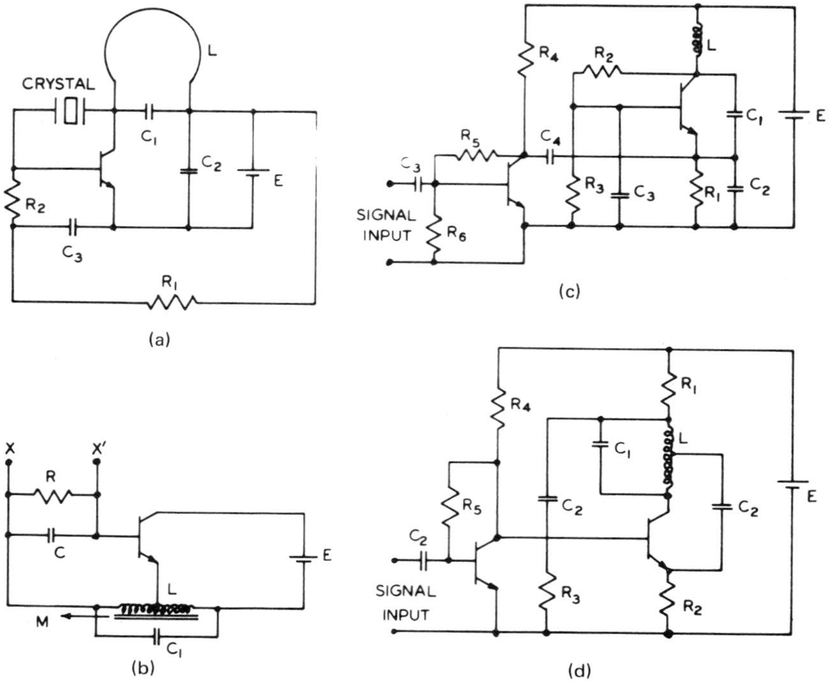


**FIGURE 2.6 Four popular telemetry transmitted circuits currently in use. (See text for further explanation.)**

Figure 2.6b is a common-emitter Hartley oscillator circuit; the transmitter is modulated by varying the rate of pulses of radio-frequency oscillations. It is often used to transmit temperature data by replacement of the resistor R with a thermistor; to transmit pressure data by allowing pressure changes to move the core M; to record acidity by attaching suitable electrodes across the voltage input X; and to transmit other variables, by means of suitable transducers. Continuous-wave operation can be obtained by reducing the value of R.

The remaining two circuits are common-base oscillators: Circuit 2.6c is a Colpitts circuit; circuit 2.6d is a Hartley circuit. Both use the voltage-sensitive emitter-to-base capacitance of the transistor to modulate the frequency of the carrier. These circuits, designed to transmit electrical signals such as electrocardiograms (EKG), should have a frequency response up to about 500 Hz. Electromyograms (EMG) require relatively wide band-widths of several kilohertz. The signal amplitude may vary from 50 $\mu$V to several millivolts, and the transmission range is usually from a few meters up to 30 meters at a power consumption of a few milliwatts.

The circuit of Figure 2.6b represents the most desirable design approach; the transducer and conditioner are integrated into the components of the oscillator. The simplicity of the circuit and the small power drain (average current of a few microamperes in the pulsed mode) explain its success and popularity. However, one should note certain limitations or disadvantages for the pulsed mode of operation: (1) the large error that can be induced by voltage variations in the power supply; (2) the interference generated over a wide frequency band because of the self-blocking pulsed-carrier mode of operation; and (3) the relatively low carrier frequency resulting from the low collector current and common-emitter configuration if self-starting is required. These limitations may be overcome when more complex circuits are used; the application of microelectronic techniques allows the small size and power requirements to be maintained even with the more complex circuits.

## RECORDERS

Recorders may preserve information on an event, elapsed time, voltage, current resistance, voltage ratios, acceleration, and so on. Most generally, the sensed data are transferred to the recorder pneumatically, or as a voltage or as a resistance. Recorders can be purchased separately or in combination with a potentiometer. The capabilities of both are constantly being improved.

### Chart Recorders

Chart or graphical recorders may be considered analog recorders since the distance between a point and some reference point on a chart is proportional to the variable being measured. The charts can be circular or in a long strip. Circular chart recorders ordinarily require charts to be changed daily or weekly; the paper roll on a strip chart may last several weeks. In either case, the variable is entered by the lateral movement of a pen, stylus, or print wheel. Several variables may be recorded on one chart by use of inks of different colors or printing the point for each variable with a different symbol or identifying number.

Chart recorders can be classified into two types: mechanical and electrical. Mechanical chart recorders use a mechanical or pneumatic linkage between the element that responds to the variable being measured and the chart measuring device. The chart-moving mechanism is also frequently of a mechanical nature (clock-driven) so that a source of electric power is unnecessary. An example is the hygrothermograph, wherein the bimetallic temperature-measuring element and the hair hygrometer provide the movement. Another example is the filled-system thermometer in which the pressure change in the sensing bulb is transmitted through a capillary tube to a sealed metallic spiral. A pressure increase causes the spiral to unwind (the free end of which is connected to the recording pen). Disadvantages of these recorders are that they must be close to the point of measurement and only a few variables can be recorded on one chart. They are relatively simple and inexpensive.

It has been indicated in previous sections that quantities representing most bioclimatic factors can be reduced to electrical forms, i.e., voltage, resistance, or current. Recorders that convert an electrical input to a mechanical displacement are of two basic types: galvanometric and potentiometric. In its simplest form the galvanometric or oscillographic recorder uses a d'Arsonval-type galvanometer. The signal to be recorded is applied to a coil that rotates on its axis in the presence of a magnetic field. When the signal is too small to operate the coil directly, an amplifier is required. A pen or stylus attached to a coil marks a moving chart. Inkless systems are available that utilize a stylus and carbon-coated paper, a heated stylus and heat-sensitive paper, and a metal stylus and electrosensitive paper. There are optical systems which use a small mirror attached to the moving coil. These have a very high frequency response since the mirror is extremely small compared with a pen. The mirror reflects light from a fixed source onto photosensitive paper. Galvanometric recorders have high frequency responses; up to 200 Hz for pen-type instruments and up to 10,000 Hz for light-beam systems. Of course, the recorder must be capable of high paper transport speed for satisfactory resolution of small intervals of time, and speeds of up to 160 in./sec are available.

There is limited need for high-frequency response for most environmental investigations, so potentiometer-type recorders with a maximum response of 1–2 Hz are used to a greater extent (Figure 2.7). These instruments have the advantage of high sensitivity (down to microvolt signals) and ability to have the recorder at great distance from the measuring point. Most use a "null-balance" system in which an emf (generated by a thermocouple, for example) is compared with an emf of known value. When a difference of "unbalance" occurs, the system acts to counterbalance it through a servo-type motor, which changes a resistance (slidewire) and at the same time positions a pen or print wheel (Figure 2.8). Resistance may also be measured by using a Wheatstone bridge. The current from an unbalanced bridge is used to drive the servomotor (Figure 2.9).

Potentiometer-type recorders are well adapted to measuring several variables consecutively (multipoint recorders). The frequency of repe-

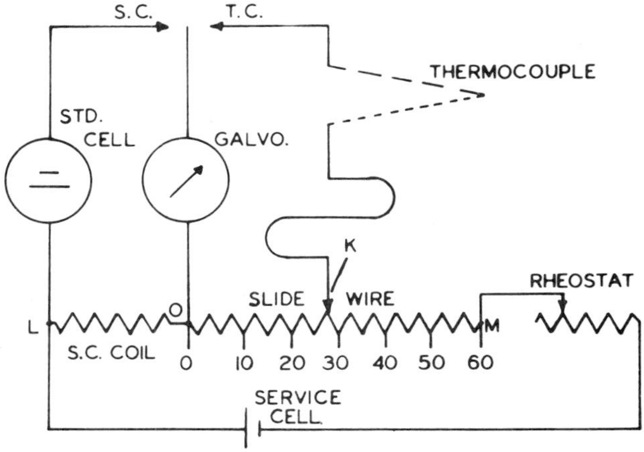


FIGURE 2.7 Millivolt potentiometer.

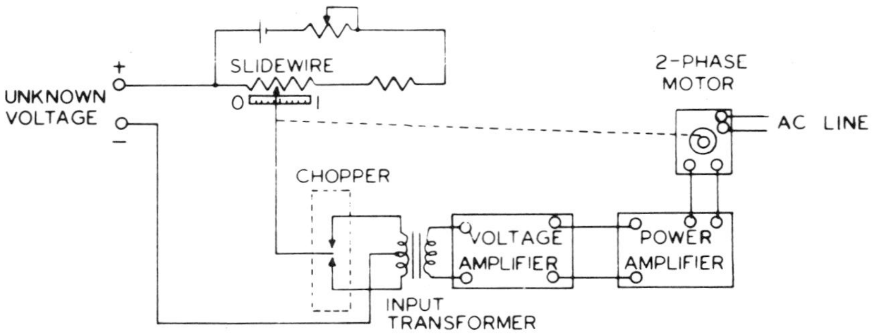


**FIGURE 2.8 Electronic self-balancing potentiometer.**

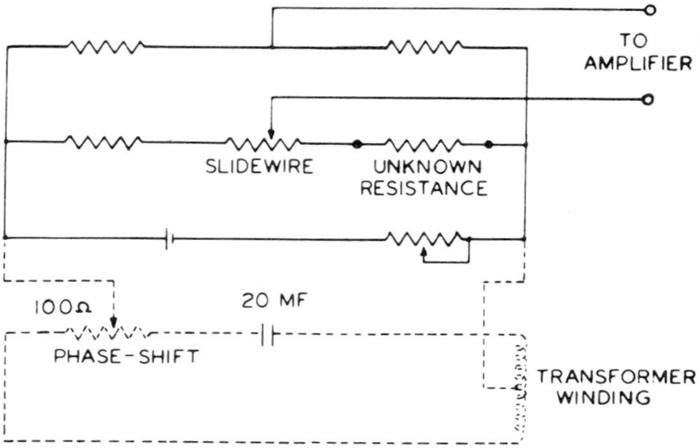


**FIGURE 2.9 Resistance-measuring circuit.**

tition of the cycle is limited by the time required for the recorder to move from minimum to maximum readings and the number of points involved. Desirable features in a recorder are adjustable range or span, adjustable zero, and adjustable sensitivity.

The discussion thus far has been limited to recording variables against one independent variable—time. The servo principle used in potentiometer recorders can be used to record two variables simultaneously with one pen. One signal moves the pen in one direction ($Y$-axis) and the other the pen or chart in a perpendicular direction ($X$-axis). This type of recorder, commonly called an $X$-$Y$ recorder or plotter, has many applications in function plotting but limited potential in bioclimatic data recording.

## Digital Recorders

Analysis of bioclimatic data usually requires data to be in a numerical or digital form. In the case of chart recorders, conversion of selected points to digital forms is necessary (chart reading). This may be done completely manually, partially automatically by using a manually operated curve follower that automatically converts the selected point to a digital form and records it on punched cards, or completely automatically with automatic chart readers and computers. The latter devices may also perform certain operations such as addition, multiplication, and integration.

There are advantages to recording data originally in digital form, and there are several systems available that do this. It is beyond the scope of this report to give the details of their operation. However, the system must include a scanner that accepts analog inputs (voltage, resistance, and so on) and sequentially connects them to a measuring instrument, a signal converter that changes the analog signal to a form acceptable to the analog-to-digital converter, and an analog-to-digital converter that converts the signal to its equivalent digital form. A digital clock is usually included as auxiliary equipment to initiate measurement at predetermined intervals and to supply a time signal for the output. There must be an output coupler that translates the information from the analog-to-digital converter to the proper form for the digital recorder. The recorder may place the data on punched cards, perforated paper tape, magnetic tape, or typewriter paper. Information in these forms is then suitable for computer processing. Even when digital data acquisition systems are used, it is desirable to use indicators or chart recorders for monitoring purposes and backup records.

## REFERENCES

Altman, P. L., and D. S. Dittman (Ed.). 1966. Environmental biology. Fed. Am. Soc. Exp. Biol., Bethesda, Maryland.

American Institute of Physics. 1941. Temperature, its measurement and control in science and industry. Reinhold Publishing Co., New York.

American Society of Heating, Refrigerating, and Air Conditioning Engineers. 1966. Fundamentals and equipment. *In* ASHRAE guide and data book. Soc. of Heating, Refrigerating, and Air Conditioning Engrs., New York.

American Society of Heating, Refrigerating, and Air Conditioning Engineers. 1967. Psychrometrics. *In* Handbook of fundamentals. ASHRAE, New York.

Annals of the International Geophysical Year. 1958. Vol. V. Part III. Pergamon Press, New York.

Bainbridge, R., C. G. Evans, and O. Rackham. 1966. Light as an ecological factor. John Wiley & Sons, New York.

Baker, H. D., E. A. Ryder, and N. H. Baker. 1953. Temperature measurement in engineering. Vol. I. John Wiley & Sons, New York.

Baker, H. D., E. A. Ryder, and N. H. Baker. 1961. Temperature measurement in engineering. Vol. II. John Wiley & Sons, New York.

Beaumont, R. T. 1965. Mt. Hood pressure pillow snow gage. J. Appl. Met. 4:626–631.

Bedford, T., and C. G. Warner. 1934. The globe thermometer in studies of heating and ventilation. J. Hyg. 34:458–473.

Behar, M. F. 1954. Handbook of measurement and control. Part II. Instr. Automat. 27:12.

Benzinger, T. H., R. G. Huebscher, D. Minard, and C. Kitzinger. 1958. Human calorimetry by means of the gradient principle. J. Appl. Physiol. 12:S1–S28.

Beranek, L. L. 1949. Acoustic measurements. John Wiley & Sons, New York.

Bianca, W. 1965. Sweating in dehydrated steers. Vet. Sci. 6:33–37.

Biswas, A. K. 1967. Development of rain gages. J. Irrig. Drain. Div., Proc. ASCE, 93:99–124.

Blinn, B. G. 1965. Properties and uses of color change humidity indicators. Vol. I, p. 602. *In* Humidity and moisture-measurement and control in science and industry. Reinhold Publishing Co., New York.

Bond, J. and C. F. Winchester. 1963. Effects of loud sounds on the physiology and behavior of swine. Tech. Bull. No. 1280, Agr. Res. Ser., U.S. Dep. Agr. Government Printing Office, Washington, D.C.

Bond, T. E., C. F. Kelly, and H. Heitman. 1952. Heat and moisture loss from swine. Agr. Eng. 33:148–152.

Bond, T. E., and C. F. Kelly. 1955. The globe thermometer in agricultural research. Agr. Eng. 36:251–255.

Bond, T. E., H. Heitman, and C. F. Kelly. 1965. Effect of increased air velocities on heat and moisture loss and growth of swing. Trans. Am. Soc. Agr. Eng. 8:167–169.

Bond, T. E., S. R. Morrison, and C. F. Kelly. 1967a. Measurement of incoming short and long wavelength radiation with directional thermopile radiometers. Trans. Am. Soc. Agr. Eng. 10:462.

Bond, T. E., C. F. Kelly, S. R. Morrison, and N. Pereira. 1967b. Solar, atmospheric and terrestrial radiation received by shaded and unshaded animals. Trans. Am. Soc. Agr. Eng. 10:662.

Bowen, H. D. 1961. Parallel thermocouple circuits in agricultural engineering applications. Trans. Am. Soc. Agr. Eng. 4:58–61.

Braude, R., K. G. Mitchell, P. Finn-Kelcey, and V. M. Owen. 1958. The effect of light on pigs. Proc. Nutr. Soc. 17:38.

Brewer, A. W. 1965. The dew-or-frost-point hygrometer. Vol. I, p.135. *In* Humidity and moisture-measurement and control in science and industry. Reinhold Publishing Co., New York.

Brody, S. 1945. Bioenergetics and growth. Waverly Press, Baltimore, Md.

Brooks, C. F. 1945. Impracticability of precipitation gages that are kept pointed into the wind. Am. Met. Soc. Bull. 26:98.

Brooks, F. S. 1959. An introduction to physical microclimatology. Syllabus 397. Assoc. Students Store, Univ. of Calif., Davis, Calif.
Brown, M. J., and E. L. Peck. 1962. Reliability of precipitation measurements as related to exposure. J. Appl. Met. 1:203–207.
Caceras, C. A. (Ed.). 1965. Biomedical telemetry. Academic Press, New York.
Chow, V. T. (Ed.). 1964. Handbook of applied hydrology. McGraw-Hill, San Francisco, Calif.
Clayton, G. A., and A. Robertson. 1960. Light induction of out-of-season reproduction in the turkey. Brit. Poult. Sci. 1:17–23.
Conover, J. H. 1950. Tests and adaptation of the foxboro dew-point recorder for weather observatory use. Bull. Am. Meteor. Soc. 31:13.
Courvoisier, P., and H. Wierzejewski. 1954. Dos Kugelpyranometer Bellani. Arch. Meteorol. Wien 5:43.
Dal Nogare, S., and R. S. Juvet. 1962. Gas-liquid chromatography. Interscience, New York.
Daubenmire, R. F. 1964. Plants and environment. John Wiley & Sons, New York.
Davey, F. K. 1965. Hair humidity elements. Vol. 1, p. 571. *In* Humidity and moisture-measurement and control in science and industry. Reinhold Publishing Co., New York.
Day, D. L., E. L. Hansen, and S. Anderson. 1965. Gases and odors in confinement swine buildings. Trans. Am. Soc. Agr. Eng. 8:118–121.
Deacon, D. L. 1951. The overestimation error of cup enemometers in fluctuating wind. J. Sci. Instr. 28:231–234.
Dike, P. H. 1954. Thermoelectric thermometry. Leeds and Northrup Co., Philadelphia, Pa.
Ditchburn, R. W. 1963. Light. John Wiley & Sons, New York.
Dobie, J. B., T. E. Bond, R. L. Givens, H. Heitman, Jr., C. F. Jacob, C. F. Kelly, C. M. Sprock, and W. O. Wilson. 1966. Effects of air ions on swine and poultry. Trans. Am. Soc. Agr. Eng. 9:883–886.
Drury, L. N. 1966. Air velocity and broiler growth in a diurnally cycled hot environment. Trans. Am. Soc. Agr. Eng. 9:329–332.
Duncan, A. D. 1967. The measurement of shower rainfall using an airborne foil impactor. J. Appl. Met. 5:198–204.
Dutt, R. H. 1960. Temperature and light as factors in reproduction among farm animals. J. Dairy Sci., Suppl. 43:123–144.
Duvdevani, S. 1947. An optical of dew estimation. Quart. J. Roy. Meteorol. Soc. 73:282.
Eckert, E. R. G., and R. M. Drake. 1959. Heat and mass transfer. McGraw-Hill, New York.
Eichmeier, A. H., R. F. Wheaton, and E. H. Kidder. 1965. Local variation in precipitation induced by minor topographical differences. Quart. Bull. Michigan Agr. Exp. Sta. 47:533–541.
Fritschen, L. J. 1963. Construction and evaluation of a miniature net radiometer. Appl. Met. 2:165–172.
Garrett, W. H., T. E. Bond, and N. Pereira. 1967. Influence of shade height on physiological responses of cattle during hot weather. Trans. Am. Soc. Agr. Eng. 10:433.

Gates, D. M. 1962. Energy exchange in the biosphere. Harper and Row, New York.
Geiger, R. 1959. The climate near the ground. Harvard Univ. Press, Cambridge, Mass.
Glorig, A. 1958. Noise and your ear. Green and Stratton, New York.
Hahn, L., T. E. Bond, and C. F. Kelly. 1961. Use of models in thermal studies of livestock housing. Trans. Am. Soc. Agr. Eng. 4:45–51.
Hamilton, E. L. 1954. Rainfall sampling on rugged terrain. Tech. Bull. No. 1096. U.S. Dept. Agr. Gov't Printing Office, Washington, D.C.
Harris, C. L., and P. B. Siegel. 1967. An implantable telemeter for determining body temperature and heart rate. J. Appl. Physiol. 22:846–849.
Henderson, S. M. 1952. A constant feed all-temperature wet bulb. Agr. Eng. 33:644.
Hicks, W. W., and J. C. Beckett. 1957. The control of air ionization and its biological effects, AIEE Trans. 76:108.
Hienton, T. E., D. E. Wiant, and C. A. Brown. 1958. Electricity in agricultural engineering. John Wiley & Sons, New York.
Hindman, E. E., and R. L. Rinker. 1967. Continuous snowfall replicator. J. Appl. Met. 6:126–133.
Horton, R. E. 1919. Measurement of rainfall and snow. Monthly Weather Rev. 47:294–295.
Hudson, N. W. 1963a. Advisory notes No. 44 and 45. Fed. Dept. Conserv. Ext. Rhodesia and Nyasaland.
Hudson, N. W. 1963b. Raindrop size distribution in high intensity storms. Rhodesian J. Agr. Res. 1:6–11.
Hungerford, K. E. 1967. An acetate dew gage. J. Appl. Met. 6:936–940.
Ittner, N. R., C. F. Kelly, and T. E. Bond. 1957. Cooling cattle by mechanically increasing air velocity. J. Anim. Sci. 16:732–738.
Ittner, N. R., T. E. Bond, and C. F. Kelly. 1958. Methods of increasing beef production in hot climates. Calif. Agr. Exp. Sta. Bull. 761.
Jacob, F. D., T. E. Bond, H. Heitman, Jr., and C. F. Kelly. 1965. Air ion studies with swine. Paper presented at Rural Electric Conference, Davis, Calif.
James, J. W. 1964. Effect of wind on precipitation catch over a small hill. J. Geophys. Res. 69:2521–2524.
Kebbon, E. R. 1961. Bi-metal thermometers, instruments, and control systems. Instr. Contr. Sys. 34:841–845.
Kelly, C. F., T. E. Bond, and C. Lorenzen, Jr. 1949. Instrumentation for animal shelter research. Agr. Eng. 30:297–300 (304).
Kelly, C. F., and N. R. Ittner. 1948. Artificial shades for livestock in hot climates. Agr. Eng. 29:239.
Kelly, C. F., H. Heitman, and J. R. Morris. 1948. Effect of environment on heat loss from swine. Agr. Eng. 29:525–529.
Kelly, C. F., T. E. Bond, and N. R. Ittner. 1950. Thermal design of livestock shades. Agr. Eng. 31:601–606.
Kelly, C. F., T. E. Bond, and H. Heitman. 1954. The role of thermal radiation in animal ecology. Ecology 25:562–569.
Kelly, C. F., T. E. Bond, and H. Heitman. 1963. Direct air calorimetry for livestock. Trans. Am. Soc. Agr. Eng. 6:126–128.

Kelly, C. F., T. E. Bond, and W. Garrett. 1964. Heat transfer from swine to a cold slab. Trans. Am. Soc. Agr. Eng. 7:34–37.

Kleiber, M. 1961. The fire of life. John Wiley & Sons, New York.

Koller, L. R. 1932. Ionization of the atmosphere and its biological effects. J. Franklin Inst. 214:543–568.

Kornblueh, I. H., and J. E. Griffin. 1955. Artificial air ionization in physical medicine. Am. J. Phys. Med. 24:618–631.

Kreuger, A. P., and R. F. Smith. 1957. Effects of air ions on isolated rabbit rachea. Proc. Soc. Exp. Biol. Med. 96:807.

Larson, W. E., W. C. Burrows, and T. E. Hazen. 1959. Equipment and methods of measurement of soil temperature. USDA Bull. ARS 41–27.

Lee, D. H. K. 1953. Manual of field studies on heat tolerance of domestic animals. FAO Develop. Paper No. 38.

Leighton, A. T., P. B. Siegel, and H. S. Siegel. 1966. Body weight and surface area of chickens. Growth 30:229–238.

Lewis, J. E., Jr., S. E. Curtis, and V. A. Garwood. 1968. Relation between atmospheric pressure fluctuations and swine gestation length variation. Intern. J. Biometeorol. 12:159–162.

Linsley, R. K., Jr., M. A. Kohler, and J. L. H. Paulhus. 1958. Hydrology for engineers. Maple Press Co., York, Pa.

Littlewood, A. B. 1962. Gas chromatography. Academic Press, New York.

Lomas, J. 1965. Note on dew-duration recorders under semi-arid conditions. Agr. Met. 2:351–359.

Lorenzen, C., Jr. 1949. The thermocouple in agricultural research. Agr. Eng. 30:275–279.

Lourenze, F. J. 1964. Investigation of factors influencing the origin and quantity of dew which forms on a turf crop. Unpublished Masters Thesis, Univ. of California, Davis.

Mackay, R. S. 1968. Bio-medical telemetry. John Wiley & Sons, New York.

Mathews, D. A. 1965. Review of the lithium chloride radiosonde hygrometer. Vol. 1, p. 219. *In* Humidity and moisture-measurement and control in science and industry. Reinhold Publishing Co., New York.

McArthur, J. M., and J. E. Miltimore. 1961. Rumen gas analysis by gas solid chromatography. Can. J. Anim. Sci. 41:187–196.

McConnell, W. J., and C. P. Yagloglou. 1924. The Kata-thermometer—Its value and defects. Reprint No. 953, U.S. Public Health Service Report, p. 2293–2337.

Middleton, W. E. K., and A. F. Spilhaus. 1953. Meteorological instruments. Univ. of Toronto Press, Toronto, Canada.

Monge, M. C., and C. C. Monge. 1966. High-altitude diseases: Mechanism and management. Charles C Thomas, Springfield, Illinois.

Moon, P. 1940. Proposed standard curves for engineering use. J. Franklin Inst. 240:583–617.

Morgan, D. L., and F. J. Lourenze. 1969. Comparison between rain-gauge and lysimeter measurements. Water Resour. Res. 5(3):724–728.

Morris, T. R., and S. F. Fox. 1960. The use of lights to delay sexual maturity in feedlots. Brit. Poult. Sci. 1:25–36.

Morrison, S. R., T. E. Bond, and H. Heitman. 1967. Skin and lung moisture loss from swine. Trans. Am. Soc. Agr. Eng. 10:691–692.

Mount, L. E. 1962. Evaporative heat loss in the new-born pig. J. Physiol. 164:274–281.

Mount, L. E. 1964. Radiant and convective heat loss from the newborn pig. J. Physiol. 173:96–113.

Neuberger, H. 1951. Introduction to physical meteorology. Mineral Industries Extension Service, State College of Pa.

Nixon, P. R., and G. P. Lawless. 1965. Annual research report. Lompoc, California, U.S. Dept. Agr., ARS, Soil Water Conserv. Res. Div., SW Br., A11-A13.

Ohme, W. E. 1967. Loudness evaluation. Hewlett-Packard J. 19:2–11.

Ower, E. 1949. The measurement of air flow. Chapman and Hall, Ltd., London.

Palmatier, E. P. 1963. Psychrometric chart. p. 55. Am. Soc. Heat., Refrig., Air Cond. Eng. J. May.

Parr, J. A. 1966. Congressional record, 89th Cong., 2d sess., 21 April 1966, p. 8343.

Penton, V. E., and A. C. Robertson. 1967. Experience with the pressure pillow as a snow measuring device. Water Resour. Res. 3:405–408.

Pereira, N., T. E. Bond, and S. R. Morrison. 1967. "Ping pong" ball into black globe thermometer. Agr. Eng. 48:341.

Peterson, A. P. G., and L. L. Beranek. 1963. Handbook of noise measurement, 5th ed. General Radio Co., Cambridge, Mass.

Platt, R. B., and J. Griffiths. 1964. Environmental measurement and interpretation. Reinhold Publishing Co., New York.

Pruitt, W. O., and D. E. Angus. 1960. Large weighing lysimeter for measuring evapotranspiration. Trans. Am. Soc. Agr. Eng. 3:13.

Reifsnyder, W. E., and H. W. Lull. 1965. Environmental measurement and interpretation. Reinhold Publishing Co., New York.

Reser, W. F. 1940. Thermoelectric thermometry. J. Appl. Physiol. 11:388–407.

Richards, C. H., A. M. Stoll, and J. D. Hardy. 1951. The panradiometer: An absolute measuring instrument for environmental radiation. Rev. Sci. Instr. 22:925–934.

Richardson, L. 1965. A thermocouple recording psychrometer for measurement of relative humidity in hot arid atmosphere. Vol. 1, p. 101. *In* Humidity and moisture-measurement and control in science and industry. Reinhold Publishing Co., New York.

Riemerschmid, G., and J. S. Elder. 1945. The absorptivity for solar radiation of different colored hair coats of cattle. Onderstepoort J. Vet. Sci. 20:223–234.

Robertson, C. E. 1965. An easy-to-use rain drop sensor. J. Appl. Met. 4:642–644.

Ruskin, R. E. 1965. Humidity and moisture, measurement and control in science and industry. Vol. 1. Reinhold, New York.

Saul, R. A. 1956. A continuous recording wet-bulb apparatus. Agr. Eng. 37:448.

Schleusner, R. A., and P. C. Jennings. 1960. An energy method for relative estimates of hail intensity. Am. Metereol. Soc. Bull. 41(7):372–376.

Scott, W. W. 1939. Standard methods of chemical analysis. Van Nostrand Co., New York.

Semplak, R. A. 1966. Gauge for continuously measuring rate of rainfall. Rev. Sci. Instr. 37:1554–1558.

Silverman, B. A., B. J. Thompson, and J. H. Ward. 1964. A laser fog discometer. J. Appl. Met. 3:792–801.

Slater, L. E. (Ed.). 1963. Bio-telemetry. Pergamon Press, New York.

Smith, A. H., H. Ablanalp, L. M. Harwood, and C. F. Kelly. 1959. Poultry at high altitudes. Calif. Agr. 13:8–9.

Smith, C. F. 1964. A quantitative relationship between environment comfort and animal productivity. Agr. Met. 1:249–270.

Snow, W. B. 1959. Significance of readings of acoustical instrumentation. Noise Contr. 5:304–307.

Sprock, C. M., W. E. Howard, and F. C. Jacob. 1967. Sound as a deterrent to rats and mice. J. Wildlife Management 1:729–741.

Stattlemen, A., and W. Buck. 1965. A transmitter for telemetering electrophysiological data. Proc. Soc. Exp. Biol. Med. 119:352–356.

Stewart, R. E. 1953. Absorption of solar radiation by the hair of cattle. Agr. Eng. 35:235–238.

Stoll, A. M., and J. D. Hardy. 1952. A method of measuring radiant temperatures of the environment. J. Appl. Physiol. 5:117–124.

Story, H. C., and H. G. Wilson. 1944. A comparison of vertical and tilted rain gages in estimated precipitation on mountain watersheds. Am. Geophys. Union Trans. 24:518–525.

Suomi, V. E., and P. M. Kuhn. 1958. An economical net radiometer. Tellus 10:160–163.

Swan, J. B., C. A. Federer, and C. B. Tanner. 1961. Economical radiometer performance, construction and theory. Univ. of Wisconsin Soils Bull. 4.

Swindells, J. F. 1959. Calibration of liquid-in-glass thermometers. Natl. Bur. Stand. Cir. 600. U.S. Dept. Commer. U.S. Govt. Printing Office, Washington, D.C.

Tanner, C. B. 1963. Basic instrumentation and measurements for plant environment and micrometeorology. Univ. of Wisconsin Soils Bull. 6.

Thompson, H. J., R. M. McCroskey, and S. Brody. 1949. Influence of ambient temperature, 0 to 105°F on insensible weight loss and moisture vaporization in Holstein and Jersey cattle. Mo. Agr. Exp. Sta. Res. Bull. 451.

Thompson, H. J., D. M. Worstell, and S. Brody. 1951. Influence of environmental temperatures, 0 to 105°F on hair and skin temperatures and on partition of heat dissipation between evaporative and non-evaporative cooling in Jersey and Holstein cattle. Mo. Agr. Exp. Sta. Bull. 481.

Timiras, P. S. 1964. Comparison of growth and development of the rat at high altitude and at sea level, p. 21–31. *In* Physiological effects of high altitude. Pergamon Press Lt., distributed by the Macmillan Co., New York.

Turk, A. 1963. Approaches to sensory odor measurement. *In* Conference on recent advances in odor: Theory measurement and control. Ann. N.Y. Acad. Sci. 116:564–566.

Vernon, H. M. 1930. The measurement of radiant heat in relation to human comfort. Proc. J. Physiol. 70:15–17.

Wallin, J. R. 1967. Agrometeorological aspects of dew. Agr. Met. 4:85–102.

Weiss, L. L. 1963. Securing more nearly true precipitation measurements. J. Hydraul. Div., Am. Soc. Civil Eng. Vol. 89, No. HY 2, Proc. Paper 3446.

Weiss, L. L., and W. T. Wilson. 1957. Precipitation gage shields. Extrait des Comptes Rendus et Rapports. Assemblee Generale de Toronto, pp. 462–484.

Wiersma, F., and G. L. Nelson. 1967. Nonevaporative convective heat transfer from the surface of a bovine. Trans. Am. Soc. Agr. Eng. 10:733–737.

Willard, A. C., A. P. Kratz, and M. K. Fahenstock. 1933. The application of the eupatheoscope for measuring the performance of direct radiators and convectors in terms of equivalent temperature. ASHVE Trans. 39:303.

Winslow, C. E. A., and L. Greenburg. 1935. The thermo-integrator—A new instrument for the observation of thermal interchanges. ASHVE Trans. 41:149.

Wylie, R. C., D. K. Davies, and W. A. Caw. 1965. The basic process of the dew-point hygrometer. Vol. 1, p. 125. *In* Humidity and moisture-measurement and control in science and industry. Reinhold Publishing Co., New York.

Yeck, R. G., and H. H. Kibler. 1956. Moisture vaporization by Jersey and Holstein cows during diurnal temperature cycles as measured with a hygrometric tent. Mo. Agr. Exp. Sta. Res. Bull. 600.

# 3

# PHYSIOLOGIC FUNCTIONS AND MEASUREMENT TECHNIQUES

The results of biometeorologic research often are expressed in terms of well-being or as increased production as measured by growth or by quantity of milk, eggs, or wool produced. However, physiologic measurements also provide information about the impact of unfavorable bioclimatic conditions. For example, the effect of high temperature may be expressed in the altered composition of milk or in the decreased thickness of egg shells.

Functions of organs or of systems can be measured by standardized techniques, and new ones are constantly being developed. Standardized tests in addition to those reviewed here are described in the textbooks referenced in this chapter. Research and development in this field is extremely active, however, and it was possible to review only a fraction of the literature as this chapter was prepared. Furthermore, only responses affected by environmental factors are considered. Data logging, reduction, and analysis are not discussed at all.

Physiologic measurements may be made in several degrees of refinement. These include (1) field measurements that are aided by portable instruments or that rely on a classification system, such as a pictorial scale of coat characteristics; (2) laboratory measurements of specimens representative of the population, such as blood samples;

(3) use of delicate instruments in a laboratory on intact or surgically altered animals, e.g., to determine blood pressure using a pressure transducer in conjunction with a recorder; (4) use of biotelemetry in field or laboratory studies.

The Bioinstrumentation Advisory Council of the American Institute of Biological Sciences, Washington, D.C., collects and distributes information to biologists and engineers who have indicated a continuing interest in such information. This council was established to initiate development and dissemination of new techniques and technology for bioinstrumentation and to increase information flow between the biological and the physical sciences.

For an extensive survey of the techniques and methods used to measure a large number of physiologic phenomena, see Geddes and Baker (1968). The accuracy of measurement desired will influence the choice of technique. For example, if one uses a scoring system which has only five classes to characterize haircoat, the error could be as high as 20 percent. Simularly, body weight of a bird is changed by the laying of an egg. As a result, the weight of a chicken may vary as much as 1 or 2 percent. However, if a Japanese quail (*Coturnix coturnix japonica*) were being studied, the error might be as great as 8 percent. Obviously, to cite a third case, it is not logical to compare the weights of fed and starved animals.

## CLIMATIC CONDITIONS AFFECTING ANIMALS

The effect of altitude or barometric pressure may be studied through observation of such physiologic parameters as the cardiovascular system, blood constituents, respiration, production, behavior, endocrine function, and metabolism. High altitude affects all animals and is generally related to oxygen requirements. Animals differ in their response to changes in altitude. The critical elevation for some animals is approximately 1500 meters (Tromp, 1967).

Photoperiods affect some species of animals more than others. The reproductive cycles of poultry and sheep are influenced by light; the effect in cows and horses is more subtle as they respond to light by change in their pelage. Other parameters to measure in some animals include productive traits and changes in activity in respect to changes in light and dark cycles. Endocrine function, metabolic behavior, production, reproduction, and activity all respond to changes in photoperiod.

All animals are affected by thermal stimuli. In addition to temperature, related environmental factors are air movement, humidity, and radiation, i.e., factors affecting heat loss or gain (see Chapter 8). A number of physiological measurements closely associated with these bioclimatic factors may be studied. Changes in metabolism are the means by which animals cope with extreme changes in thermal conditions. The central nervous system is often investigated, as it is related to perception of the environment as well as to changes in behavior related to thermoregulation. Shivering and sweating are responses to the thermal environment. Body composition may reflect both short- and long-term changes resulting from temperature. The function of certain organs of the body, such as the liver and kidneys, are affected by temperature. Endocrine function is associated with changes in environmental temperature. Blood constituents, cardiovascular changes, respiratory activity, reproduction, and production are all influenced by the thermal environment.

A limited amount of information is available concerning the physiologic functions affected by some of the other environmental factors. These include sound (communication), pollution, cyclic changes, and gravity. Doubtless other factors of the environment, e.g., ions and changes in barometric pressure, as yet not documented, will also be found to affect physiologic responses.

The detection of deviation from homeostasis is perhaps the best evidence that the environment is affecting the animal. These effects may be observed and reported on the basis of either qualitative or quantitative data. Procedures such as bioassay, gravimetric analysis, electrical, chemical, and linear measurements, and tracer techniques are examples of procedures for taking quantitative data.

The use of transducers permits various changes in states of energy prior to recording. The transduction and measurement of physiologic events may employ transducers that are resistive, mechanoelectronic, inductive, capacitive, photoelectric, piezoelectric, thermoelectric, or chemical (Geddes and Baker, 1968).

## CENTRAL NERVOUS SYSTEM

Environmental physiologists interested in the workings of the nervous system should become acquainted with the three volumes on neurophysiology that are part of the series of the *Handbook of Physiology* published by the American Physiological Society.

The sensory mechanism may act either in proximity or at a distance. For reception at distance, an animal relies on hearing or sight. Sensory perception involving proximity could include touch, taste, and perhaps smell.

## Behavioral Patterns

Measurement of behavioral patterns is discussed by Stebbins (1966). This work defines behavior broadly to include, for example, feeding and drinking, which can regulate body temperature considerably. Reproduction may also be included in a definition of behavior.

Shivering can be measured, or indicated, by electromyographic (EMG) activity. Action potentials from shivering muscles can be recorded from either skin electrodes (Davis and Mayer, 1954) or intramuscular electrodes (Heroux *et al.*, 1956). Recitified EMG voltage recorded in arbitrary units has been used as a measure of shivering intensity (Davis and Mayer, 1954). Peak amplitudes of the nonrectifield EMG potentials have also been used as measures of shivering intensity. In another technique, the potential peaks of the EMG waves are grouped according to their amplitudes, and the amplitudes are multiplied by their numbers per second (Goepfert *et al.*, 1953).

Denenberg and Banks (1962) describe and evaluate stimulus characteristics. A quantitative description of behavior is given, as well as a discussion the role of major factors in the evocation of behavior. The tables are helpful in evaluating several techniques for studying ingestive, sexual, parental, and social behavior.

The operant conditioning technique (Hafez, 1968) may be useful in studying thermoclimes (Mount, 1968).

## Electrophysiologic Studies

The role of the central nervous system in eye movement, cardiovascular control, respiration and digestive function is discussed in the *Handbook of Neurophysiology* of the American Physiological Society (1959, 1960). Also included are neurophysiology, brain potential and rhythm, technique for recording activity of individual neurons, and the recording of gross action potentials from unanesthetized animals. Electrical stimulation of neural tissue is described by Ranck (1966), providing the principles of electrical recording and the limitations of available methods. The electrical recordings from the nervous system include the use of extracellular microelectrodes, intracellular microelectrodes, and recording from large electrodes (Vol. 9, Section 4).

Other references on the electronic measurement of physiologic phenomena in experimental animals are to be found in Gay (1965). Discussion of the techniques of analysis of electrical signals may also be found in Kay (1964). In the first few chapters, detection of electrical activity, amplification of biological signals, display and recording of biologic electrical signals and stimulation are discussed.

Devices that can be permanently implanted permit control of the central nervous system activity so that the system can be manipulated by electric stimulation and the activity measured. Such techniques are described by John (1966).

### Hypothalamic Control of Endocrine Releasing Factors

Neurosecretion and central control of certain functions, as well as secretions from the pituitary, are discussed in the *Handbook of Neurophysiology* of the American Physiological Society (1959, 1960).

The hypothalamus exerts a control over the hypophysis, and thus over the entire endocrine system. As to the nature of the ultimate information from the hypothalamus to the hypophysis, it is now generally accepted to be neurohumoral; neuroendocrine cells in various, and probably discrete, areas of the hypothalamus apparently have the ability to secrete substances that will reach the hypophysis.

Two major specialized systems link the hypothalamus with the hypophysis: (1) nerve fibers coming from the paraventricular, supraoptic, tuberal, and periventricular nuclei; and (2) the hypothalamo-hypophyseal portal system (Fortier, 1962; Nalbandov, 1963). Electrical stimulation of particular regions of the hypothalamus has been found to excite gonadotrophic hormone secretions, provided the hypothalamo-hypophyseal portal system is intact. Harris (1960) reported that direct electrical stimulation of the anterior lobe of the pituitary gland has such an effect. He also found that sexual atrophy occurred following lesion in the tuber cinercum. Taubenhaus and Soskin (1941) and Nalbandov (1963) did much of the early work in this field.

The central nervous system plays a role in regulating the secretion of follicle-stimulating hormone (FSH). This regulatory control may be exerted by some neurohumoral mechanism (Nalbandov, 1963).

There is no evidence that the hypothalamus is concerned with either the release or control of growth hormones (Hertz, 1950), although this possibility has been suggested.

The concept that the hypothalamus produces a material that can cause release of adrenocorticotropic hormone (ACTH) from the adenohypophysis appears to be supported by observations that hypothalamic lesions prevent the usual adrenal responses to conditions of stress. Hume (1958) reported that lesions in the anterior part of the median eminence abolish ACTH release.

The matter of origin and storage of antidiuretic hormone (ADH) is not fully settled as yet, but the posterior lobe, the supraoptic-hypophyseal tract fibers, and the cells of the supraoptic and paraventricular nuclei are involved. The neurosecretions from the cells in the hypothalamus, and their relationship to ADH, have been reviewed (Archer and Fromageot, 1957).

Physiologic and endocrine changes induced by environmental factors may markedly alter drug activity; similarly, chemicals may alter the response of the organism to his environment. The interaction of drugs with autonomic nervous functions and thermoregulation has been cited by Maickel (1970). A more general discussion of pharmacologically active substances and chemical activity has been assembled by Sourkes (1961). Having determined the function of the anatomic units, the researcher must often go on to enzymic studies.

Hammel *et al.* (1963) describe the hypothalamic control of temperature regulation. Benzinger *et al.* (1963) discuss methods of studying the physiology of thermoregulation and methods of measuring the skin temperature at the tympanic site (to best approximate that of the interior hypothalamus); measurements of the responses include sweating, vasodilatation, and increased metabolic heat production measured indirectly from oxygen consumption.

Andersson and Larsson (1961) have reported on the influence of local temperature changes in the preoptic area and rostral hypothalamus on regulation of feed and water intake of goats. Their studies were related to the feed intake regulation through the satiety center of the ventral medial hypothalamus and appetite center at the lateral hypothalamus described by Brobeck (1955).

In contrast to the previous report, in a study with rats (Spector *et al.*, 1968), it was observed that selectively cooling the area preoptica to 24° ± 1°C resulted in hyperthermia and a significant decrease in blood intake. Heating the same area to 43° ± 1°C resulted in the opposite effect.

## CARDIOVASCULAR SYSTEM

All cells need a constant supply of nutrients and oxygen. Cells also need to constantly dispose of waste products, some of which are extremely poisonous. In all vertebrates, the cardiovascular system satisfies these requirements. Blood is exposed to fresh air supplied by the lungs, has nutrients added to it by the gut and liver, and has waste products removed from it by the kidney.

The level of blood pressure is regulated by both the nervous system and the endocrine system. Any increase in the flow of blood through any portion of the body is compensated by an increase in the amount of blood pumped by the heart. An increase in cardiac output is normally produced by an increase in heart rate, not usually by an increase in the amount of blood pumped per stroke. The cardiovascular system is especially sensitive to changes in temperature and barometric pressure.

### Heart Rate and Electrocardiogram

From the voltage recorded from the heart as electrocardiograms, the degree of rhythm, or lack of it, and the heart rate may be readily determined. The electrocardiogram yields information on cardiac refractoriness.

Cain *et al.* (1967) report heart rate determinations in intact chicken embryos. A highly sensitive ballistocardiograph was used to measure heart rate (Deboo and Jenkins, 1965). For comprehensive reading, see Katz and Pick (1956) and Elson (1963).

### Arterial and Venous Pulse

The increased filling of the aorta due to the ejection of blood during systole causes a distension of the vessel and a rise in pressure. The pressure wave travels rapidly away from the heart and can be felt, after a brief interval, at any superficial peripheral artery, such as the radial artery at the wrist. Thus, the heart rate can be counted easily by feeling the pulse. The venous pulse is seen best in the low neck region.

### Blood Pressure

Arterial pressure may be measured by a transducer. Strain gauge transducers in manometer systems are the most common. They use fluid linkage for the conversion of membrane deflection to electric signals. Other methods for the translation of membrane displacement into electric signals include inductive, capacitive, piezoresistive, and optical (Franke, 1966).

The indirect measurement of arterial blood involves the determination of the amount of air pressure that is required, when applied to a limb, to cause collapse of the artery and obstruction of the blood flow.

Two indices may be used to detect the peripheral pulse. The first is auscultatory. The second method is the oscillatory criterion developed by Erlanger (1916).

### Flow Dynamics

Blood flow may be considered to be the product of heart rate and stroke volume. Cardiovascular flow measurements have been classified by Fry and Ross (1966) as follows:

1. Positive volumetric displacements: Drop counters, flow meter, displacement, plethysmograph gas spirometer, and reciprocating bellows gas meter.
2. Hydrodynamic techniques: Orifice meters, venturi meter, and pressure gradients.
3. Linear displacement or velocity-sensing techniques: Bubble flow meter, indicator transit time techniques, chemical and thermal convection techniques, pulsed sonic flow meter, and Doppler shift technique. The principle of the last depends on the change in frequency of an ultrasonic wave as it is reflected back toward the sending direction from the moving particles in the flowing liquid. In principle, ultrasound systems could be miniaturized to catheter tip dimension. Franklin *et al.* have recently developed an instrument that can be applied directly to the skin overlying a superficial blood vessel, to obtain qualitative information on blood flow, including pressure.
4. Dilution of indicator substance as inert gases, nondiffusible dyes, and natural metabolites. The direct Fick method assumes that, with the animal at rest, the pulmonary oxygen uptake is equal to that used by the tissues. Oxygen is the indicator, and cardiac output is equated with blood flow through the lungs. In practice, arterial and venous blood samples are obtained simultaneously, whereas oxygen consumption is measured by spirometry and subsequent chemical analysis of the expired gas.

Selectively excreted indicators include bromsulphalein (BSP) for hepatic blood and *p*-aminohippurate, which may be used for estimating renal blood flow.

5. Other measures include the ballistocardiograph and pneumograph.

Pressure is one of the most commonly studied variables in physiological systems and is important in a great variety of applications. The measurement of fluid pressure by use of physiological pressure transducers is discussed by Franke (1966), who presents characteristics of some pressure transducers that are commercially available. Cannulation and catheterization procedures have also been utilized (Thompson and McIntosh, 1966).

The general types of dimensional transducers are discussed in a section edited by Peterson (1966) in *Methods in Medical Research*. For a treatment of blood volume and various cardiovascular responses to the environment, see the three volumes of *Handbook of Physiology of Circulation* of the American Physiological Society (1962, 1963, 1965).

## BLOOD CONSTITUENTS

Blood constituents and other blood characteristics may reflect the effect of changes in the environment. For example, the environment may alter either the number of erythrocytes and leukocytes or the differential cell count.

The normal values in blood morphology for many species of animals are found in Schalm (1961), who describes the materials and techniques necessary for the study of blood and also staining techniques. For information on the blood of avian species, see Lucas and Jamroz (1961).

### Cells

Hematocrit values or packed cell volumes are obtained by centrifuging a quantity of blood and determining the volume of packed red cells. The reading is expressed as a percentage of total blood. When the red cell count or hemoglobin estimation falls below normal values, the hematocrit reading also drops below normal. In dehydration the body loses fluid and the erythrocytes become more concentrated; consequently the hematocrit reading rises above normal values. Two methods for obtaining hematocrit values are described by Seiverd (1964).

Reeve (1960) describes measurement of red cell volume with radioactive phosphate. Whole blood is incubated with $^{32}P$ for 30–60 min at 32°C until 4–6 $\mu c$ have entered the cell. The red cell volume is determined by the number of counts in the $^{32}P$ injected, divided by the number of counts per milliliter of venous red cells.

Erythrocyte and leukocyte counts may be made by using the hemacytometer counting chamber. Standard references on the general topic of hematology are available (Schalm, 1961; Lucas and Jamroz, 1961; Mark and Zimmer, 1967). Electronic methods of cell counting have been described (Seiverd, 1964). These counters have the advantage of completely eliminating subjective counting errors, reducing by a third the scatter of replicated counts. A main advantage of the electronic counter is its speed: The maximum count reaches 7000 particles/sec with a mean error of 0.5 percent as against 8 percent in the counting chamber. This method is therefore suitable for mass studies.

To observe the types of white cells and to make differential counts, stained smears are prepared. Wright's stain is probably the most frequently used. The peroxidase stain (Graham, 1918) is helpful for distinguishing between monocytic and lymphoid forms.

Hemoglobin determination (Wintrobe, 1961; Gradwohl, 1956) may be divided into two groups of procedures: visual and photoelectric. The photoelectric procedure offers the utmost in accuracy. Some photoelectric procedures convert the hemoglobin to acid heme (hematin), oxyhemoglobin, and cyanmethemoglobin. Another procedure is colorimetric: Sahli's hemoglobinometer provides fairly accurate results. The cyanhematin method is similar.

Erythrocyte indices correlate the data from hemoglobin determinations, erythrocyte counts, and the hematocrit, in order to explore functional problems of erythrocytes.

The color index method is used to indicate the hemoglobin content of each cell relative to a normal one. The mean corpuscular hemoglobin (MCH) is a similar comparison. It is found from the volume of the red cells as determined by the hematocrit (Schalm, 1961).

## Serum

Total lipids of the blood usually are determined from the serum. The lipids are extracted by a mixture of alcohol and ether, which precipitates the protein. After the solution has been evaporated to dryness, the lipids in the residue are reextracted with petroleum ether. The ether is evaporated, and the remaining residue, containing the lipids, is weighed (Mark and Zimmer, 1967).

To assay the metabolism of lipids, plasma–lipid determinations are often made. Reinhold *et al.* (1963) describe a colorimetric method for measuring of total fatty acids and glyceride.

Total protein and the albumin/globulin ratio are also determined from blood serum. The protein undergoes a biuret reaction with copper sulfate in an alkaline medium to form a blue-violet compound. Total protein is determined on a diluted sample of serum. The globulin is then precipitated with a solution of sodium sulfate, and the albumin remaining in the solution is determined. The globulin is determined by difference to obtain the albumin/globulin ratio.

The Kjeldahl method is used to determine serum nitrogen (Archibald, 1957; Hiller *et al.*, 1948).

Alkaline or acid phosphatase may be determined from the serum or from heparinized plasma. The procedure is based on the colorimetric determination of the amount of *p*-nitrophenol formed in the hydrolysis of *p*-nitrophenol phosphate by the phosphatase of the serum sample at pH 10.3 (or 4.8) (Mark and Zimmer, 1967).

Calcium in the serum is determined by precipitation as calcium oxalate. The oxalate is separated, dissolved in acid, and titrated with potassium permanganate.

Protein-bound iodine determination involves precipitating the protein from the serum with zinc hydroxide and drying with sodium carbonate. The dried mixture is ashed at approximately 600° C to convert the organically bound iodine to inorganic sodium iodide. The iodide is extracted with dilute hydrochloric acid to catalyze the reaction of ceric sulfate with the arsenite ion. The rate of the reaction is proportional to the amount of iodide present and is determined colorimetrically by following the change in the ceric ion concentration (Mark and Zimmer, 1967).

Blood chloride may be determined from the serum. The chloride is titrated with mercuric ions (from mercuric nitrate) to form soluble but undissociated mercuric chloride. The end point of the titration is the appearance of mercuric ions in solution. These combine with the indicator diphenylcarbazone to form a lilac-blue complex.

Sodium, potassium, and calcium may be determined in the serum or heparinized plasma by use of a flame photometer. The specimen is diluted with water, and the solution is atomized directly into a flame. The sodium, potassium, and calcium ions are energized to give off light of characteristic wavelengths, which may be isolated by means of filters and other monochromators. The intensity of the light is measured by a photometer, and the concentration of an ion is determined by comparison with the reading of the photometer using either a standard sodium, potassium, or calcium solution (Mark and Zimmer, 1967).

Analysis of the blood serum or plasma for creatinine and creatine may give some indication of renal function. Creatinine, a protein-free filtrate, is treated with picric acid in alkaline solution to produce a deep red tautomer of creatinine picrate. Creatine is converted to creatinine in an acid medium by heating in an autoclave or boiling water bath (Mark and Zimmer, 1967).

## Blood Volume Determination

The plasma volume or corpuscular components of the total blood volume may vary independently although there is a homeostasis of both. Direct determination of whole blood volume entails the sacrifice of the animal; therefore, indirect methods are more often used. The principal indirect method is to introduce a known amount of a substance into the circulation and to allow it to become uniformly distributed throughout the blood of the body. The concentration of the indicator in a blood sample is then determined. Suitable tagging materials may be used for plasma or red cells independently. It is best to study both components and to combine the determination with hematocrit measurements.

Plasma tagging was originally done with vital dyes which bind firmly to plasma protein. Evans blue (T-1824) gained wide acceptance (Gregersen *et al.*, 1935). In this procedure a series of blood samples is obtained at intervals after the injection, and the photocolorimetric analyses are plotted graphically against the time to give a dye disappearance curve. This can be extrapolated back to zero time. Plasma volume multiplied by 100/plasma hematocrit gives the total blood volume.

Red cell tagging may also be used: A measured volume of carbon monoxide (CO) is introduced into the blood of the animal. Also used are radioisotope tracers such as radioactive iron (Hahn and Rouser, 1950). In the latter method, tagged cells are injected intravenously. The degree of dilution in the nonradioactive cells is determined by the radioactivity analysis of the sample. The recipient cells give a measurement of the total circulating red cell volume. The computed total erythrocyte volume multiplied by a 100/red blood cell hematocrit gives the total blood volume. Many investigators prefer to use $^{32}P$ or $^{51}Cr$, which have a shorter half-life (Gray and Sterling, 1950). One of the main advantages is that phosphates or chromates can be equilibrated with red cells *in vitro* so that the animal's own blood can be used by bleeding, tagging, and reinjection.

### Blood pH

Determination of hydrogen ion concentration of the blood has been outlined by Gradwohl (1956). The indicator method used provides a means of separation of protein by dialysis, using collodion sacks. The dialysate is adapted to the use of indicators. Another method is the LaMotte blood pH (Cullen, 1922). This method is more rapid than the former and can be used to estimate the reaction of the blood in acidosis and alkalosis. A third method is by the capillary glass pH electrode introduced by Sanz (1957). This micromethod for measuring pH requires some time to reach equilibrium. Severinghaus (1965) gives the criteria required for accurate determination of pH in the blood by the capillary glass pH electrodes.

### Coagulation Time

The coagulation time is determined on blood in a container, the capillary tube method, or the Lee-White method. For a detailed description of these and of tests associated with blood coagulation—for example, bleeding time, platelet count, prothrombin time, and thromboplastin generation tests—see Seiverd (1964).

### Sedimentation Rate

When red cells are allowed to settle out from the plasma, the speed of their fall is known as the sedimentation rate, the number of millimeters of sedimentation after the blood has stood undisturbed for 1 hr. Seiverd (1964) gives the details of the Westergren, Culter, and Wintrobe methods.

### Blood Gas Composition and Analysis

Oxygen content and capacity and percent saturation may be determined in whole blood which has been protected from the air. Oxygen and carbon dioxide are released from hemoglobin by hemolyzing with saponin and ferricyanide. The $CO_2$ is absorbed in KOH; and the pressure, which represents the pressure of oxygen and dissolved inert gases, is measured. The oxygen is then absorbed in sodium hydrosulfite, and the pressure is again measured. The difference in the two measured pressures is the pressure due to oxygen. To obtain percent saturation, a sample of blood is oxygenated and its oxygen content is determined on a second sample. Percent saturation is calculated from the two values obtained (Mark and Zimmer, 1967).

Carbon dioxide in the blood plasma is liberated by lactic acid under controlled conditions in an airtight manometric apparatus (Natelson microgasometer). The gas is shaken free from the liquid under reduced pressure and is then brought to a constant volume to measure the pressure, $P_1$. The $CO_2$ is absorbed in alkali and the pressure of the residual gas, $P_2$, is measured at the original volume. The pressure drop ($P_1 - P_2$), in millimeters, is multiplied by a factor to convert the pressure into an equivalent amount of $CO_2$.

The intimate involvement of oxygen and carbon dioxide in respiratory exchange under any bioclimatic condition makes their measurement a problem of recurring nature. Currently most measurements are made with a physiologic gas analyzer, such as the Beckman Model 160 and Instrumentation Laboratory Model 113. These instruments also measure blood pH. The $CO_2$ electrode is essentially a pH glass electrode arranged to measure the pH of a very thin film of aqueous sodium bicarbonate solution, which is separated from the blood sample by a membrane permeable to $CO_2$ gas molecules but not to ions that might alter the pH. The pH is a linear function of the logarithm of $P_{CO_2}$. The $O_2$ electrode consists of a platinum cathode and a silver anode in electrolyte solution separated from the blood sample by a permeable membrane. When $O_2$ diffuses across the membrane, reduction of oxygen at the cathode and oxidation of silver at the anode allow a direct current to flow which is directly proportional to $P_{a_{O_2}}$.

Blood gases can also be measured manometrically or volumetrically, but these procedures are laborious (Oser, 1965).

## RESPIRATION

Several physiologic processes are collectively or separately termed "respiration." Measurements of ventilation, respiratory exchange, respiratory evaporative heat loss, and pulmonary function tests may be included under discussions of external respiration; transport of blood gases and various oxidative aspects of metabolism may be termed internal respiration.

Before any measuring technique is used, the limits of its static and dynamic accuracy must be assessed in terms of the desired tolerance. If only total respiratory exchange is required, for example, to determine metabolic rate or the sources of a metabolite, a consideration of potential errors is relatively simple, particularly if extended periods of measurement are used (see the section on metabolism). If a detailed analysis of respiratory exchange is necessary for careful comparison of different environments, or if dynamic changes are needed for a description of diurnal variations, for example, then it is essential to consider the potential interactions between the animal and the environment. This is particularly true if chambers are to be used to determine respiratory exchange.

### Measurements of Respiratory Exchange in a Chamber

A general description of various chambers is given in the section on metabolism; see also Kleiber (1961).

Carbon dioxide concentrations should always be kept as low as feasible. Concentrations above 0.5 percent involve disturbances in ventilation and in whole-body stores of carbonates. Whole-body carbonate stores can also undergo alteration due to disturbances in ventilation, which are brought on by behavioral responses or body temperature alterations (including diurnal variations). Most masks increase dead space and therefore the level of inhaled carbon dioxide. The smaller the animal, the less important becomes the error involved in whole-body carbonate stores because of the ratio of chamber content to the intensity of metabolism (Kleiber, 1961).

High levels of oxygen tension sometimes used in indirect calorimeters alter the proportionality of ventilation to metabolism because of altered peripheral chemoceptor reactivity and cerebral blood flow. Long-term exposure to one atmosphere of oxygen can give rise to "oxygen toxicity." In both cases, alterations occur in carbonate stores and metabolism.

High levels of chamber ventilation can increase heat loss, which cannot be totally corrected by adjustment of environmental temperature. Additionally, most animals show strong aversion to cool air at high rates of movement. The latter can significantly alter metabolic rate.

Complications arise when metabolism is altered. All chambers have a delayed response to alteration of the steady state. This results in reduction of the response to short-term alterations and averages metabolism, shifting the phase of observed metabolic variation. An appreciation of the importance of apparently insignificant events is the observation that metabolic rate in man doubles when the position is changed from supine to sitting. Arm movements increase metabolic rate over basal by 300 percent; short periods of moderate exercise increase metabolism significantly for hours.

In an empty closed respirometer the approach to equilibrium following a sudden withdrawal of gas can be observed to follow an exponential curve. For an open system, Kleiber (1940) gives a solution to the transient alteration of chamber gas for $CO_2$. The static error in carbon dioxide or oxygen analysis must be balanced against the dynamic error estimated from the time constant as it is impossible to reduce chamber volume to a small ratio to body weight without use of a mask.

A closed system has additional complicating factors when an animal is introduced into the chamber. The absorption of carbon dioxide and equilibrium of water vapor are generally concentration dependent. In many closed systems, the loss in dynamic response due to long time constants in humidity and carbon dioxide equilibration is probably larger than that for supply of oxygen.

Alteration in heat flux of an animal in a closed chamber observably gives rise to the largest and quickest fluctuations in chamber pressure. The largest source of heat flux alterations occur when the animal moves, releasing trapped air heated considerably above chamber environmental temperatures.

An appreciation may be gained for the importance of temperature if an alteration in metabolism occurs. Kleiber (1940) defines a "relative chamber volume" to calculate the error involved in a change in chamber temperature relative to the animal's consumption of oxygen. His formula for error of 1° is $0.58V/O$ in percent of total consumption, where $V$ is chamber volume and $O$ is the volume of oxygen consumed over the period of test. If temperature does change, the test may be extended for long periods of time; conversely, the large relative error in transient response can be seen for small alterations in temperature over short periods of time. The use of masks for collection of the expired air simplifies analysis and reduces analytical errors in those species that tolerate a mask. Tracheotomy procedures have been used, but rarely, to bypass behavioral responses to mask or to alter processes occurring in the upper air passageways. Generally, tidal volume and rate, or total inspiratory or expiratory volume, is determined. In large animals, a sampling procedure is often used to reduce analyzed gas volume.

### Measurement of Gas Concentration

Expired gas consists of dead air volume and an "alveolar" volume, which often must be mixed for analysis (in a Douglas bag or spirometer). However, in a gas meter or flow meter, the total sample is not collected, and the sampling procedure must accurately reflect the gas concentration in mixed, expired air. The sample cannot be continuously drawn by pump since air will be drawn back during inspiration. The physical principle of the flow meter is again utilized to advantage by allowing the pressure drop in the flow-measuring device to force air through the sampling or analyzing chamber (e.g., Pauling meter).

If determination of metabolic rate is the purpose of respiratory exchange measurements, oxygen concentration in expired air is often the important parameter. In certain cases it is desirable to measure carbon dioxide concentrations as well. Chemical methods of analyzing for carbon dioxide and oxygen have been described in detail (Consoliazio *et al.*, 1951). Particular notice should be taken of the Scholander micrometer gas analyzer, which permits determination of oxygen, carbon dioxide, and nitrogen in 0.5-ml samples, with an accuracy of ±0.015 percent. This analysis requires 6–8 min. Also deserving of consideration is the Fry (1949) analyzer, which is economical and simple to handle. It analyzes a 2–3-ml sample with an accuracy of ±0.5 percent.

Rein (1933) combined the chemical method with a flow-sensitive device to obtain continuous recording of gas concentration. If a component of a gas flowing through two chambers is absorbed in one chamber, the rate of flow becomes unbalanced. The Rein analyzer measures this imbalance.

There are several convenient and accurate physical methods for gas analysis. For oxygen measurement, the Beckman Model C oxygen analyzer has become standard in many laboratories. Its operation depends on the paramagnetic properties of oxygen. This instrument can be obtained with a full scale range of 60–760 mm with an accuracy of ±1 percent. The response time is of the order of 10 sec if the gas sample is passed through the pole pieces, requiring close flow rate control, and approximately 50 sec if the analyzer is protected by a fine porous diffusion plate. An oxygen analyzer, utilizing the ionization of $O_2$, with a response (70 percent) time of 0.07 sec and which is suitable for use in respiratory measurements, is available commercially (from Westinghouse).

For carbon dioxide measurement, conductivity, light refraction, and infrared absorption methods have been used. The Cambridge Instrument Company manufactures a thermal conductivity cell. Clamans describes a refractometer (1962). A number of infrared analyzers are described in the literature (Blinn and Noell, 1949; Fowler, 1949; Luft, 1943; Young, 1952), and the Liston Becker Company manufactures two commercial models with a wide response range (Baird Instrument Company and Leeds and Northrup manufacture instruments. Accuracy is ±0.1 percent carbon dioxide, and speed of response (90 percent) is 0.1 or 0.3 sec, depending on the model. Spoor (1948), in addition to describing the use of a Leeds and Northrup infrared analyzer, reports another method for determining metabolism (see Figure 3.1). A constant volume of air exceeding in amount the maximal respiratory volume is drawn past the subject by a pumping system. An analyzer sample is drawn off by means of a small, constant-speed rotary pump. Other investigators using this principle introduce a rubber bag of 2–3-liter capacity between the subject and the measuring devices to "buffer" the peaks in flow velocity.

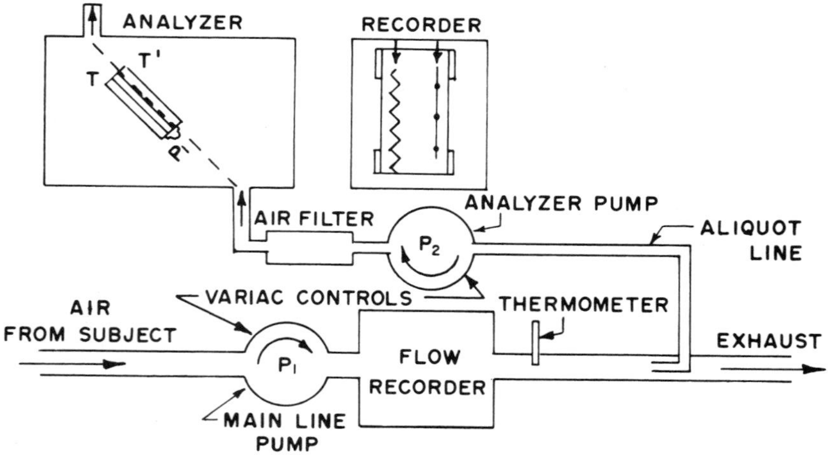


FIGURE 3.1 Schematic drawing of air pumping system and air analyzing system used by Spoor (1948).

### Pulmonary Function Tests

Of increasing importance to studies of the environmental responses of man and other animals is the analysis of pulmonary function. Lungs are exposed to ever-increasing levels of gaseous and particulate pollutants in the atmosphere. A discussion of clinical pulmonary function tests, i.e., lung volumes, flow rates, compliance, air-way resistance, and diffusing capacity, can be found in Comroe *et al.* (1963). Many standard clinical tests are difficult to perform on noncooperating subjects. Consideration should be given to adapting the methods (Avery, 1964) used for the human infant to animals.

## METABOLISM AND BODY TEMPERATURE

Energy metabolism is a measure of the regulated release and storage of chemical energy by organisms, i.e., the sum of energy transformation taking place in an organism per unit of time.

Animals are dependent exclusively on food for their source of energy. The quantity of energy used is normally reflected by either a gain in body weight as energy is stored or a loss in body weight as the body consumes itself for fuel.

The units for energy are the same units used to express work. Whenever calories are discussed without qualification, the large calorie (kilocalorie) is meant. It is defined as the amount of heat required to raise the temperature of 1 kg of water 1°C. Another term used for expressing the energy equivalent is the kilogrammeter, which is simply the amount of work performed when a body of given weight is moved against gravity through a known distance. For example, if an individual raises a 3-kg weight through 3 meters, then he has performed 9 kg-meters of work.

The measurement of energy expenditure is referred to as calorimetry, in which energy is measured as heat; heat is one of the most conveniently handled forms of energy. Calorimetry is based on the law of the conservation of energy. This law states that energy can neither be created nor destroyed, and, therefore, the energy content of any system can be increased or decreased only by the amount of energy that is added to or subtracted from the system.

Potential energy is defined as either the energy of position of a body (a book on a table possesses energy by reason of its position and it can perform work if it falls off the table) or the potential chemical energy of certain substances.

Kinetic energy, the other general type of energy, has various forms, such as heat, light, and motion. It is the energy of movement, and it is the work that can be performed by moving bodies.

The principal calorimetry measures are of two types. *Direct* calorimetry is the measurement of energy expenditure (radiation, conduction, convection, and evaporation) in the form of heat (see Chapter 2); all types of energy are converted to heat and then measured. Since energy is utilized by means of chemical reactions, it is possible to evaluate energy utilization from the measurements of the substances consumed and the products formed. Indirect calorimetry, the second type, determines energy expenditure from the amounts of oxygen consumed and carbon dioxide and methane produced.

### Direct Calorimetry

Direct calorimetric measurements of metabolism involve placing an animal in an enclosure that permits the estimation of the number of calories lost as heat per unit time. In principle, direct calorimetry of the whole animal is a measure of heat of combustion of compounds catabolized within the body. The heats of combustion of the various materials excreted are not measured.

In recent years, partitional thermoelectric or gradient layer calorimetry has increased the precision of estimating heat production and thus our knowledge of the subject. Partitional thermoelectric calorimeters designed for domestic animals are in operation at several research centers. The only disadvantages are rather stringent restrictions on the environmental and behavioral states of the animals being studied.

The Missouri thermoelectric partitional calorimeter (Figure 3.2) measures electrically all of the heat added to the air by the subject and, in addition, partitions the radiation fraction from the total sensible heat. The walls of the calorimeter act as a constant temperature heat sink, and the thermal gradient across the wall, proportional to the amount of heat flowing through the wall, is measured with two thermopiles: radiation and total sensible. The total sensible gradient layer lies next to the chamber well and is covered by the radiation thermopile, which consists of a series of thermojunctions serially connected in one plane. This layer is covered with a radiation-absorbing layer of Mylar plastic over which is placed a reflective aluminum tape covering alternate junctions. The alternating reflection and absorption permits thermoelectric separation of radiation as a component of the

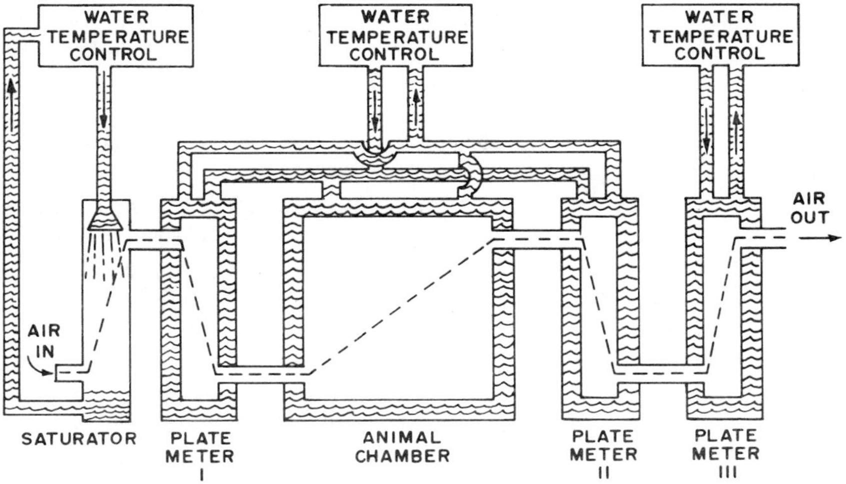


**FIGURE 3.2 Schematic diagram of the Missouri partitional calorimeter. Fresh air is conditioned to the desired dew point temperature by a precisely controlled constant temperature water spray in the saturator. The air is then heated to the desired dry bulb temperature at plate meter I. Sensible heat gain from the animal is removed at the animal chamber and plate meter II. The moisture gain is removed at plate meter III. The necessary heat transfers are attained by circulating water around the plate meters and the animal chamber.**

total heat loss of the subject. Insensible heat loss, vaporization from skin and lungs, is measured from the amount of water added to the ventilating air as it passes through the chamber.

Output of the radiation thermopile is recorded with a Leeds and Northrup Speedomax G, Model S, series 6000 recorder; the outputs from the total sensible thermopile and from the plate meters used to measure latent heat are recorded with Leeds and Northrup Speedomax H, Model S, series 60 recorders.

Calibration data, periodically collected, reveal that the accuracy of the direct system is ±0.2523 Btu per hour for the gradient layer (total sensible heat), ±0.050098 Btu per hour for the radiant layer, and ±0.0002611 Btu per hour for evaporation or insensible heat loss.

## Indirect Calorimetry

In measuring aerobic energy metabolism indirectly, the known proportionality between the oxygen consumption or the carbon dioxide production and the total energy production is used. This method for measuring energy expenditure is also referred to as respiratory calorimetry. It is performed in one of two ways. (1) In the open circuit method the animal is permitted to breathe air from the "outside," while its expired air volume is measured or collected for volumetric gas analysis. This gas volume is corrected for standard conditions and is analyzed for its oxygen and carbon dioxide content; subsequently oxygen consumption and carbon dioxide production will be calculated. (2) In the closed circuit method the animal is completely cut off from the outside air and breathes through a closed system. The respirometer originally contains pure oxygen. As the gas is expired by the subject, the carbon dioxide is continually removed as it passes through soda lime. The decrease in the gas volume in this closed system is related to the rate of oxygen consumption, from which the metabolic rate is then calculated.

Since the indirect calorimetry methods do not measure heat production directly, the heat equivalents of the oxygen consumed and the carbon dioxide produced must be determined. The heat produced when one consume 1 liter of oxygen varies with the foodstuff burned as fuel. The ratio of the volume of carbon dioxide produced to the volume of oxygen consumed is called the respiratory quotient, RQ.

## OPEN CIRCUIT

Open systems for indirect calorimetry avoid some of the problems of having an animal breathe pure oxygen and of absorbing the $CO_2$. The open circuit principle is to move the air through a chamber in which the animal's head, face, or body is enclosed, measuring the differences in quality of the air between the input and output parts of the system.

## MASK METHOD

Figures 3.3 and 3.4 illustrate the measuring of metabolism with the mobile open circuit respiratory exchange apparatus used for short-term measurements in the Missouri Climatic and Shelter Engineering Laboratory (Kibler, 1960). Exhaled air from the animal is directed by one-way valves from the mask to a 100-liter rubberized mixing bag. A small vacuum pump (Figure 3.4, upper right corner) draws a small part of the air from the bag (about 3 liters/min) and passes it over a dew point indicator and through a gas drier, a collection tube, and a small gas meter. This pathway provides a gas sample for laboratory analysis and data that can be used in determining respiratory vaporization rate. The major part of the exhaled air that enters the mixing bag follows

**FIGURE 3.3 Measuring metabolism with the mobile open circuit respiratory exchange apparatus. (From Kibler, 1960.)**

another pathway through the large gas meter and is expelled back into the room by a larger pump. The sum of the volumes recorded by the two gas meters during a timed interval measures the volume rate of exhaled air or the ventilation rate. All air volumes are corrected to dry air volumes at standard temperature and pressure (0°C, 760 mm Hg) by use of a psychrometric chart or table. The samples of exhaled air are analyzed in the gas analysis laboratory in Figure 3.5.

The gas analyzers used with the open circuit metabolism apparatus are also shown in Figure 3.5. A magnetic susceptibility analyzer indicates the percentage oxygen in inspired (room) and expired air in the ranges 0–25 and 16–21 percent. Infrared-type analyzers indicate the carbon dioxide (0–5 percent) and methane (0–1 percent). The analyzers are connected in series so that values for all three gases can be read at one time from one gas sample. The analyzers are first evacuated by means of the high-vacuum pump shown behind the analyzers. Gas is then admitted from the 250-ml collection tube attached to the oxygen analyzer. Additional gas is displaced as needed from the collection tube and passed through the analyzers by raising the mercury bulb attached to the bottom of the collection tube. Readings are then made on all

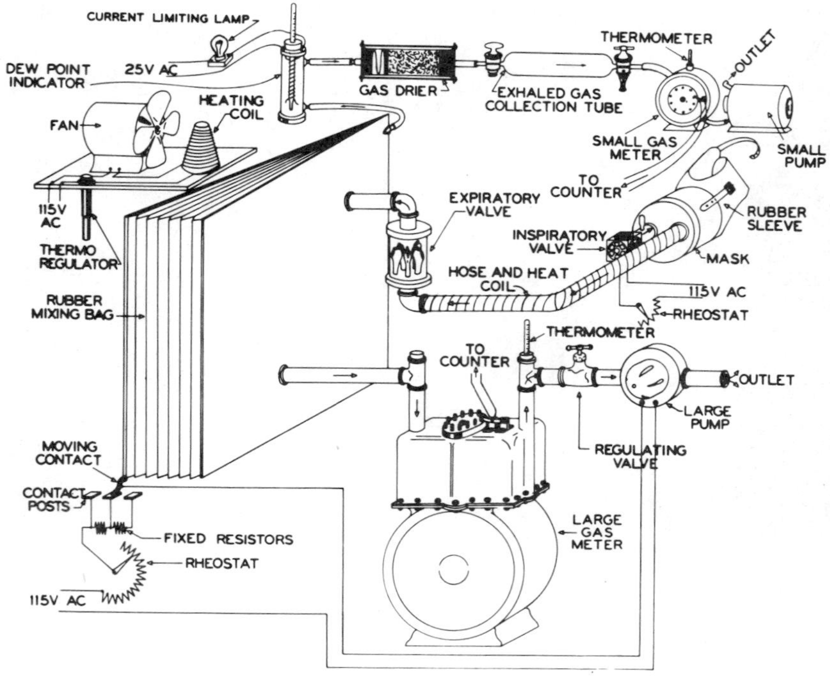


**FIGURE 3.4 Energy metabolism and related thermoregulatory reactions in Brown Swiss, Holstein, and Jersey calves during growth at 50° and 80°F temperatures. (From Kibler, 1960.)**

analyzers. Gases of known concentrations are used to calibrate the analyzers. For an alternative method of transferring a gas sample from tube to analyzer, see Kibler (1965).

The thermal equivalent (TE) calories per liter of $O_2$ at the computed RQ can be obtained from a table in Brody (1945). This may be used to calculate energy metabolism as follows: energy metabolism, kcal/hr = $O_2$ consumption liter/min (STP) × 60 × TE.

The working with rats, hamsters, and other small animals, enclosures for the whole animal are used in a closed or open circuit apparatus. For short-term measurements of oxygen consumption, a closed circuit apparatus is convenient. For continuous measurements, an open circuit apparatus with recording analyzers is used.

**FIGURE 3.5 Electronic gas analyzer used at the University of Missouri. (From Kibler, 1960.)**

### CLOSED CIRCUIT

The closed circuit mask method has been used for many years for laboratory animals and still is applicable because of its simplicity. However, it is gradually being replaced by more elaborate and easily operated open and closed circuit systems.

Figure 3.6 describes the principal type of closed circuit mask system for cattle and other species ranging from rabbits to elephants. It may be adapted to any trainable domestic animal.

In measuring energy metabolism by the closed circuit, spirographic mask method, an oxygen spirometer is filled from a cylinder of oxygen. As the spirometer fills, the counterweighted bell or lid rises. An animal is connected to the spirometer by means of a mask and hose. As the animal uses oxygen, the bell falls. The respiratory air flow is directed by unidirectional valves through soda lime, a process that removes the carbon dioxide produced by the animal. The fall of the bell is recorded graphically on a kymograph drum which also has time marks every minute. The rate of oxygen consumption is computed from the slope

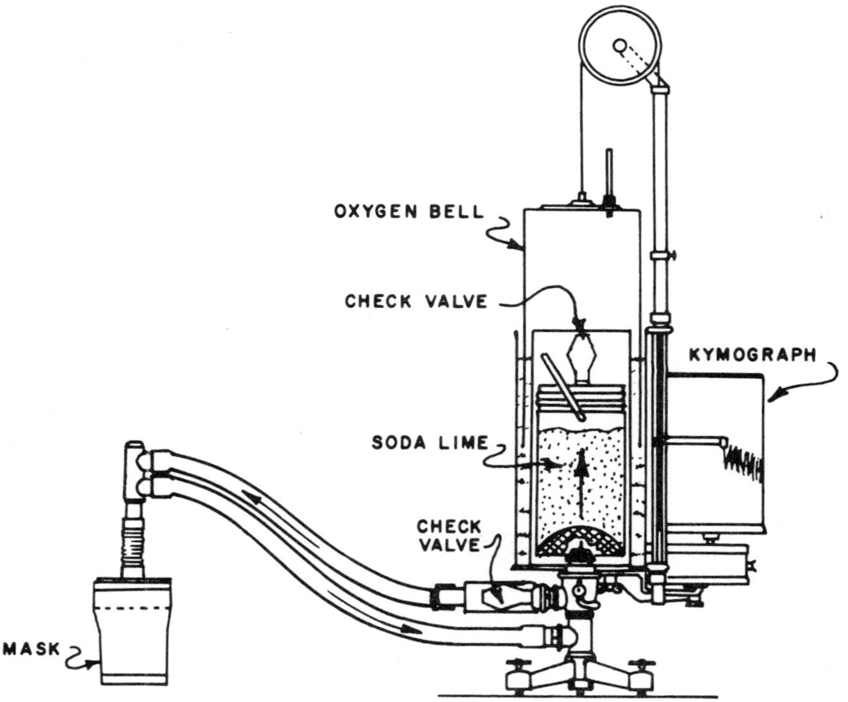


**FIGURE 3.6 Clinical metabolism apparatus used by Benedict-Roth-Collins. (From Brody, 1945.)**

of the graphic record and the volume calibration factor of the spirometer employed. It is assumed that the soda lime maintains a relative humidity of 80 percent. The temperature is indicated by a thermometer, and the atmospheric pressure is measured by a barometer. The measured rate of oxygen consumption is adjusted to STP conditions by the use of a psychrometric chart or table (Brody, 1945). As the RQ is unknown, the thermal equivalent (TE) of oxygen in kcal/liter $O_2$ is assumed to be approximately 4.7 for a fasting animal or 4.8 for a normally fed animal.

The *continuous* measurement of respiratory gas exchange from an animal is the most effective way to determine instantaneous changes in the metabolic pattern. The respiratory exchange chamber (Figures 3.7 and 3.8) is supplied with air through a distribution plenum by a low-volume, high-pressure fan.

The advantages of a continuous system include (1) the potential to

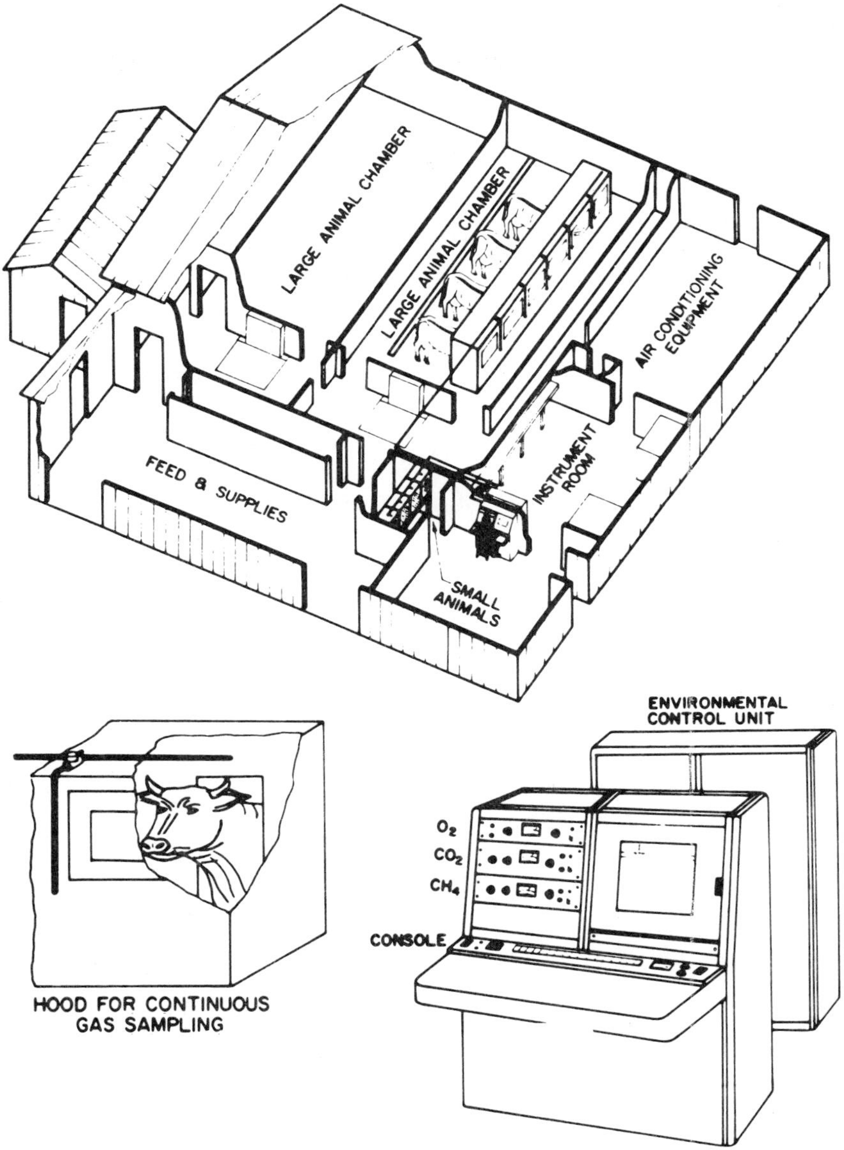


**FIGURE 3.7** Floor plans of University of Missouri Climatic Laboratory with inserts of head box and respiratory exchange apparatus.

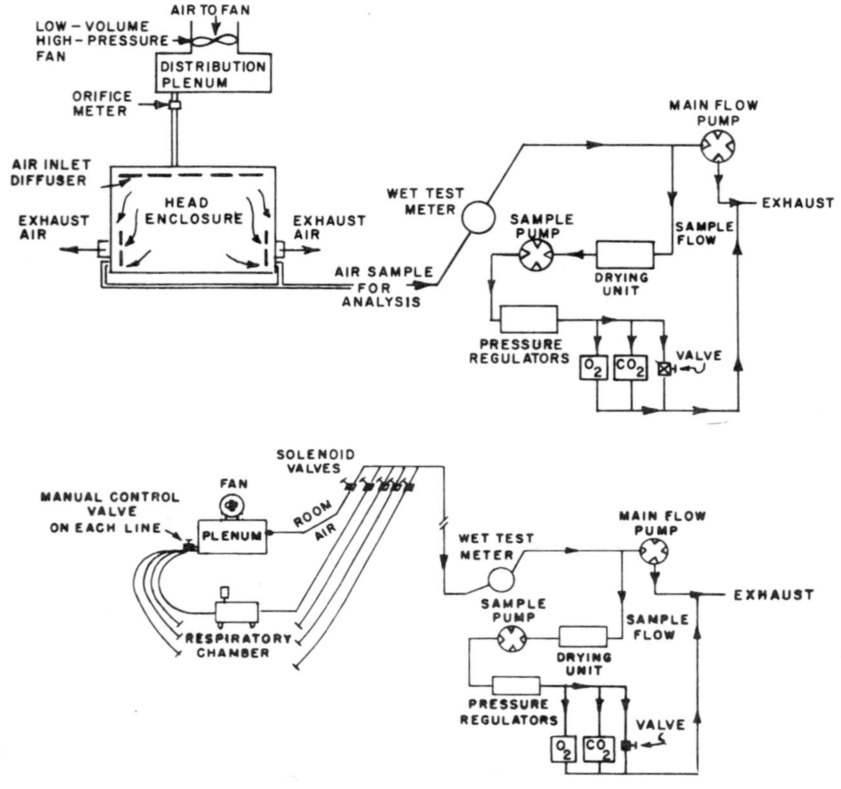


**FIGURE 3.8 Schematic diagrams showing component parts of respiratory change apparatus for continuous analyses.**

determine reaction times of hormones, metabolic stimulants or depressants, and environmental influence on heat production; (2) estimation of circadian rhythms of heat production; and (3) ability to monitor the continuous analyses that are uniquely applicable to the time phasing of biological reactions that occur during environmental changes.

The California respiration apparatus described by Kleiber (1958) is also of the open circuit type.

### COMPARISON OF THE OPEN AND CLOSED CIRCUIT MASK METHODS

An obvious advantage of the closed circuit mask method is its simplicity and economy—no expensive gas analysis equipment is needed. Several disadvantages are inherent in its use, however. (1) It furnishes

$O_2$ consumption only—not $CO_2$, $CH_4$, or $H_2O$ vapor. (2) The $CH_4$ exhaled into the spirometer by ruminants is not absorbed, and its accumulation introduces an error in the $O_2$ consumption reading. (3) The temperature of the spirometer may rise several degrees during a test, causing increased discomfort to the animal at warm environmental temperatures. (4) The efficiency of the soda lime as a $CO_2$ and $H_2O$ absorbent may change with temperature, causing a change in the humidity of the inspired air. The open circuit method, in addition to furnishing complete respiratory exchange data, also allows the animal to breathe normal room air.

The mask method, whether closed or open circuit, of course, is not appropriate for continuous measurements. The animals require freedom to eat and drink, and tend to become irritable if restricted by a mask for a long period. The mask method is not appropriate either for small laboratory animals such as mice.

Many laboratories throughout the world employ closed or open circuit procedures for cattle (Blaxter, 1966).

A description of open circuit and closed circuit respiration chambers and of gas analysis apparatus for calves and sheep has been given by Blaxter *et al.* (1954; Blaxter, 1962).

At the Rowett Institute a gradient-layer calorimeter suitable for sheep and swine has been in use for at least 10 yr (Pullar, 1958).

Calorimetric studies with sheep at the Australian Council of Scientific and Industrial Research Organization (CSIRO) Sheep Biology Laboratory has been reported by Alexander (1958).

For more detail on pigs refer to Mount (1968).

In the United States the respiration calorimeter in the USDA's Dairy Section at Beltsville, Maryland, is currently the most elaborate open circuit indirect system in operation for nutritional studies. The indirect, open circuit calorimetry recording system for cattle and smaller animals in the Environmental Physiology Section in the College of Agriculture at the University of Missouri is currently actively investigating environmental metabolic processes. The dynamic state of a metabolizing individual's caloric intake in relation to body weight becomes important under certain environmental conditions. Computation of the caloric content of the ration and measuring the quantity of feed consumed in relation to body weight is desirable.

## Measurement of Body and Surface Temperatures

Body temperature is measured at different locations. For rectal temperature, a clinical thermometer, a resistance thermometer, or thermocouple may be used. Findlay (1960) has discussed their use, together with errors that must be guarded against (see also Chapter 2).

Indwelling thermistors or telemeters (Slater, 1962) are also useful for determining deep body temperature. They have the advantage over mercury in that they may be used for continuous recording. Methods of monitoring temperature are discussed by Calatayud *et al.* (1965).

For temperature of extremities and skin, usually either a very fine thermocouple or a thermopile is used. A radiometer is one such instrument for determining skin temperatures of man that was adapted for use with animals (Kelly *et al.*, 1949) (see also Chapter 2). Another method of determining temperature differences of parts of the bodies is the use of infrared radiometry (Veghte and Solli, 1962).

Maintenance of "normal" body temperature is a characteristic of homeothermic animals. The measurement of temperature is subject to certain experimental errors and interpretation. Normal values for body temperature have been compiled by Altman and Dittmer (1966). In some animals such as birds there is a diurnal change in temperature of several degrees from being at rest and active. Thus, hypothermy and hyperthermy must be related to the species under study.

Simple thermometers, such as mercury rectal thermometers, were used to a greater extent in the past. Thermocouples, thermistors, and resistance thermometers are in general use. Because of the possibility of hyperthermy being produced in the animal by excessive muscular activity, as in struggling, or of producing hypothermy as with pigeons and rabbits, the use of transmitters for body temperature measurements is now popular. Radio telemetry permits the animals to be unrestrained and not conscious of being under observation.

The topography of temperature gradients has been shown to vary from the cooler extremities and less vascular areas, such as hooves and beaks, to the deep body or core temperature. Differences in internal temperature of indwelling transmitters may be related to their location. The temperature of the hypothalamus reflects changes in environmental temperature faster than the core temperature. The blood supplying the hypothalamus or thermostat may rapidly reflect a change such as being in hot water or drinking ice water.

Surface temperatures may be measured with dermal radiometers; see Chapter 2. Equipment for measuring the temperature of different parts of the body by means of infrared radiometry is available (Veghte and Solli, 1962).

Indices have been proposed for assaying heat tolerance of animals under both field and laboratory conditions. In a way this parallels the integration of multivariables discussed in Chapter 2. The idea of "equivalent" environment that has been used includes effective temperature, mean radiant temperature, and operative temperature. In addition, McDowell (1966) has included indices of adaptability. In this area he has discussed the following: a heat tolerance test based on rectal temperature, proposed by Lee and Phillips (1948); the lines of equal effect based on rectal temperature, respiration rate, and respiration volume, proposed by Barrada (1957); ratios of evaporation based on skin and respiratory evaporation of two temperatures, proposed by Yeck and Kibler (1956); and a 6-hr hot room test studying rectal temperatures and respiration rate, proposed by McDowell *et al.* (1953). In addition, McDowell (1966) discusses field tests used in assessing the adaptability of cattle to hot environments.

Guidry and McDowell (1966) describe an adaptation for cattle of the Benzinger tympanic temperatures. By the use of semiflexible transistor probes, they sensed the temperature near the tympanic membrane and the rectal temperature and then compared the two. This work shows that, in cattle, temperature sensed at the tympanic membrane may be more suitable than rectal temperature for determining the speed of response to both internal and external temperature changes.

Bianca (1961, 1965) reviews research on temperature measurements as indicators of domestic animal responses to thermal stress. His references are recommended as information sources.

## BODY COMPOSITION AND VAPORIZATION

The techniques used for measuring body composition were reviewed, critically and systematically, at a conference held in 1959 at the Quartermaster Research and Engineering Center, Natick, Mass. (Brozek and Henschel, 1961). Participants described three approaches in detail: (1) direct body measurements and radiographic analysis; (2) determinations of body volume and calculations of body density; and (3) measurement of fluid volumes, electrolyte concentrations, and metabolic balances. A more recent reference (Reid *et al.*, 1968) gives some peculiarities in the body composition of animals.

### Fluid Volumes

The measurement of fluid volume is based on a dilution principle and utilizes a substance that distributes itself throughout the cells or extracellular areas to be measured.

## TOTAL BODY WATER

Total body water is measured by determining the distribution of an appropriate indicator introduced by a single intravenous injection. The substance must be able to distribute itself uniformly in all fluid spaces. The most reliable and common indicators in use today seem to be antipyrene and the heavy isotopes of water, deuterium oxide and tritium oxide. Both sulfanilamide and thiourea have been tried (Danowski, 1944; Painter, 1940), but they were unsatisfactory because of unequal distribution.

## ISOTOPE TECHNIQUES

Pace *et al.* (1947) describe the use of tritiated water. For tritium determination, they devised a counting assembly that involves the introduction into a Geiger tube of tritiated water as water vapor at low pressure, the emission being counted directly. Total body water can be calculated by dividing the total amount of labeled water administered (after correction for urinary excretion) by the equilibrium concentration in plasma water.

A procedure described in detail by Deane (1952) uses a radioactive substance injected intravenously.

For deuterium concentration analysis, Lilly (1950) gives details.

## ANTIPYRENE METHOD

The antipyrene method is a valuable contribution for the determination of total body water because of the simplicity of analytical procedure and the fact that multiple studies are economically feasible (Brodie, 1951; Soberman *et al.*, 1949).

## EXTRACELLULAR FLUID

Sucrose, thiosulfate, and inulin are the preferred indicators. Some disagreement exists over exactly what space these substances measure. A method of continuous infusion for measurement of extracellular fluid is described by Schwartz *et al.* (1949). Inulin determination in plasma and urine (Schreiner, 1950) depends on the reaction of resorcinol with inulin. A method of calibrated infusion described by Deane *et al.* (1951) uses a constant intravenous infusion of sucrose without a priming injection.

#### INTRACELLULAR FLUID

The intracellular water can be estimated as the difference between total body water and the total extracellular water.

### Vaporization

The dissipation of heat by respiration, surface diffusion, or sweating has been measured rather extensively in cattle and, to a lesser extent, in poultry, swine, and other domestic animals. Moisture is vaporized from the exterior surface and from the respiratory tract. Two methods (with modifications) have been used to measure total, or whole animal, vaporization: (1) insensible loss measurements, and (2) psychrometric measurements. Respiratory vaporization is measured by using a mask to collect the expired air. Surface vaporization from the whole animal can be obtained by subtracting the respiratory vaporization from the total vaporization. Vaporization from small areas of the surface is measured by some form of capsule technique.

#### TOTAL VAPORIZATION

Insensible weight loss is a measurement of the difference between the weight of gases and vapors given off by the body and those absorbed by the body. It is measured by observing the rate of loss of body weight on a scale. Insensible loss = $H_2O + [(CO_2 + CH_4) - O_2]$, where the bracketed term represents the metabolic weight loss factor. Total vaporization = insensible loss – metabolic weight loss.

The insensible weight loss method of measuring total vaporization (Thompson *et al.*, 1949), because of technical problems when large animals are involved, is not currently being used.

#### PSYCHROMETRIC MEASUREMENTS

Yeck and Kibler (1956) report a method of measuring total vaporization with a hygrometric tent. The animal is placed inside a polyethylene enclosure through which air is circulated. The volume of air passing through the enclosure is measured by determining the pressure drop across an orifice. The moisture content of the air entering and leaving the tent is measured by lithium chloride dewcells. The change in moisture content of a measured volume of air in passing through the tent in a specified period of time measures the total vaporization rate of the animal. Separation of respiratory evaporation and total surface evaporation in sheep may be done by the method of Guidry and Hofmeyr (1968).

Respiratory vaporization is determined by an apparatus designed to measure the volume rate and moisture content of the expired and inspired air (Kibler, 1960). As indicated in Figure 3.4, the exhaled air is directed by one-way valves into a rubberized mixing bag. A small part of the exhaled air (3 liters/min) is drawn over a dewpoint indicator and through a gas drier, a collection tube, and a gas meter. A larger meter measures the major part of the exhaled air. All volumes are corrected to standard temperature and pressure. The respiratory vaporization rate is computed as the difference in moisture content, in grams per hour, between the expired and inspired air.

In the apparatus described above, the moisture content of air is determined by dewcells, and the drier was used only to obtain dry air samples for metabolic measurements. However, in earlier work, gravimetric moisture determinations were made by using a separate drier tube which was weighed before and after each measurement (Kibler and Brody, 1950).

Surface (or cutaneous) vaporization for the whole animals can be determined as the difference between the total vaporization and the respiratory vaporization. Both measurements should be made under the same environmental conditions and as close together in time as possible.

## MEASUREMENT OF SWEATING

Six standard methods have been described (Robinson and Robinson, 1954) for measuring sweating: (1) determination of weight loss of the subject; (2) collection of water vapor from evaporative sweat; (3) direct collection of liquid sweat; (4) direct observation by microscope of droplets from sweat pores; (5) use of color indicators on the skin or by imprint of an absorbent paper; and (6) measurement of changes in skin resistance to galvanic current.

Cutaneous vaporization measurements for small local areas on the animal are reported by several workers. McDowell *et al.* (1954) employed a closed circuit system using a salt solution for absorbing water vapor from air circulated through an inverted capsule cemented to the shaved skin of the animal. McLean (1963a,b) has called attention to the discrepancy between using skin capsules and indirect method for determining moisture vaporization from the skin of cattle. He believes the values of McDowell *et al.* (1954) to be too high for technical reasons. He discusses the geometry of the capsule, the physics of humidity in air, and ventilation rate as related to the rate of evaporation from the skin. Berman (1957) determined the color change in chemically treated strips of paper placed on the skin under inverted transparent strips.

Brown and Motasem (1965) obtained a significant correlation between the salts deposited under a covered area on the skin and the vaporization rate from the same area.

Kibler and Yeck (1959) used an open circuit system employing an inverted cup, a dewcell, a meter, and a vacuum pump (Figure 3.4). As shown, the open-bottomed collector is placed over an animal surface. Lithium chloride dewpoint indicators measure the difference in moisture content of the air before and after it passes over the enclosed area. Two readings are made at each position. The difference between these two readings is then used in computing the evaporation rate from a known area of surface.

Ohara (1964) describes a gravimetric method for determining weight increases in desiccant material in an absorption capsule and a more refined method of automatic recording of electrical conductance.

## LIVER AND LIVER FUNCTION

Basically, the function of the liver is to help regulate the addition, removal, and alteration of the individual constituents of the blood by means of metabolic and excretory products. Numerous function tests are available to indicate hepatic dysfunction. The selection of one as a reliable index of function under all conditions is subject to severe limitations; consequently, tests must be selected after consideration of the problem in question (Howe, 1960).

Methods of measuring liver function by indirect blood flow have been reviewed in *The Handbook of Physiology, Circulation* by Bradley (1963).

Tests for liver function in domestic animals have been reviewed and the results of various tests and methods summarized by Benjamin (1961).

### BROMSULPHALEIN (BSP) DYE EXCRETION TEST

The principle of the bromsulphalein test is based on the removal of dye from the blood by the reticuloendothelial cells and subsequent excretion of this dye by the parenchymal cells of the liver. The amount of dye retention in the blood serves as a measure of the degree of hepatic disease and blood flow.

In the BSP liver function tests the dye removal from the plasma is expressed as a percentage of the injected BSP present in the plasma, i.e., the percentage retention at a given time interval, which is usually 30–40 min after a single intravenous injection of 2–5 mg of dye per kilogram of body weight. After complete mixing of BSP with plasma, the rate of disappearance of the dye from the plasma with time is a function of the original concentration. By plotting the log of the plasma BSP concentration against time, a straight line can be fitted against the initial points. The slope of this line gives the BSP disappearing constant. For detailed procedures, see Benjamin (1961).

A discussion of the modifications necessary for liver function tests before they can be used in domestic animals has been given by Cornelius and Kaneko (1963). A review on liver circulation and function by Brauer (1963) is an excellent source of information on this topic.

### GLYCOGEN DETERMINATION

Tissue glycogen is commonly determined by either the anthrone reagent or by the acid hydrolysis method. The anthrone method is simple because the hydrolysis of glycogen and the development of color, which is measured photometrically, occur simultaneously (Seifter *et al.*, 1950).

In acid hydrolysis, tissue glycogen is hydrolyzed to glucose, which is subsequently determined by any of a variety of methods after the neutralization of the hydrolysate. Highly reliable results have been obtained by mild hydrolysis followed by determination of the resultant glucose with glucose oxidase (Johnson, 1963).

Kemp and Van Heijningen (1954) report a simple method using concentrated acid to produce color, which is measured directly.

Glucose originally present in the tissue will contribute to the color reactions; thus the glycogen content will be overestimated. As glucose concentrations in tissue are normally relatively low, this overestimate can either be disregarded or it can be determined separately by extra steps in the procedures.

### METABOLISM AND EXCRETION OF BILIARY PIGMENTS

The theory of bile pigment metabolism, briefly, is as given in the following steps: (1) Hemoglobin is released when erythrocytes are destroyed in the reticuloendothelial system; (2) hemoglobin is broken down to iron, globin, and bilirubin; (3) bilirubin is released into the blood and is called free bilirubin or "indirect-reacting" bilirubin; (4) free bilirubin is conjugated in the liver with glucuronic acid to form bilirubin glucuronide (conjugated bilirubin, "direct-reacting" bilirubin), which is excreted with the bile into the small intestine; (5) bacterial action in the small intestine reduces bilirubin glucuronide to urobilinogen; (6) some of the urobilinogen is oxidized to urobilin, which, with some other bilirubin pigments, is responsible for the color of normal feces. Part of the urobilinogen is absorbed into the portal blood system and conveyed to the liver, where most of it is excreted into the bile, thus completing the enterohepatic circulation of the bile pigments. A small amount of absorbed urobilinogen that is not extracted from the plasma by the liver reaches the kidney and is excreted in the urine.

Measures of serum bilirubin are the Icterus index, Van den Bergh test of urine bilirubin (azo-pigment conjugate of glucuronic acid), and the Icotest. For reliable detailed procedures, see Benjamin (1961).

### ALKALINE PHOSPHATASE

Serum alkaline phosphatase is found in appreciable amounts and probably forms in the osteoblasts, which apparently release the enzyme into the blood. Some alkaline phosphatase is normally excreted in the bile, and therefore the serum activity of alkaline phosphatase is said to increase as a result of interference with bile flow. There is evidence that, under abnormal circumstances, especially biliary stasis, the liver forms excessive alkaline phosphatase, or one of its activators, and releases it into the blood.

The principle underlying the Phosphatab test (Warner-Chilcott) is the use of a buffered phenolphthalein phosphate substrate that remains colorless until decomposed by the phosphatase enzyme in the serum. The amount of phenolphthalein liberated is determined by a color change and is a measure of alkaline phosphatase activity (Benjamin, 1961).

### LIVER PROTEIN METABOLISM

Albumin, prothrombin, and fibrinogen are probably formed only in the liver. Globulin is formed in the liver, but to a far greater extent in extrahepatic tissue. For total protein, albumin, globulin, and prothrombin procedures, see Benjamin (1961).

### CHOLESTEROL

Liver is the main source of plasma cholesterol, although most tissues are able to synthesize it. Cholesterol is secreted with bile, but is also destroyed in the liver by conversion to cholic acid (Benjamin, 1961).

### LIVER BIOPSY

A technique for obtaining liver biopsies in cattle, by use of a simplified instrument, has been described by Hughes (1962). The instrument employed incorporates a neoprene (O) ring near the base of the cannula. The trocar fits snugly within the ring of the cannula, thus eliminating the need for disengaging the trocar in order to attach a syringe to obtain the necessary vacuum for aspiration (Hughes, 1962).

### FUNCTIONAL HEPATECTOMY

A procedure for producing functional hepatectomy in unanesthetized birds has been described by Ranney *et al.* (1951). This consists of tightening two exteriorized ligatures previously placed around the afferent hepatic blood vessel. Since the ligated liver was left *in situ* in the bird, its functional activity was tested. It was found that the circulation of the ligated liver was reduced 99 percent. This method for functional hepatectomy of the relatively untraumatized, unanesthetized fowl is useful in studying blood glucose and other metabolites.

In contrast to ligation of the hepatic afferent blood vessels, permanent constriction of the posterior vena cava caudal to the hepatic vein allows collection of portal and hepatic venous blood in conscious sheep (Moodie *et al.*, 1963).

### LIVER PERFUSION

Procedures for rat liver perfusion have been described by Hems *et al.* (1966). The rate of gluconeogenesis from most substrates can be tested in livers so perfused.

## KIDNEY FUNCTION

In studying reabsorption or secretion of any substance, it is necessary to know the rates at which it is filtered and excreted. Several methods are available for measuring the clearance of these variables. Earle and Berliner (1946) describe an infusion technique utilizing inulin for measuring renal function without either bladder catheterization or collection of urine. (To avoid pyrogenic reactions, the inulin must be specially processed.) See Shannon and Fisher (1938) for determination of glucose reabsorption.

### Quantification of Tubular Reabsorption

- Reabsorption rate equals filtration rate minus excretion rate.
- Secretion rate equals excretion rate minus filtration rate.
- Filtration rate equals glomerular filtration rate times plasma concentration.
- Excretion rate equals urine concentration times urine flow.
- All variables must be measured simultaneously.

Many, but not all, substances secreted are partly bound to plasma proteins and, consequently, are incompletely filterable through glomerular capillaries. For a substance bound to protein, the rate of filtration is the product of the freely filterable fraction of the substance present in plasma and the rate of glomerular filtration. The concentration in the ultrafiltrate then should be related to the water in the plasma. Protein binding of secreted substances varies widely: phenol red is about 80 percent bound, 20 percent free; PAH (*p*-aminohippurate) is 10–30 percent bound and 70–90 percent free (Shannon, 1939).

If the substance is filtered, reabsorbed, and in part secreted, exact quantification of the rate of secretion is impossible. One can only calculate net secretion or net reabsorption. However, in some instances it is possible to block either reabsorption or secretion with a transport inhibitor and gain some idea of the time magnitude of the process.

### Measurement of the Renal Blood Flow

The use of the PAH (*p*-aminohippurate) clearance in measuring minimum effective renal plasma flow, described by Barker and Clark (1960), is an extension of the well-known Fick principle. The difficulty in employing the Fick method for the measurement of renal plasma flow is that of obtaining renal venous plasma. In the dog, one kidney can be explanted subcutaneously in the flank and a skin tube constructed around the renal vein. Renal venous blood can then be obtained by venipuncture with minimum disturbance to the animal.

Arterial blood can, of course, be obtained by puncture of any peripheral artery, for its composition is the same throughout the body. Obviously such methods are inapplicable to man. However, it is possible to introduce a radiopaque cardia catheter into a peripheral vein, maneuver it into the inferior vena cava, and then direct it into a renal vein under fluoroscopic observation. The technique, though, is unsuited for routine clinical studies.

Techniques involved in direct measurement of renal blood flow are by Selkurt (1948). The measurement of venous outflow and arterial inflow, utilizing direct cannulization and the use of flow meters, e.g., the rotameter, orifice meter, or bubble flow meter, are described in this publication. One of the desirable attributes of a directly recording flow meter is that it offers minimal resistance to blood flow so that nearly normal pressure relationships would be obtained whether it is in series with an artery (arterial inflow method) or a vein (venous outflow method). Although the optically recording bubble flow meter is unsurpassed for simplicity and precision, it suffers from the disadvantage of not continuously recording and, in the range of renal blood flow, causes considerable pressure drop because of its resistance. The latter makes it unsuitable for venous outflow measurement. To obtain continuous recording of renal blood flow with minimal pressure loss, use has been made of a larger optically recording rotameter.

The intensity of urinary acidity is a striking indication of the magnitude of one of the fundamental regulating activities of the organism. The pH of the urine can be determined by indicators; on the average, it is 6.00 and varies from 4.82 to 7.45, according to Henderson and Palmer (1912). They estimated the urinary pH after the addition of indicators by matching colors of urinary samples with those of standard solutions ($NaH_2PO_4$, $CH_3COOH$, and $CH_3COONa$) of known reactions.

In determining the titratable acid of the urine, it is usual to adhere to one indicator, e.g., phenolphthalein, and to give the acidity in terms of decinormal acids naming the indicator used. The total acid set free in the body per day is equal to about 700 ml 0.1 *N* acid, but some is neutralized and excreted as ammonium salts (= 400 ml 0.1 *N* acid daily).

The rest, the titratable acid, amounts to about 300 ml 0.1 *N* acid daily. The titratable acidity of urine is always much greater than the hydrogen ion concentration would indicate. By titrating, first with methyl red and an acid, and then with phenolphthalein and an alkali, a useful measure of the urine buffers can be obtained.

## ENDOCRINE FUNCTION

The animal response to a particular environment, via the central nervous system, is usually rapid and is brought about by behavioral, cardiovascular, and respiratory activity. Changes in endocrine function are slower in response, involve enzymic and metabolic changes, and may indeed take days or weeks before the full response to the environment is evident. If an animal touched a hot object, a neuromuscular reflex would follow. On the other hand, if day lengths became noticeably shorter, circulating gonadotropic hormones would become less, and eventually a chicken would respond. If the animal were a sheep, the circulating hormones would be increased and the breeding "season" would begin.

Because of the interaction of several hormones in relation to physiologic responses to environmental factors, it is not always easy to state which endocrine should be investigated. However, it is fairly obvious that, in the study of water exchange, the concern should be with the releasing factor of the brain affected by osmolarity; but the antidiuretic hormone of the posterior pituitary vasopressin is related as well as ACTH and the adrenal's aldosterone.

Reproduction involves studies of the central nervous system, hypothalamic thermoregulation centers, and hypothalamic hormonal releasing factor, as well as the endocrine glands: thyroid, pancreas, adrenals, and pituitary. Adequate food intake necessary for reproduction involves the hypothalamic control of appetite and thirst, and proper enzymes and substrates for adequate metabolizing of nutrients.

Thermoregulation can become very complex and involves metabolism and the endocrines: the sympathetic nervous system's 5-HT; norepinephrine; the adrenal's epinephrine; the $\alpha$-cells of the pancreas and the glucagon that they produce; the hypothalamic releasing factors; and the tropins TSH (thyroid stimulating hormone), STH (somatotropic hormone), and ACTH and their effect on the secretion of thyroxin and the corticosteroids. Acclimatization to cold is associated with hypertrophy of the adrenals, thyroid, liver, lungs, kidney, intestines, and heart. These and other endocrine-associated changes in metabolism and cellular functions are reviewed by Smith and Hoijer (1962).

The hormones may be defined as specific substances secreted by particular organs into the general circulation, which carries these substances to their sites of action elsewhere in the body. Here they regulate the rates of specific processes without contributing significant amounts of energy or matter to the tissues.

The final stage in the development of the modern view of endocrine function was its linkage to Claude Bernard's concept that the internal environment is maintained relatively constant or under extreme environmental conditions may compensate for maintenance of the organism.

## Assay Methods

The general principles of assay methods for hormones and their metabolites can be divided into three maingroups—biologic, chemical, and immunologic. The use of immunologic techniques is a relatively recent development. Bioassays can often be more specific than chemical assays. However, the latter are generally more precise, more sensitive, less laborious and expensive, and therefore more suitable for routine use. In the case of the steroid hormones and their metabolites, the past 10 yr have witnessed the virtual replacement of bioassays by chemical procedures for quantitative determinations of body fluids. Immunologic assays are gradually replacing bioassays because of their greater practicality in the measurement of protein and polypeptide hormones, epinephrine, norepinephrine, and thyroid hormones. This is a rapidly progressing field of research.

At the time of writing, biologic or immunologic methods must be used for the estimation of pituitary gonadotropin (PG), chorinonic gonadotropin (CG), thyroid-stimulating hormone (TSH), adrenocorticotropic hormone (ACTH), growth hormone (GH), prolactin, thyrocalcitonin, posterior pituitary hormones, parathyroid hormone, insulin, and glucagon. The determination of these hormones in blood and urine by chemical methods is either impossible or impractical at present. Chemical assay methods should now be employed for the determination of estrogens, progesterone and its metabolites, corticosteroids, and 17-oxosteroids.

The essential requirements in hormone assay procedures have been reviewed by various workers (Gray and Bacharach, 1967; Loraine and Bell, 1966; Zarrow *et al.*, 1964). It is generally agreed that the efficiency of a given method depends on two main factors—*reliability* and *practicability*.

One of the main difficulties inherent in the estimation of many hormones in body fluids stems from the fact that available assay methods are not sufficiently sensitive to detect very small amounts of material. One of the most pressing needs in the general field of hormone assay is the development of techniques of sufficient sensitivity to permit quantitative determinations. This need is being satisfied by immunologic and radiochemical methods in the case of most of the protein and steroid hormones, respectively.

The two main types of immunologic methods are reactions depending on hemagglutination or hemolysis inhibition and radioimmunologic assays. Reactions of the first type usually lack sensitivity and specificity (Hays *et al.*, 1967; Selenhow *et al.*, 1967). The double antibody radioimmunoassay has been reviewed (Berson and Yalow, 1964; Hays *et al.*, 1967; Loraine and Bell, 1966; Selenhow *et al.*, 1967; Wright and Taylor, 1967). The principle depends on the competition between radioactively labeled and unlabeled hormones for the specific binding sites on the antibody. The higher the concentration of the unlabeled hormone (Ag), the less radioactive labeled hormone ($^{125}$I-Ag) will be bound to the antibody. Thus, radioimmunoassay depends on the ability of labeled hormone (as a standard, as serum or urine samples, or pituitary homogenates) to inhibit competitively the binding of radiolabeled hormones by a specific antibody. The percentage of total radioactivity present and bound to the antibody (precipitate count) is expressed as percent radioactivity in antibody-bound hormone. A standard curve is prepared by plotting this variable (the ordinate) against the concentration of the hormone in a standard preparation. The concentration of hormones in unknown samples can be read off the standard curve directly, and the sensitivity of the assay can be high enough to detect as minute a quantity as 50 pg ($10^{-12}$ g) of hormone per milliliter of the material to be assayed.

With certain hormones, for example, corticotropin (Berson and Yalow, 1968), oxytocin (Glick *et al.*, 1968), and prolactin (Bryant and Greenwood, 1968), hormone concentration in terms of a few picograms or a fraction thereof has been measured.

In radioimmunoassay it has been assumed that (1) radioactively labeled and unlabeled hormones behave identically; (2) the circulating hormone is chemically and immunologically identical with the standard hormone preparation; (3) the amount of hormone bound by the antibody is dependent on the concentration of the hormone; and (4) different antisera made against a particular antigen will give a similar quantitative response in the immunoassay system. An outline of the method is shown in Table 3.1.

The purpose of mentioning the periods of incubation (Table 3.1) with first and second antibody is to indicate that the first antibody invariably needs a longer incubation than the second antibody. For example, Garcia and Geschwind (1968) incubated rabbit growth hormone antiserum generated in monkeys for 6 days, while incubation with the

**TABLE 3.1 General Radioimmunoassay Procedure for Measuring Protein and Peptide Hormone Concentrations in Blood**

0.05–0.1 ml unlabeled hormone (standard or serum)
0.05–0.1 ml labeled hormone (0.05–0.2 ng)
0.05–0.1 ml antiserum (diluted so as to get 50–60% binding of labeled hormone)

↓

First reaction: Incubate at 4°C for 3–7 days

↓

0.05–0.1 ml anti-γ-globulin antiserum
0.05–0.1 ml normal serum (both of these sera should be from the same animal in which the antiserum against the hormone is produced)

↓

Second reaction: Incubate at 4°C for 1–2 days
Centrifuge at 3200 rpm for 50 min at 4°C

↓

| Supernatant count | Precipitate count (PPT) |
|---|---|

Calculations: $\%\text{PPT} = \dfrac{\text{CPM}^a\ \text{PPT}}{\text{Corrected CPM}^a\ \text{super} + \text{CPM}^a\ \text{PPT}}$

[a] = counts per minute.

second antibody (anti-monkey γ-globulin goat serum) was for 2 hr only. Special features of the technique are described with the discussions about the individual hormones, below.

Recommended procedures for hormone assays follow. For the latest techniques see the method section of the *Endocrinology Index*.*

## Thyroid-Stimulating Hormone (Thyrotropin, TSH)

The thyroid-stimulating hormone (thyrotropin) is a glycoprotein secreted by the basophil cells of the anterior pituitary gland, the physiologic function of which is stimulation of the development of the

*A bi-monthly publication of National Institute of Arthritis and Metabolic Diseases, available from Superintendent of Documents, U.S. Government Printing Office, Washington, D.C.

thyroid gland and the secretion of the thyroid hormones. The molecular weight of TSH is about 30,000, and it contains approximately 8 percent carbohydrate.

Many of the methods of assay of TSH (McKenzie, 1960; Kirkham, 1966; Condliffe and Robbins, 1967) described in the literature depend on histologic changes in the thyroid gland and on alterations in thyroid weight. Most such procedures are too insensitive and laborious for routine use. Some of the *in vivo*, radiometric, and immunologic assay methods considered follow:

Manley *et al.* (1969) utilized the depression of $^{131}I$ concentrating activity in the propylthiouracil (PTU) blocked guinea pig slices for *in vitro* bioassay of TSH in human serum. The dose range for optimal precision extends from 1 to 64 micro units of international standard TSH and 2 to 16 micro units of human TSH research standard A (final concentration/milliliter in the incubation medium). This method, as applied by Bouke *et al.* (1969) to measure TSH in rat serum, yields mean values of 6.5 and 4.7 international mμ/100 ml for males and females, respectively. These values compare favorably with 4.1 mμ/100 ml as reported by Wilber and Utiger (1967) by radioimmunoassay.

Utiger (1965) described an immunochromatoelectrophoretic assay for humans (HTSH), which is capable of detecting as little as 1.5 mμg HTSH per milliliter of plasma. A somewhat similar method has been developed by Odell *et al.* (1965) and is capable of detecting 2.5 ng/ml of TSH in human serum. This topic has been reviewed by Odell *et al.* (1967).

An accepted double antibody immunoassay technique in rats is given by Wilber and Utiger (1967). They developed a sensitive and specific assay for TSH in rat serum utilizing $^{125}I$ mouse thyrotropic tumor TSH, antibovine TSH and serum, and USP bovine TSH as a standard.

## Corticotropin (Adrenocorticotropic Hormone, ACTH)

Corticotropin is a peptide hormone secreted by the anterior pituitary gland. Its physiologic function is stimulation and secretion of adrenocortical hormones. Corticotropin from human, porcine, ovine, and bovine pituitary glands consists of 39 amino acids.

Evans *et al.* (1966) and Hodges *et al.* (1968) reviewed assay methods for ACTH. Sayers' test (1967) has been used most widely. However, at the present time the procedure of Ney *et al.* (1963) is the method of choice and is based on *in vivo* steroidogenic effects of ACTH. The material to be assayed is injected into the femoral vein of male rats hypophysectomized at approximately 200 g body weight. Two hours later the blood is sampled continuously (7–10 min) from the left adrenal vein by the technique of Munson and Toepel (1958). Corticosterone content of the adrenal vein blood was determined by the method of Sibler *et al.* (1958). Since the circulating levels of ACTH are very small in the blood of any species, it is necessary to extract and concentrate the sample by one of the following methods: (1) acid–acetone; (2) oxycellulose; or (3) amberlite absorption, as recently reviewed by Sayers *et al.* (1944). Purves and Sirett (1968) discussed the stability of endogenous and exogenous ACTH in rat plasma as established by bioassay in intact rats.

Donald (1968) measured ACTH in extracted blood plasma and was able to identify as little as 0–25 pg/ml. Berson and Yalow (1968), in a radioimmunoassay of unextracted human plasma, sensed a level of 1 pg/ml (equivalent to 0.014 m$\mu$/100 ml).

### Growth Hormone (Somatropin, GH)

Growth hormone (GH) is a simple protein hormone secreted by the anterior pituitary gland. Its action stimulates a large variety of physiologic processes that result in increased tissue and skeletal growth, nitrogen retention, and fatty acid mobilization. Bovine and ovine growth hormones have two *n*- terminal amino acids, while those of other species such as monkey and porcine animals have only one. Molecular weights of the growth hormones of ox, sheep, pig, and whale range from 40,000 to 48,000, of monkey and man from 21,000 and of rat 28,000.

Of the reported bioassay techniques (Loraine and Bell, 1966; Papkoff and Li, 1962; Zarrow *et al.*, 1964), the principal methods are based on increase in weight in normal plateaued rats, increase in weight or tail length in hypophysectomized rats, and the tibia test in hypophysectomized rats. A method utilizing lead acetate is also reported. Of these methods, the tibia test (Greenspan *et al.*, 1949) is the most widely used. It is sufficiently sensitive to detect GH in blood of calves and young pigs and in cases of human gigantism and acromegaly but not in normal plasma (Turner, 1966).

Although the test is reasonably sensitive, its specificity to measure GH in pituitary homogenates or in blood plasma is questionable since thyroxine and prolactin stimulate cartilage growth, while ACTH depresses it (Cargill -Thompson and Green, 1963). However, the tibia method is more reliable in measuring the growth hormone releasing factor (GHRF) or the growth hormone inhibiting factor (GHIF); neither thyroidectomy, thiouracil feeding, nor the injection of thyroxine equivalent to thyroxine secretion rate had any effect at the hypothalamic level in altering the releasing factor for GH (Meites and Fiel, 1965). Also, several central nervous system depressant drugs failed to release hypothalamic GHRF. Recently, Yonaga (1967) claimed that the lead acetate method is more sensitive than the tibia test.

Special problems concerned with radioimmunoassay are described by Yalow and Berson (1967). Bovine and ovine growth hormones crossreact with each other; bovine GH also crossreacts with porcine GH. The ready availability of these hormones from the National Pituitary Agency, ease of antibody production, and less vulnerability to isoptopic damage during labeling (due in part to their higher molecular weight) make immunoassays with these hormones relatively easy. Machlin *et al.* (1967) measured plasma growth hormone levels in sheep in response to intravenous injection of stalk median eminence (SME) extract. Hartog *et al.* (1967) assayed growth hormone in dog plasma using antiporcine serum. The sensitivity of the assay was 10 ng/ml.

Birge *et al.* (1967) and Daughaday *et al.* (1967) describe assay procedure for measuring rat growth hormone in plasma and pituitary extracts, respectively. Garcia and Geschwind (1968) discuss a double antibody radioimmunoassay procedure for plasma GH in monkeys, and for the concentration of GH in plasma and pituitary homogenates in rabbit, rat, and mouse. Plasma levels in humans, in a period less than 24 hr, have been recently measured by Mitchell *et al.* (1969), using the proteolytic enzyme facin to separate antibody-bound and free radiohormone $^{125}I$ (HGH). The precision and the sensitivity is comparable to results with a technique incorporating a longer time and with a double antibody. The production of antiserum against rat growth hormone (RGH) in chickens has proved economical and rapid (Martin, 1968). The growth hormone in unextracted urine has been measured by Sakuma *et al.* (1968).

## Insulin

Insulin is a peptide hormone formed in the β-cells of the Islets of Langerhans and is secreted in response to hyperglycemic stimuli. Although the effects of insulin are many, its primary physiologic role is to facilitate utilization of glucose, presumably by promoting its rate of entry into the body cells.

Insulin activity is expressed in terms of international units (IU), which are defined as the amounts of insulin activity contained in 0.04082 mg of the international standard preparation. The molecular weight of ox insulin is 5734.

Earlier bioassay methods (Heroux *et al.*, 1956; Randle and Taylor, 1960; Rieser, 1967) depended on the fall in blood sugar response. These are now rarely employed because they lack sensitivity and as such will not be discussed here. *In vivo* hypoglycemic methods have been superseded by *in vitro* rat or mouse diaphragm methods. The latter are based on increased glucose uptake by the diaphragm with increasing insulin concentration and have been subsequently replaced by immunoassays because the relationship between glucose uptake and insulin concentration is not always linear. Assays based on the use of isolated adipose tissue also lack specificity and are affected by other hormones, such as growth hormone, glucagon, prolactin, and corticotropin, if they are present in concentrations not physiologic. For this reason, the results obtained by this method are usually termed "insulin-like activity" (Steinke *et al.*, 1962).

Radioimmunoassay employing double antibody technique is the current method of choice for insulin assay. Insulin was the first hormone to be measured by radioimmunoassay and is probably the hormone most widely assayed by this means. Problems of iodination seem to be less with this molecule than with other hormones. The double antibody procedure is outlined in the introduction to this section (Assay Procedures).

### Glucagon

Glucagon is a peptide hormone produced by the $\alpha$-cells of the Islets of Langerhans in the pancreas. It has a hyperglycemic effect, presumably due to stimulation of hepatic glycogenolysis. Glucagon is a straight-chain polypeptide with a molecular weight of 3485.

Glucagon in blood is determined by bioassay (*in vivo* and *in vitro*), radioimmunochemistry, and radioimmunoassay.

*In vivo* bioassay methods are based on the hyperglycemic activity and are only sufficiently sensitive to measure large quantities of glucagon present in pancreatic extracts. They are not capable of detecting the minute quantities of glucagon in the circulation (Staub and Behrens, 1954).

*In vitro* methods are based on the glycogenolytic action of glucagon and measure various products in the glycogenolytic pathway, such as glucose production from liver slices, and phosphorylase and adenyl cyclase activity of liver homogenates. Since adrenalin and free or proteinbound calcium interfere with the glycogenolytic response of glucagon, this method is nonspecific for glucagon assay (Unger and Eisentraut, 1967a).

Unlike that for insulin, glucagon radioimmunoassay faces difficulty in production of suitable antisera and damage to the labeled hormone during its incubation with plasma, particularly of human origin. Detailed procedures for the immunization and the assay has been outlined by Unger and Eisentraut (1967b). Although the precision and sensitivity are satisfactory, the specificity of the assay is questioned because of $\gamma$-globulin complications.

Using the radioimminoassay technique, Unger and Eisentraut (1967a) measured human plasma levels using beef–pork glucagon (available commercially as a standard $^{131}I$ glucagon) and antiglucagon serum from rabbits. The unknown samples can be assayed with a standard deviation of ±1.0 percent, and the sensitivity is 0.2 ng/ml. It may be concluded that, of all the available assay methods, radioimmunoassay can be regarded as precise, sensitive, and semiquantitative for detection of relative changes in plasma glucagon in physiologic and pathologic states.

## Prolactin (Lactogenic Hormone, Luteotropin, LTH)

Prolactin is a simple protein hormone secreted by the adidophils in the anterior pituitary gland. Its physiologic function is stimulation of the mammary glands in mammals, and maintenance of the corpus luteum in some species, including the rat and the mouse. Both sheep and ox prolactins have molecular weights of approximately 25,000.

By definition, one international unit (IU) is the active amount contained in 0.045 mg of sheep pituitary tissue (Bangham *et al.*, 1963). Results should now be expressed in terms of the standard unit.

Many methods for the assay of prolactin (Forsyth, 1967) have been proposed but most of them are too insensitive. These methods can be divided into three main groups: (1) assays in pigeons; (2) assays in mammals; and (3) immunological assays.

### ASSAYS EMPLOYING PIGEONS

Prolactin preparations stimulate the crop glands of pigeons by causing rapid proliferation of the epithelial lining. In addition, the hormone increases the production of "crop milk," which is a caseous fluid consisting mainly of desquamated epithelial cells. Assay methods in pigeons are generally more precise than those in mammals, but require rigid standardization of technique during their performance. Among the many factors that have been shown to influence the response are season of the year, body weight of the birds, strain and race of pigeon, environmental temperature, route of inoculation, and volume of solution administered.

The crop sac methods depend on measuring two end points: increased weight of the crop gland, and increase in diameter of the crop gland. The method of Riddle and Bates (1939), as modified by Coppedge and Segaloff (1951) for urine, and the methods of Clark and Folley (1953), Hall (1944), and Bates *et al.* (1963) may be considered. A sensitive objective bioassay method for prolactin based on $^{3}H$-methyl-thymidine uptake by the pigeon crop mucosal epthelium has been described by Ben-Daird (1967). The use of a balanced incomplete block design was recommended by Kanematsu and Sawyer (1963).

#### ASSAYS EMPLOYING MAMMALS

The Kovacic (1968) method is practical, precise, and sensitive. It uses nonparous mice 3–4 months old and is based on hyperaemia of corpora lutea of ovulation. The method is specific for prolactin and is sufficiently sensitive to detect traces of contaminants in gonadotropic hormones.

#### RADIOIMMUNOASSAY

Recently the double-antibody procedure has been proposed by Arai and Lee (1967) for ovine prolactin and by Kwa and Verhafstad (1967) for rat prolactin, although there is some doubt as to whether all the material measured immunologically is biologically active. Radioimmunoassay for sheep, goat, and cattle prolactin having sensitivities equal to 5.9 and 0.2 ng/ml of sheep plasma and tissue extract has been described (Bryant and Greenwood, 1968).

### Gonadotropins

The gonadotropins are protein hormones secreted by the anterior pituitary and by the placenta. Those of pituitary origin are follicle stimulating hormone (FSH); luteinizing hormone (LH), also referred to as interstitial cell stimulating hormone ICSH; and prolactin (LTH), also called luteotrophic hormone and lactogenic hormone (see above).

The molecular weights of LTH, LH, and FSH are approximately 26,000, 28,000, and 30,000, respectively.

Both FSH and LH are glycoproteins containing 10 and 25 percent carbohydrate, respectively. Sialic acid is required for the biological activity of both FSH and LH. Luteinizing hormone is characterized by a low content of tryptophan and a relatively high content of proline and cystine (Damm *et al.*, 1966).

The luteinizing hormone causes luteinizing of the ovarian follicles in the female and stimulates testicular interstitial cells in the male. The function of FSH in the female is the development of the ovarian follicle, and in the male, stimulation of spermatogenisis. Prolactin stimulates the mammary gland in mammals, maintains corpus luteum in some species, and stimulates the crop sac in pigeons.

A gonadotropin of placental origin appears in pregnant mare's serum (PMSG), and it acts like FSH. (Another is found in the blood and urine of pregnant humans and is secreted particularly between the eighth and twelfth weeks of pregnancy. It is called human chônionic gonadotropin, HCG, and it acts like LH.)

Both PMSG and HCG are used in the bioassay of the pituitary gonadotropins. The international unit of HCG is equivalent to 0.1 mg of the standard preparation. Pregnant mare's serum assay uses end points similar to those for HCG. The purity is very good: 1 IU is equivalent to 0.25 mg of the standard preparation. Reference standards are available from the National Institutes of Health and from some pharmaceutical concerns.

Bioassay methods for FSH are based on follicular growth in hypophysectomized immature rats, augmentation of ovarian weight in rats or mice (a routine method of choice), increase in testicular weight in hypophysectomized immature rats treated with an excess of HCG, and augmentation of uterine weight in mice (Igarashi and McCann, 1969) and by Steelman and Pohley (1953) in rats. The augmentation tests are the methods of choice and depend on the ability of HCG to augment the action of FSH on the ovary. Paesi *et al.* (1955) and Gans and Rees (1966) have reported modifications of Steelman and Pohley's test.

Assays of LH and FSH are now conducted in intact rather than hypophysectomized animals. The ovarian ascorbic acid depletion (OAAD) test is Parlow's (1961) widely used bioassay method for LH. This method depends on the depletion of ascorbic acid in the ovaries of immature rats pretreated with PMSG and HCG. The chief advantage of the OAAD method is its relatively high degree of sensitivity. However, Crighton (1968) reported nonspecificity of rat OAAD test with unextracted serum from sows; depletion of ovarian ascorbic acid occurred and found not to be due to LH nor to any other hormone of pituitary orgin. Another disadvantage of OAAD test is its low degree of precision in the hands of some investigators. The ovarian cholesterol depletion test (OCD) of Zarrow and Clark (1969) is far more precise than the OAAD test. It is specific, practical, and reproducible.

The ability of HCG to cause ovulation or sperm release in a number of amphibians has provided a useful assay method. Early diagnosis of pregnancy in man is based upon the ability of urinary concentrations of HCG to effect ova or sperm release in the frog, ovulation in the rabbit, hyperaemia of the ovary of immature rat, and other phenomena indicative of the action of LH.

The radioimmunoassay (double-antibody) procedure is the same as that described above under "Assay Methods." For a review of bioassay and radioimmunoassay of human gonadotropins, see Albert (1968). The subject has also been treated by Hunter (1967), Raiti *et al.* (1968), Midgley (1966), and Odell *et al.* (1967).

Bioassay and radioimmunoassay results were found by Kulin *et al.* (1968), Monroe *et al.* (1968), and Niswender *et al.* (1968, 1969) to agree well.

## Melanocyte Stimulating Hormone (Intermedin, MSH)

Intermedin is a peptide hormone secreted by the intermediate zone of the pituitary gland. Its physiologic function is associated with the dispersion of melanin in the epidermal melanophores. Two types of MSH are known—$\alpha$-MSH with 13 amino acid residues and $\beta$-MSH with 18 amino acid residues. Melanin dispersion in mammals has been noted (Lerner *et al.*, 1954); MSH also causes an increase in melanin synthesis (Freiden and Bozer, 1951). The melanocyte stimulating hormones are remarkably similar to a portion of the ACTH molecule (Harris and Roos, 1956; Harris and Lerner, 1957). The structure of $\beta$-MSH varies slightly in different species.

The assay of MSH is based upon its principal action, the dispersion of melanin in the melanocytes of amphibians (Landgrebe and Waring, 1944). The frog (*Rana pipiens*) or the South African toad (*Xenopus laevis*) is commonly used. Assay procedures vary from a subjective grading of skin color or microscopic observation of melanin dispersion to measurement of degree of darkening (under constant illumination) by a photoelectric reflectometer (Shizume *et al.*, 1954).

The international unit of MSH is equivalent to the activity present in 1 μg of lyophilized standard posterior pituitary powder.

### Oxytocin and Vasopressin

Oxytocin is a peptide hormone of the neurohypophysis, the physiologic effect of which is to increase the rate and force of contraction of a smooth muscle, especially of the uterus, as well as stimulation of milk ejection during lactation and the transport of sperm.

The structure of oxytocin prepared from human, bovine, and porcine sources is identical. Oxytocin is chemically related to vasopressin, its structure differing only in positions 3 and 8. The chicken's neurohypophysis contains arginine vasotocin instead of vasopressin.

These hormones appear to be synthesized in the supraoptic and paraventricular nuclei of the hypothalamus. They travel down the axons of the hypothalamo–hypophyseal tract bound in some way to the inert protein neurophysin, and are stored in the termination of these axons in neurosecretory vesicles (Douglas and Poisner, 1964; Sloper, 1966; Walker, 1967). Most stimuli liberate both hormones, but oxytocin is released in greater amounts than vasopressin. However, hemorrhage in the cat causes the release of vasopressin without oxytocin (Beleslin *et al.*, 1967). The hormones may circulate as the "free" form or perhaps bound to plasma proteins (Hipsley and McKellar, 1959).

Several bioassays have been developed for the measure of oxytocin and vasopressin (Van Dyke *et al.*, 1955). For oxytocic activity, measures such as blood pressure depression in the chicken, uterine contractility in the guinea pig, mammary gland contraction, milk ejection, and parturition have been used. Vasopressin activity has been measured by elevation of blood pressure in cats and dogs and by its antidiuretic action in rats and mice (Walker, 1967). Since methods for a high degree of purification and for artificial synthesis of the hormones have been developed, an accurate estimation of the potency of the pure hormones is now possible. The international unit is equivalent to 0.5 mg of the standard powdered posterior pituitary gland, assayed by the avian blood pressure depressor method for oxytocin, but applicable to antidiuretic and depressor activity also. Lysine vasopressin is about half as active as arginine vasopressin. Gupta (1968) measured vasopressin with bioassay procedure in unextracted human plasma and attained a sensitivity of 0.1–0.25 μU/ml.

Radioimmunoassay for lysine vasopressin is available (Permutt *et al.*,

1966); however, the sensitivity is not sufficient to replace the current bioassay methods.

### Parathyroid Hormone

The parathyroid hormone is a hypercalcemic factor secreted by the parathyroid gland in response to low levels of ionized calcium in serum. Its physiologic role is to maintain blood levels of calcium within narrow limits. About 10 mg calcium/100 ml plasma is maintained by direct effects on bone, where it promotes dissolution of bone salts, and on kidney, where it acts to prevent losses of calcium into the urine. Parathyroid hormone is of major importance in the maintenance of homeostasis by virtue of its regulatory role in calcium (and also phosphorus) metabolism.

Parathyroid hormone from bovine animals is a single-chain polypeptide consisting of 83 amino acids and is approximately three times the size of thyrocalcitonin (Potts *et al.*, 1967).

Although several methods based on hypercalcemic and phosphaturic activity of parathyroid hormone are available, these bioassays have proved to be insufficiently sensitive to measure the amounts of the hormone present in body fluids.

Since no international standard has been prepared for parathyroid hormone, an absolute measure of potency is not available. However, the biological response of the dog has been utilized as a basis for comparison of extracts. The USP unit is defined as the amount of an extract required to raise the serum calcium of a dog weighing 20 kg by 1 mg/100 ml within 16–18 hr after the administration of the preparation.

Amer (1968) described bioassay for parathyroid activity (PTA) using thyroparathyroidectomized rats maintained on normal diet supplemented with thyroid extract. The animals were most sensitive to PTH (5–20 USP units/100-g body wt.) 3–5 days postoperative. Since percentage increase, absolute increase, and postinjection calcium levels were linear with the logarithm of the dose, any of the three criteria may be used for the assay.

Sherwood *et al.* (1966) developed a sensitive radioimmunoassay using bovine parathyroid hormone. The high degree of cross-reactivity (the bovine parathyroid hormone cross-reacts with human, goat, and sheep) is an asset. However, preparation of the hormone and adherence of the iodinatied bovine parathyroid hormone (Yalow and Benson, 1966) to laboratory glassware at neutral or alkaline pH are two of the limiting factors precluding the wide applicability of radioimmunoassay to this hormone.

### Thyrocalcitonin or Calcitonin

Thyrocalcitonin from pigs is a single chain polypeptide hormone having 32 amino acids (Potts *et al.*, 1967). In mammals it is secreted by the *c* cells of the thyroid (Pearse, 1966); in birds and fishes, the ultimobranchial body is its source (Copp *et al.*, 1967). Thyrocalcitonin caused hypocalcaemic and hypophosphatemic conditions.

Hypocalcemic activity of the hormone forms the basis of the bioassay. The fall in plasma calcium level is measured 60 min after subcutaneous injection (Cooper *et al.*, 1967). The plasma calcium level can be measured by atomic absorption spectrometry; it requires 2 weeks to complete one estimation. For these reasons, a short method is preferred to increase sensitivity and speed; such measures are, however, available.

The Medical Research Council of England supplies a standard preparation, "A," which is prepared from acetone-dried pig thyroid by the method of Baghdiantz *et al.* (1964). Crude acetone-dried preparations can also be obtained commercially from Pel Freeze Biologicals, Inc., Rogers, Arkansas. Recently, standard "B," which is four times as potent as standard "A," has become available. The calcitonin measure is the "Hammersmith unit," defined as one tenth the amount expected to cause a 10 percent fall in plasma calcium concentration in a 150-g rat.

The use of paper chromatoelectrophoresis for separating bound and free plasma has been reported (Arnaud and Littledike, 1966). The method is sufficiently sensitive to detect 0.5 ng of thyrocalcitonin per milliliter of plasma from pigs. A radioimmunoassay for calcitonin has been reported by Arnaud *et al.* (1968) and is exceedingly sensitive.

### Thyroid Hormones

L-Thyroxine ($T_4$) is an amino acid secreted by the thyroid gland in response to stimulation by thyrotrophin (TSH). It is stored in the thyroid gland in the iodized glycoprotein, thyroglobulin (molecular weight, 680,000), from which it is released, presumably through the action of a thyroid proteinase. The corresponding 3,3′,5′-triiodothyronine ($T_3$), also produced by the thyroid gland has in most studies had a higher biological activity than $T_4$.

A widely used method for determining the thyroxine secretion rate (TSR) in cattle was reported (Lodge *et al.*, 1957; Pipes *et al.*, 1957). Although an excellent method, this procedure requires special equipment for restraining the animals for radiation detection. Also, it requires a rather long period of time, about 2 weeks, to complete one estimation. For these reasons, a short method is preferred to evaluate rapidly the physiologic alterations that may be induced by environmental changes. Yousef and Johnson (1967) describe a simple and more rapid method for the estimation of TSR in cattle, in which TSR was calculated as the product of the product of the plasma thyroxine (PT), the thyroxine disappearance rate ($K$), and the thyroxine distribution space (TSD).

To simplify the method Post and Mixner (1961) estimated $K$ in 2 rather than 4 days after inoculation of $T_4$-$^{131}I$. Recycling of $^{131}I$ was not blocked; therefore they concluded that $K$ is not influenced significantly by the recycling of $^{131}I$ through the thyroid gland during the first 96 hr after radiothyroxine inoculation.

The TSR method that Yousef and Johnson (1963) followed is similar to that reported by Post and Mixner (1961) except that radioactive L-thyroxine was used instead of a high dose of L-thyroxine. This probably simplifies the estimation of the fractional $K$ value. Reineke and Lorscheider (1967) described a quantitative direct output method for TSR in the rat.

## Catecholamines

Epinephrine and norepinephrine are in a group known as catecholamines. Both hormones contain an asymmetric carbon and therefore each can occur as two optically active isomers, the L-form (levorotatory) and the D-form (dextrorotatory). The naturally occurring isomer of the medulla is the L-form, which is approximately 15 times more active than the D-isomer in the elevation of blood pressure.

Epinephrine and norepinephrine are similar in structure and differ only by the presence of a methyl group. The side chain of epinephrine, containing a terminal methyl group, is a secondary amine. The hormones are rapidly inactivated *in vivo* or oxidized to inactive forms. They give distinctive color reactions with both ferric and chromium salts because of their catechol structure. Oxidation of these hormones leads to the formation of adrenochromes, which condense with ethylenediamine to form fluorescent compounds. This occurrence forms the basis of a sensitive and relatively specific chemical assay for the adrenomedullary hormones (Alvarez, 1968).

Epinephrine is 5 to 10 times as active as norepinephrine in increasing oxygen consumption. Alvarez *et al.* (1969) compared the effects of epinephrine and norepinephrine on heat production and dissipation in cattle. They present data on the effects of environmental temperature on plasma catecholamine and describe the assay procedure for their measurement in bovines. [The double-isotope derivative method using *S*-adenosyl L-methionine-methyl-$^{14}C$ and 7-$^{3}H$-norepinephrine to measure total catecholamines (epinephrine plus norepinephrine) in human plasma and urine has been shown (Engelman *et al.*, 1968) to have a sensitivity of 0.12–0.52 mg/liter.]

## Steroid Hormones

The procedures for assaying various steroid hormones in blood are provided by Dixon *et al.* (1967) for corticosteroids; van der Molen and Aakvaag (1967) for progesterone; O'Donnell and Preedy (1967) for estrogens; Eberlein *et al.* (1967) for androgens; and by Coghlan and Blair-West (1967) for aldosterone. Butt (1967) has written an excellent source book on hormone chemistry dealing with chemical and physical characteristics, methods of separation, and quantitative estimations. For gas chromatographic analysis of steroid hormones, the work of Wotiz and Clark (1966) and Neher (1964) may be referred to. Loraine and Bell (1966) have critically evaluated the assay methods. Separation procedures and various methods of quantitative estimation have been reviewed (Gray and Bacharach, 1967; Butt, 1967; Carstensen, 1967).

### ADRENAL CORTICOIDS

The secretions of the adrenal cortex are steroidal and include the sex hormones (progesterone, estrogens, and androgens) in addition to the adrenocorticoids, such as corticosterone (compound B), 17-hydroxycorticosterone (cortisol, hydrocortisone, compound F), and aldosterone. In addition to the active steroid hormones, a number of inactive steroids have been extracted from the adrenal gland and identified primarily as intermediates in the biosynthesis of the adrenocorticoids. The adrenocorticoids have been classified into two categories, based on the physiologic activity. The glucocorticoids, characterized by an 11-oxygen moiety (oxycorticoids), bring about an increase in deposition of liver glycogen and other responses associated with adaption to stress. The mineralocorticoids are usually devoid of an oxygen grouping on carbon-11; however, the potent, naturally secreted mineralalocorticoid aldosterone has a hydroxyl group on carbon-11 and, in addition, has an aldehyde group on carbon-18, through which the internal condensation is possible. The main physiologic response to the mineralocorticoids is increased retention of sodium ions by the kidney. Since the chemical structures of the glucocorticoids and mineralocorticoids are quite similar, each class of compounds possesses some physiologic action of the other group. While 17-hydroxy-corticoids (17-OHCS), cortisol, and cortisone are the glucocorticoids of major importance in most species, corticosterone is the principal corticoid in the rat, rabbit, mouse, and bird. Aldosterone appears to be the true adrenal hormone in all species for the regulation of electrolytes.

The biologic responses to stressing agents initially served as the end points for assay of corticoids (Zarrow *et al.*, 1964). A widely used technique for the assay of glucocorticoids is based on the increased deposition of liver glycogen (Greenman and Zarrow, 1961). This technique has been used to evaluate the activity of the individual corticoids as well as the activity of corticoid material in blood and urine. Various colorimetric reactions such as Porter-Silber reaction, blue tetrazolium reaction, and oxidative clevage of the side chain followed by Zimmerman reaction have been used, although the Porter-Silber seems to be the most commonly employed at the present. Fluorometric techniques and a number of procedures for quantitative assay of corticosteroids have been described by Silber (1966). The color reagent of Porter and Silber has been coupled with methods of chromatography and solvent partition to estimate corticoids in blood (Nelson and Samuels, 1952; Peterson *et al.*, 1957). While these methods are mainly used for 17-OHCS, the techniques can be used for corticosterone (McCarthy *et al.*, 1959, 1960). The corticoids also exhibit fluorescence in sulfuric acid. A very precise method has been developed for the separation of groups of corticoids by column chromatography and determination by fluorescence analysis (Moncola *et al.*, 1959). Fluorescence analysis has been applied widely to the study of rat and bird blood, where corticosterone is the significant corticoid (Greenman and Zarrow, 1961; Guillemin *et al.*, 1958, 1959; Moncola *et al.*, 1959). For the determination of plasma glucocorticoids in cattle, a modification of the techniques reported by Bergman (1963) was used by Alvarez *et al.* (1969). [A rapid and precise method for the estimation of the production rate of cortisol in man has been described (Kelly *et al.*, 1968). It utilizes a known dose of $^3$H-cortisol into subjects followed by collection of urine for 1 or 2 days. Those metabolites having 3-$\alpha$-hydroxy-5-$\beta$-pregnone configuration are chemically converted into 11-$\beta$-hydroxy-etiocholanolone, which is then isolated and crystallized, and its specific activity determined by counting a weighed sample.]

For the rat plasma and adrenal, the method of Guillemin *et al.* (1959) and Vernikos-Danellis *et al.* (1966) is reliable. Townsend and James (1968) described a semi-automated procedure for determination of plasma corticoids. The results obtained by this procedure gave a mean value that was 1.4 μg/100 ml higher than the true cortisol level (some part of it can be accounted for by corticosterone). The precision is approximately 1.0 μg/100 ml, and the analytical rate is 20 samples per hour.

A test by Eisenstein (1967) is a further source of information on the adrenal cortex and mechanisms thereof, and on methods analyzing its hormones.

## PROGESTOGENS

Progesterone, the most prevalent naturally occurring progestogen, is secreted by the ovary, placenta, adrenal cortex, and testis. Its systemic name is pregn-4-ene-3,20-dione, its empirical formula $C_{21}H_{30}O_2$, and its molecular weight 314.4. The hormone acts on the endometrium, previously prepared by estrogen, inducing mucus secretory activity for implantation of the embryo. If pregnancy ensues, continued secretion of progesterone is essential for completion to term in some species such as the rat, rabbit, and goat. Progesterone also contributes to the growth of the mammary gland and is thought to maintain the uterus quiescent during pregnancy.

Progesterone is present in more generous quantities than estrogens in the blood stream. During the luteal phase, concentrations of progesterone as high as 50 μg/100 ml plasma are measured in the ovarian blood, and less than 1 μg in the same amount of peripheral blood. Numerous bioassay methods using different end points have been suggested for the estimation of progesterone and have been critically reviewed (Zander, 1962). The Hooker and Forbes (1947) test is of high sensitivity (0.0002 μg). However, it lacks specificity, and several progestational steroids and estrogens may exert synergistic or antagonistic actions on the effect of progesterone, depending on the concentration of these compounds in the progesterone-containing extracts. Variability in quantitative estimation of progesterone in different tests is another serious disadvantage. The Hooker and Forbes test depends on the hypertrophy of the stromal nuclei in the endometrium of ovariectomized mice after intraluminal injection of progesterone. Chemical methods may be preferred to bioassay for less variability between assays and less cost. These methods are (1) colorimetric and fluorimetric, (2) gas chromatographic, (3) isotopic, and (4) competitive protein-binding (CPB).

Among other colorimetric and fluorimetric methods, that of Heap (1964) may be considered for its high sensitivity (3–5 ng). The plasma samples, after treatment with alkali, are extracted by ether, separated by paper chromatography, and converted enzymically to 20-β-dihydroxy-progesterone, whose fluorescence is measured.

Collins and Sommerville (1964), using gas-liquid chromatography with argon ionization detector, reported plasma progesterone levels of 0.20–0.29 mg/100 ml in males and 0.52–1.10 mg/100 ml in females. Lurie *et al.* (1966) used a flame ionization detector to measure progesterone in plasma of human females. The mean recovery was 78.6 ± 11.7 percent, and the coefficient of variation was 14.8 percent when measured at 0.1 μg. The method tends to overestimate the progesterone level in the range of 20–80 ng.

The use of radioactively labeled reagents that will react with progesterone has permitted the preparation of labeled progesterone derivatives. After purification of such derivatives, the amount of radioactivity in the final residue will be proportional to the amount of progesterone or its derivative. This is the principle of isotopic methods. An isotopic derivative of progesterone such as 20-β-dihydroprogesterone is prepared using tritium and/or $^{14}C$ or $^{35}S$. In the double isotope derivative method ($^{3}H$ and $^{14}C$ or $^{35}S$), the sensitivity and precision of the technique depend upon the efficiency for simultaneous counting of isotopes with different energy spectra. Depending on the combination of isotopes used, the sensitivity of the double-isotope dilution derivative technique ranges between 2 and 100 ng (Riondel *et al.*, 1963; Wiest, 1966; Woolever and Goldfien, 1963).

Progesterone in plasma is largely (95–98 percent) bound to protein and is the basis for a competitive protein-binding technique. Although albumin appears to have the largest capacity for progesterone binding and is important for quantitative measurements, other plasma protein fractions, such as α-acid glycoprotein (orosomucoid), β-lipoprotein, and transcortin, also have higher affinity for progesterone (Murphy, 1967). The sensitivity of the protein-binding method, like radioimmunoassay of protein hormones, depends on the binding affinity of the protein and the specific activity of the radiolabeled form of steroid hormone under investigation. Short's (1958) method of ether extraction of alkaline plasma is widely followed for its speed, efficacy, and simplicity. [Yoshimi and Lipsett (1968) used the CPB method to measure plasma level in the human male, ovariectomized female, and normal female in follicular and luteal phase of the menstrual cycle. The specificity of the method is very high. The coefficient of variation was 7 percent in duplicate samples having a progesterone level of 0.25–3.0 ng/ml. In a human plasma sample of 5 ml, 3 ng/100 ml of progesterone can be detected.]

## ESTROGENS

Estrogens are 18-carbon compounds. Ring A of the steroid nucleus in estrogens is unsaturated. Estriol is excreted in urine in the largest quantities. In the mare, estrone, equilin, and some equilenin are found.

During the ovarian cycle, estrogens (1) induce proliferation of the vaginal epithelium, (2) augment secretion of mucus by the cervical glands, and (3) cause endometric proliferation. Estrogens are responsible for the secondary sex characteristics of the female, maturation of the skin, alteration in body contour, and closure of the long-bone epiphyses. Estrogens also stimulate proliferation of the mammary gland during pregnancy. In addition they exert profound metabolic effects, such as retention of NaCl, Ca deposition, protein anabolism, fat deposition, and ketogenisis.

The international standard preparation of one international unit represents the activity contained in 0.1 mg of standard preparation.

Various bioassay methods have been used. These include measurement of cornification of vaginal epithelium (Allen-Doisy test), vaginal mitotic count of epithelial cells (Martin and Claringbold, 1960), reduction in 2,3,5-triphenyltetrazolium chloride to form a red formazon, which is measured colorimetrically (Martin, 1960), and measurement of uterine weight (Dorfman and Dorfman, 1954). All of these assay methods are unreliable, variable, and of doubtful specificity for impure extracts such as urine.

Of the colorimetric chemical methods, the one that utilizes the Kober reaction is the best known. The method consists of the initial formation of a yellow compound with a green fluorescence on heating with $H_2SO_4$ and a reducing agent such as a phenol, followed by its conversion into a pink nonfluorescent complex on reheating with more dilute $H_2SO_4$. Modifications of Kober's (1931) method by Brown and Coyle (1963) for estriol, and particularly those of Ittrich (1960), merit consideration.

The fluorimetric chemical methods are reliable since the estrogens fluoresce on heating with strong $H_3PO_4$ (Finkelstein, 1952) or $H_2SO_4$ (Preedy and Aitken, 1961). Using this principle, the density of fluorescence, which is proportional to the quantity of estrogens, is measured with a fluorometer. Gordon and Ville (1956) have claimed a high degree of specificity using enzymic reaction methods. The double isotope derivative method for measuring estrone and estradiol-17-β in human and sheep blood has been described by Baird (1968).

The gas chromatographic method of Fishman and Brown (1962) uses β-glucuronidase for hydrolysis and ether for extraction. The urine samples determined by this method and by another method of Brown (1955) yield comparable results. The methods described by Wotiz and Chattoraj (1964) and Menini (1965) are laborious; that of Yousem (1964) for estradiol is relatively simple and rapid. The most recent method of Huang (1968) may be consulted for speed and reliability in measuring urinary estriol. The literature describes related methods (Pincus *et al.*, 1966; Carstensen, 1967; Gray and Bacharach, 1967).

## ANDROGENS

Androgens are steroid hormones with 19 carbon atoms, methyl groups being attached at C-10 and C-13. Androgens are secreted by the testis, ovary, and adrenal cortex. Both androstenedione and testosterone exist in the plasma in a directly extractable form, and it is measurement of this moiety that has been reported in most of the published analytical techniques. In addition to the androgenic effects, testosterone has powerful metabolic activity, and causes retention of nitrogen, phosphorus, and potassium. It is responsible for increases in body weight, particularly in growth of bone and muscle, but growth is arrested, due to closure of the epiphyses, after an initial spurt.

The methods of assay for testosterone in blood and urine are biologic and chemical (Carstensen, 1967; Dorfman, 1962; Eberlein *et al.*, 1967; Zarrow *et al.*, 1964).

The biologic methods used are the capon comb test, chick comb test, sparrow bill test, and enlargement of prostate. The capon comb test has been used widely because of its high precision. Some researchers claim better results through intramuscular administration (Greenwood *et al.*, 1935), while others (McCullagh and Cuyler, 1936) prefer the direct application of material to the comb. The chick comb test (Munson and Sheps, 1958), although more convenient and sensitive than the capon test, lacks precision. Blackening the bill of the castrated male or intact female English sparrow, although claimed by Pfeiffer *et al.* (1944) to be very sensitive (1 μg of testosterone can be measured), has not gained wide use. Castrated rats have been used to show the dose-dependent response of testosterone on enlargement of prostate, seminal vesicles, or total accessory gland (Dorfman, 1962). Although the test is as precise as the capon test, it lacks sensitivity. All the above-mentioned tests are incapable of detecting androgens in the circulation and are being replaced by chemical methods.

The fluorimetric chemical method of Finkelstein *et al.* (1961) depends on the fact that testosterone can be converted enzymically to estrone and estradiol-17-β, obtained from human placenta (Lamb *et al.*, 1964) and extracted and purified by the method of Ryan (1959). Estradiol-17-β is measured fluorimetrically.

In the double-isotope method of Riondel *et al.* (1963) $^{3}H$-testosterone is used as an internal standard and $^{35}S$-thiosemicarbazide for preparation of a derivative. The overall recovery is about 5 percent. The only steroid likely to interfere with specificity is 17-epitestosterone. Another approach has been to use $^{3}H$-acetic anhydride for the preparation of labeled acetate, with $^{14}C$ as an internal standard (Hudson *et al.*, 1963).

Futterweit *et al.* (1963), using gas-liquid chromatography (GLC) with a flame ionization detector, attained a sensitivity of 0.1 μg. Gas-liquid chromatography with electron capture as a detector uses a chloracetate derivative of testosterone. This technique seems to be more sensitive than flame-ionization detection. These chromatography methods have distinct advantages of speed and convenience over the laborious double-isotope techniques.

Two methods for steroid analyses that have been reviewed (Caspi and Scrimshaw, 1967) are nuclear magnetic resonance (NMR), and paper chromatography (Dominguez, 1967).

The main source of 11-deoxy-17-oxosteroids is dehydroepsiandrosterone (DHA) secreted by the adrenal gland and found in urine; testosterone is the secondary source (Bush, 1952; Eberlein, 1967; Loraine and Bell, 1966). Among various methods (Vermeulen and Verplancke, 1963; Ibayashi *et al.*, 1964), that of Ismail and Harkness (1966) seems preferable because of its greater recovery, higher sensitivity (0.5 μg), and higher specificity. The method separates testosterone from its closely related isomer, 17-α-testosterone, and hence is of high specificity.

## Hypothalamic Releasing and Inhibiting Factors

Agents formed in the hypothalamus and secreted into the hypophyseal portal blood act as important regulators of anterior pituitary function and may perform key regulatory functions in relation to environmental stressors.

### CORTICOTROPIN

The existence of CRF (corticotropin releasing factor) activity has been demonstrated both *in vitro* and *in vivo* (Brodish, 1968; Guillemin, 1968). Extensive physiological studies on the control of ACTH release have been reviewed by Butt (1967) Hodges *et al.* (1968), James and London (1968), McCann and Porter (1969), and Schally *et al.* (1967).

Two distinct peptides, or families of peptides, possess CRF activity. β-CRF is related structurally to lysin vasopressin and contains cysteine but no methionine. α-CRF does contain methionine. α-CRF is related to α-MSH (melanocyte-stimulating hormone) and contains threonine, alanine, and leucine in addition to all the other amino acids in α-MSH. α-CRF has been divided further into α-I and α-II, the former being extracted from the posterior lobe of the pig, the latter from the whole pituitary. Recently Chan *et al.* (1969) compared *in vivo–in vitro* assay with a fully *in vivo* and *in vitro* method and found that the sensitivity and precision were best with *in vivo–in vitro* assay for CRA (corticotropin releasing activity).

### LUTEINIZING HORMONE

Gonadotropin releasing factors such as the luteinizing hormone releasing factor (LHRF) are extracted in acetic acid and purified by Sephadex G-25 gel filtration and by carboxy-methyl-cellulose (CMC) chromatography. The presence of LHRF is shown as distinct from vasopressin, MSH, and CRF. It appears to have a molecular weight between 1200 and 1600, as judged by its behavior on Sephadex. It contains aspartic acid, glutamic acid, glycine, alanine, lysine, histidine, arginine, threonine, proline, leucine, serine, and traces of tyrosine and phenylalanine (Schally and Bowers, 1964). LHRF activity in rats can be measured by an ovarian ascobic depletion test (Nallar and McCann, 1965).

### FOLLICLE STIMULATING HORMONE

The follicle stimulating hormone releasing factor (FSHRF) is separated from LHRF under different conditions of extraction and elution (Igarashi *et al.*, 1964). Neither vasopressin nor oxytocin is sufficiently active to account for FSHRF activity. Dhariwal *et al.* (1965) showed that FSHRF area is devoid of CRF and GHRF (growth hormone releasing factor) activities. FSHRF may be measured by depletion of pituitary FSH in castrated male rats treated with testosterone propionate according to the method of Saito *et al.* (1967).

### PROLACTIN

The prolactin inhibiting factor (PIF) of the hypothalamus exercises a tonic inhibitory effect on the release of prolactin (Talwalker *et al.*, 1963). Intraperitioneal injection of rat or bovine stalk median eminence (SME) extracts just prior to nursing will prevent the nursing-induced decline in pituitary prolactin (Grosvenor *et al.*, 1965). This is used as a method to measure PIF activity in the hypothalamus. Cerebral cortical extracts are without effect. Purified preparations of LHRF are devoid of PIF activity, indicating the presence of distinct polypeptides that specifically influence the secretion of prolactin.

Although in mammals the neurohumoral control affecting prolactin secretion is inhibitory, the existence of prolactin-releasing factor (PRF) in pigeons (Kragt and Meites, 1965) and in other species of birds (Nicoll, 1965) has been shown.

### GROWTH HORMONE

Evidence for the existence of GHRF in the hypothalamic extracts of several animal species (Muller *et al.*, 1967a, b) has been established. Extracts of SME have been shown to cause depletion of pituitary GH content in rats (Sawano *et al.*, 1967) with simultaneous increase in plasma GH, indicating increased release of GH from the pituitary into the blood (Knobil, 1966; Machlin *et al.*, 1967).

The growth hormone releasing factor is measured in terms of pituitary GH content through the tibia test (Katz *et al.*, 1967). The tibia method of Greenspan *et al.* (1949) is somewhat more reliable for measuring the physiological variation in the hypothalamic GHRA. For example, neither thyroidectomy, thiouracil feeding, nor the injection of thyroxine equivalent to thyroxine secretion had any effect on the hypothalamic GHRF content (Meites and Fiel, 1967). Similarly, various polypeptide hormones (Krulich *et al.*, 1965) and certain central nervous system depressant drugs (Ishida *et al.*, 1965), which alter GH levels in pituitary and plasma, failed to release the hypothalamic GHRF. An *in vitro* assay of GHRF has been described by Dickerman *et al.* (1969). Hypothalamic GHRA probably diminishes with age according to Pecile *et al.* (1965). Hypothalamic GHRA is increased by insulin-induced hypoglycemia (Katz *et al.*, 1966), stress of cold exposure for 1 hr (Muller *et al.*, 1967a, b), heat exposure for 1 and 24 hr (Parkhie and Johnson, 1969), and acute starvation (Meites and Fiel, 1965).

Separation and further purification of SME extract has shown GHRF as a distinct entity (Krulich *et al.*, 1967). GIF, when screened through Sephadex G-25, elutes after GRF and is clearly separable from PIF (Dhariwal *et al.*, 1969). It inhibited the release of GH to levels of about one half of that released by control glands incubated in the presence of eluting buffer (Krulich *et al.*, 1968). It also prevented increased GH release that follows addition of GRF to incubated glands, but failed to alter release of other pituitary hormones such as FSH, LH, and ACTH.

### THYROID STIMULATING HORMONE

The thyroid stimulating hormone releasing factor (TRF) has been shown to increase plasma levels of TSH in man (Bowers *et al.*, 1967; McCann and Dhariwal, 1966; Reichlin, 1966; Schally, 1968). Large doses of tri- and tetra-iodothyronine can inhibit the effect of TRF on the release of TSH *in vivo* and *in vitro* (Guillemin and Schally, 1963; Reichlin, 1964; Sinha and Meites, 1965). This inhibition can be reversed by actinomycin D (Bowers *et al.*, 1967). The induction or incorporation of radioactive amino acid into the pituitary tissue by TRF may be indicative of TSH synthesis *de novo*.

The complete structure of TRF is unknown. The thyroid stimulating hormone releasing factor activity, unlike GHRF activity, is not destroyed by proteolytic enzymes or periodate, but is lost on incubation with human serum. Porcine TRF may contain histidine, proline, and glutamic acid, which together account for 30 percent of the dry weight. TRF reacts with diazotized sulfanilic acid (Pauly's reagent), with the destruction of biologic activity, but does not react with ninhydrin (Schally *et al.*, 1966). The *in vivo* assay methods of Redding *et al.*

(1966) and Fraschini *et al.* (1966) may be followed for measuring TRF activity.

### MELANOCYTE STIMULATING HORMONE

From *in vitro* and *in vivo* experiments with the melanocyte stimulating hormone inhibiting and releasing factors (MSH-IF and MSH-RF), it was concluded that the inhibition of MSH is due to a neurohumor in the hypothalamus. Some evidence in favor of MSH-RFhas been accumulated by Kasten and Schally (1966). The activity of MSH-IF can be measured by the lightening effect on the skin of the frog whose hypothalamic lobes have been removed (Kasten and Schally, 1966).

## REPRODUCTION

Environmental factors may affect reproduction in animals, either directly of indirectly. For example, high temperature may have a direct effect upon the testes and an indirect influence on the gonads by suppressing the secretion of the thyroid gland.

Data on reproduction of animals includes many measurable parameters—age at puberty, duration of estrus, gestation time, egg number, sperm and semen quality, and offspring number, to mention a few. Reproductive terminology varies with the species in question.

Research techniques for studying the physiology of reproduction, primarily in the female animal, are described extensively and lucidly by Casida (1969). Cole and Cupps (1969) provide comparative information on reproduction in both sexes and discuss the subject comprehensively with minor emphasis upon techniques.

Hafez (1959), in discussing the relationship between climate and reproductive capacity of farm animals, describes how climate affects puberty, conception rates, and interval to postpartum estrus. Climatic effects on fertility vary with species, breed, and stage of maturity, and depend on mechanisms associated with adaptation. The inhibition of spermatogenesis, sperm migration, and function of the pituitary and the endometrium by thermal stress are well known; Hafez attributes the suppression of sexual function by environmental stress to a protective adaptation syndrome. Hafez (1961, 1968) is comprehensive on reproductive endocrinology, reproducing physiology, patterns of reproduction, and stress and diseases.

### OVULATION

Although ovulation is associated with peaks of secretion of LH by the anterior pituitary, the precise measurement of the correlation has been difficult and ambiguous (Heald *et al.*, 1967). Winget *et al.* (1965) reported a peak in body temperature in laying hens occurring at ovulation, but the procurement of data by sophisticated electronic equipment precludes adoption of this procedure by most laboratories.

### SPERMATOGENESIS

The primary hormones, FSH and testosterone, control spermatogenesis via the blood pathway; the former exists in the blood as an impure moiety and the latter in conjunction with some blood proteins. Burrows (1939) and Cole and Cupps (1969) describe semen collection procedures for poultry and for larger animals, respectively.

### FERTILITY AND HATCHABILITY

According to Yeates (1965), fertility is related to sperm motility, and its level is influenced by abnormal sperm number. In poultry, fertility in eggs is readily ascertained via incubation unless preovipositals mortality occurs. Hatchability refers to the percentage of fertile eggs that hatch or result in living young birds (Taylor, 1949).

### MATING

An index for mating behavior in cattle and other animals has resulted from observations of the sexual responses in males and females. Lee (1953) has described a scoring system that gives a rough quantitative measure for this reproductive trait.

## EXPERIMENTAL SURGICAL TECHNIQUES

Surgical techniques are basic tools of the biologist engaged in environmental, nutritional, pharmacologic, physiologic, and physiopathologic research. Various extirpation, fistulation, and cannulation techniques have made possible studies that otherwise could not have been done. The development of new nonreactive plastics (to tissue) and the use of coating materials with continuous anticoagulant activity have given great impetus to the use of experimental surgical techniques. A number of them are potentially useful in environmental studies. Breeding care, management of anesthesia, surgical aftercare, and euthanasia of a wide variety of animals, including farm animals, are thoroughly covered in the literature (Worden and Lane-Petter, 1959; Gay, 1965; Lane-Petter, 1963; Porter and Lane-Petter, 1962; Short and Woodknott, 1963; Bustad and McClellan, 1966; U.S. Public Health Service, 1968).

### CARE OF ANIMALS

Markowitz *et al.* (1964) discusses briefly the care and feeding of rabbits, cats, and dogs undergoing experimental surgery. Physiologic surgery includes many unusual and rapidly changing techniques. Aseptic techniques (Markowitz *et al.*, 1964) should be as stringent as procedures in clinical surgery.

### SURGICAL TABLES

These, of course, depend on the kind of animal and the sophistication of the surgery. Tables should tip in all directions and should be capable of height adjustments. Figure 3.9 shows the table currently being used at the National Animal Disease Laboratory (NADL). It can be used for animals of all sizes—rabbits to horses. The operating room is copper lined to reduce electrical interference so that biophysical recordings can be made during acute experiments or from experimental surgical preparations. A monitoring oscilloscope and a heart sound assembly permit the anesthetist and surgeon to be constantly aware of the condition of the animal during surgery.

### ANESTHESIA

The literature (e.g., Lumb, 1963; Hall, 1966; Croft, 1964) on anesthesia is extensive. The effective use of inhalation anesthetics depends primarily on the kind of equipment available. Safe and effective administration of inhalation anesthetics depends on a good closed system with proper "volatilization" equipment, one in which the concentration can be accurately controlled. Respiration can be controlled in respiratory collapse and open chest work with reasonably priced artificial respiration equipment. This equipment is a "must" in animal

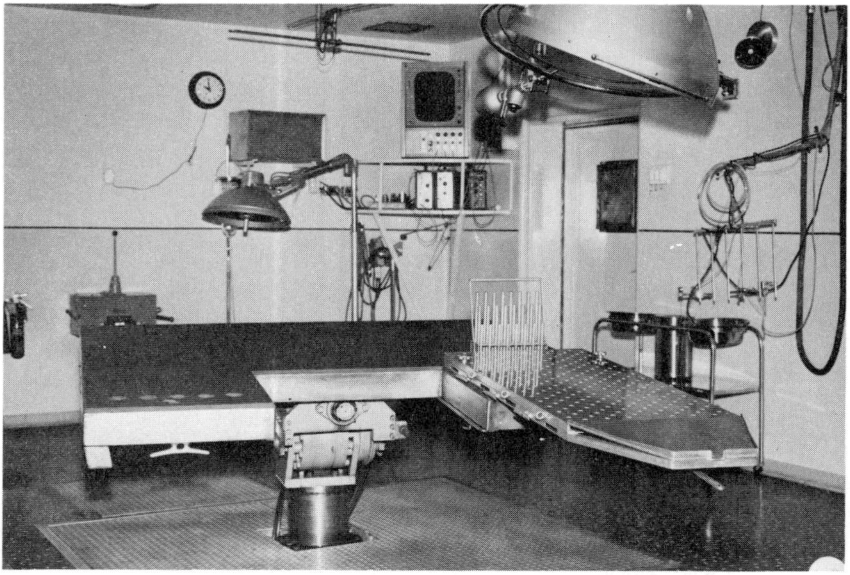

**FIGURE 3.9 Experimental surgical table used at NADL. The monitoring system is shown in the background. The table is recessed in the floor with the hydraulic motors capable of raising the table and changing the slope; on the right of the table is a small pegboard attachment.**

work because of the danger of respiratory collapse in some species and because of the inadequacy of the mediastinum in unilateral pneumothorax.

### MONITORING EQUIPMENT

Such equipment includes a transducer support that may be suspended from the ceiling or supported from the operating table, a viewing oscilloscope with appropriate preamplifiers, heart sound contact microphone and speaker, and suitable outlets and cables for making permanent recordings of physiologic events during acute experiments.

### CLEANING (SANITIZING) THE OPERATING ROOM

Proper drainage should be built into the facility so that cleaning can be quick and effective. An animal surgical facility with a full schedule needs special care in cleaning and disinfecting techniques to minimize the chances for postoperative infections.

### LIGHTING

Many operations demand maximum lighting of deep structures. This imposes two important requirements: (1) maximum mobility of the light; and (2) maximum mobility of the table and consequently of the animal undergoing surgery. Focusing surgical lamps are very helpful.

### FISTULATIONS

Many fistulations are ably described in *Experimental Surgery* (Markowitz *et al.*, 1964), especially the later editions.

### RUMEN FISTULAS

Various techniques (Dougherty, 1955; Schalk and Amadon, 1928) include use of ebonite cannulas (Quin and van der Wath, 1938). Very effective cannulas can be made in the laboratory with small vulcanizing equipment for a few cents (Dougherty, 1955).

The preparation of gastric pouches in ruminants is more difficult than in simple-stomach animals. (Hill and Gregory, 1957).

The secret of success in using open fistulas is closure with adequate and nonirritating plugs. A number of modifications are described in the literature (Dougherty, 1955; Schalk and Amaden, 1928; Quin and van der Wath, 1938).

### OTHER FISTULATIONS

Dougherty (1955) describes methods for making abomasal, intestinal, and cecal fistulas. Markowitz *et al.* (1964) describe a closure plug for the cecal fistula.

Reentry stomach and intestinal fistulations are described by several workers (Dougherty, 1955; Markowitz *et al.*, 1964; Phillipson and Innes, 1939). These permit sampling of ingesta leaving the stomach or the abomasum. Using this method, absorption areas can be studied and limited, and the rate of passage of ingesta can be estimated within limitations.

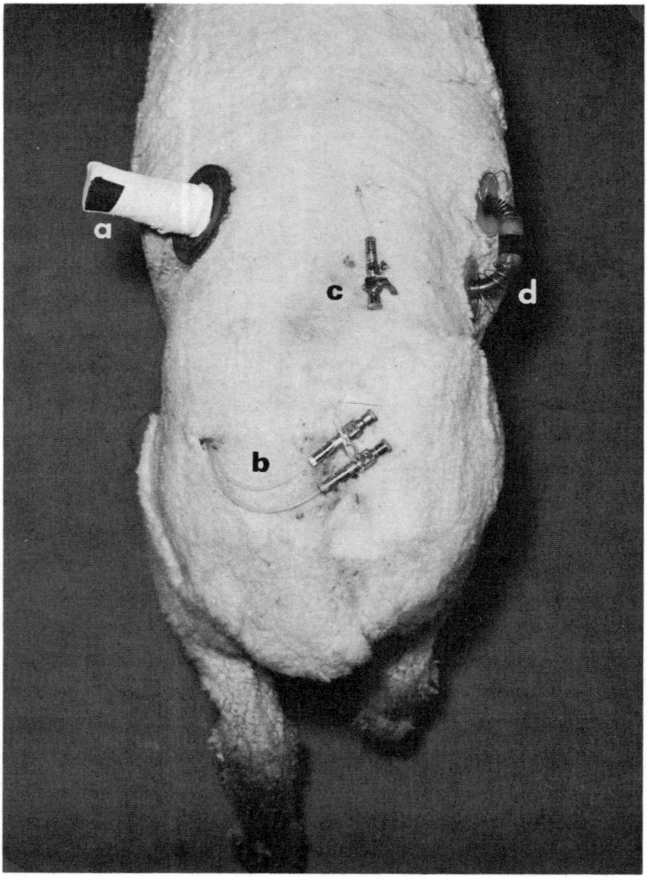


**FIGURE 3.10** Top view of sheep showing (a) rumen cannula, (b) femoral arterial and vein catheter, and (c) portal vein cannulae. The sheep also has an intestinal bypass (d) that has been functioning for about 2 months. This permits intestinal sampling and injecting material directly into the small intestine.

## CANNULATION OF THE CARIOVASCULAR SYSTEM

Popovic and Popovic (1960) describe cannulations of the cardiovascular system of laboratory animals. Other workers have applied this technique with modifications to large animals (Olsen *et al.*, 1967; Buck *et al.*, 1965).

Cardiovascular cannulations are very helpful in obtaining representative blood samples without exciting the animal. They are also quite useful in measuring blood pressure and blood pH. Olsen *et al.* (1967) describe permanent catheterization of the aorta in adult cattle. Buck *et al.* (1965) describe cannulation of the carotid artery. Lewis *et al.* (1957) have studied portal blood through cannulation. Harrison (1969), using a technique that did not seriously impair circulation has cannulized the portal and hepatic veins. Arterial catheter implanting in sheep is shown in Figures 3.10, 3.11, and 3.12 (courtesy National Animal Disease Laboratory, Ames, Iowa).

In all cases, considerable surgical care must be exercised in implanting the catheters and frequent (daily or alternate day) flushings are necessary to keep the catheters open. The catheters should be filled with an anticlotting solution such as heparin.

Procedures for exteriorizing the carotid artery in cattle have been described by McDowell *et al.* (1966). However, in some laboratories the techniques have been replaced to some extent by the use of implanted cannulas, which may be located either in the arterial, the venous part of the cardiovascular system, or both.

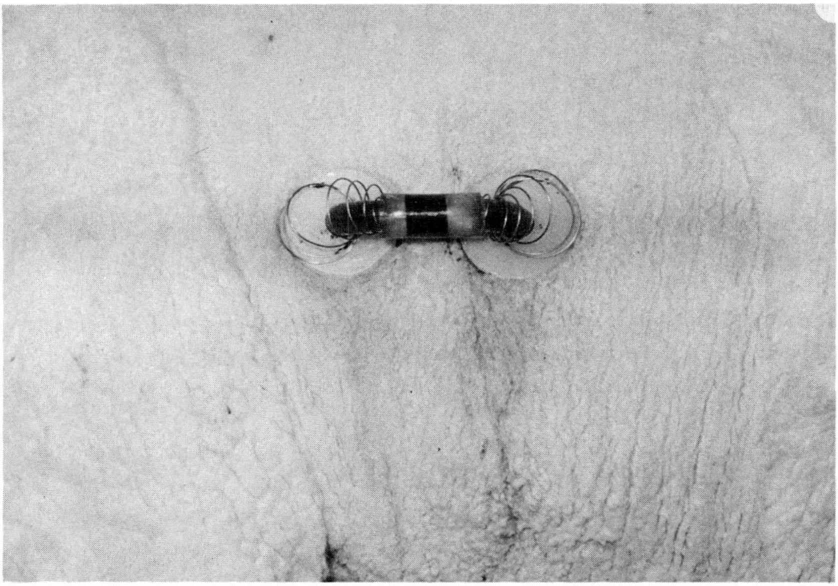

**FIGURE 3.11 Close-up view showing detail of bypass equipped with spring for maintaining pressure that permits leaking around the cannulae.**

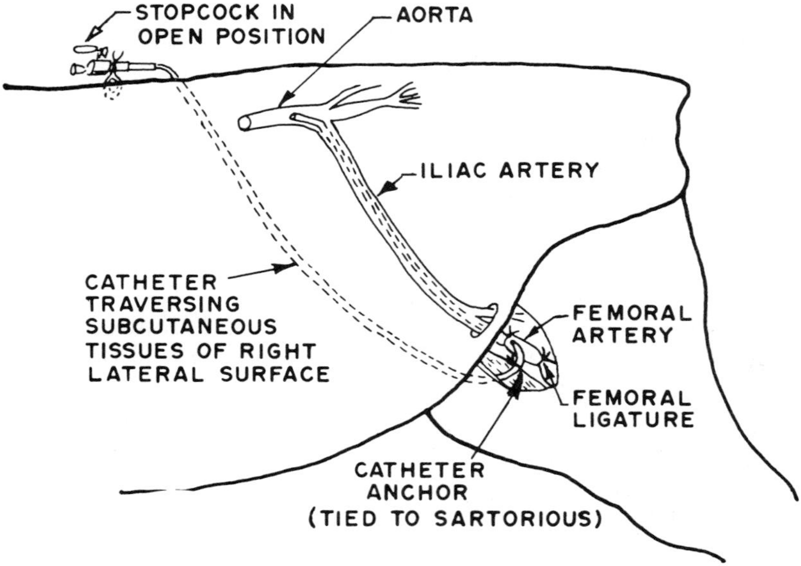


FIGURE 3.12 Chronically implanted arterial catheter for sheep.

### CANNULATION OF THE CEREBROSPINAL FLUID TRACT

The cerebrospinal fluid tract has been cannulated for fluid collection purposes (Buck, 1968). These experiments are usually of short duration because of infections and mechanical damage to the cord. Entry has been made into the cisterna magna.

### CANNULATION OF PANCREATIC DUCTS

It is sometimes desirable to collect the pancreatic juice. Pancreatic cannulations have been successful in a number of animals, e.g., sheep (Taylor, 1960) and chicken (Heatley *et al.*, 1965).

### CATHETER MATERIAL FOR CANNULATIONS

A number of materials have been used for catheters, such as polyethylene, polyvinyl, silastic, and Teflon. Judgment on the intended use, tempered with experience, must be used in selecting the best catheter material.

Gastric pouches in dogs are capably described elsewhere (Markowitz *et al.*, 1964). Gastric pouches in the ruminant have been quite successful (Hill and Gregory, 1951), and ruminal pouches have been successfully prepared (Tsuda, 1964). These are rather difficult to make, and one must be prepared for failure in about 50 percent of the preparations.

### ADRENALECTOMY

A single-stage operation for bilateral adrenalectomy of sheep, goats, and calves has been described (Buck and Bond, 1966; Whipp, 1968). It is also described for the dog and other animals (Markowitz *et al.*, 1964). Some prefer two unilateral operations, with proper supportive treatment.

### HYPOPHYSECTOMY

Hypophysectomies are fairly straightforward in some species. In cattle they are more difficult because of the deep-seated attachment of the hypophysis. A technique for hypophysectomy of calves has been described recently (Whipp, 1968). Nalbandov and Card (1943) hypophysectomized chickens.

### SPLENECTOMY

Splenectomies are not difficult, but in the horse unusual care must be exercised (Quinlan *et al.*, 1935; Witzel and Mullenax, 1964).

### THYMECTOMY—SURGICAL MANIPULATIONS

The method of surgical ablation of the thymus gland in neonatal dairy calves has been described by Coleman *et al.* (1966).

### ELECTROENCEPHALOGRAM ELECTRODES

Electroencephalograms have been made of pigs, using implanted electrodes (Mullenax and Dougherty, 1963). Holes are bored in the skull on either side of the midline using a surgical brace and a metal bit of appropriate size. Care must be taken to penetrate only the bone, exposing the dura mater. The holes are then threaded with a dye. Nylon screws with silver wire center inserts are screwed into the holes until the silver wire just touches the dura. The skin is sutured carefully around the head of the nylon screw. Healing is usually quick and effective. When in use, the electroencephalograph is connected to the exposed silver wire inserts. This method, with modifications, has proved very effective. The number and placement of the electrodes is optional (Buck, 1964).

### TRANSPLANTATION OF ORGANS

This is an immensely popular field and with surgical skill, backed by proper instrumentation and increased knowledge of antigenic and immune reactions, it may become increasingly important. Newer and more successful methods of storing and keeping organs and tissues in a dormant but living condition also have added impetus to this surgical field. Markowitz *et al.* (1964) discusses transplantation of organs in detail.

## ANIMALS POTENTIALLY USEFUL IN EXPERIMENTAL STUDIES

It is always desirable to have the right animal for a particular research goal. It is possible through proper selection of species or genetic lines, including mutations, to find one better suited than others. For example, cats are generally used for neurophysiologic experiments, dogs for circulation studies, and rats for comparative study of the lungs; monkeys and horses may be used because they have several similarities to man. For investigation of the eye, sheep, rabbits and swine are the best choice if similarity to man is desired. Featherless and scaleless chickens are suitable for hypothermic stresses. The precocious development of Japanese quail makes it a suitable pilot animal for many avian studies. It may produce eggs as early as 6 weeks after it has been hatched.

## REFERENCES

Archer, R., and C. Fromageot. 1957. The relationship of oxytocin and vasopressin to active proteins of posterior pituitary origin. Studies concerning the existence or non-existence of a single neurohypophysial hormone. p. 39. *In* H. Heller, (ed.) The neurohypophysis. Academic Press, New York.

Albert, A., E. Rosenberg, G. T. Ross, C. A. Paulsen, and R. J. Ryan. 1968. Report of the National Pituitary Agency collaborative study on the radioimmunoassay of FSH and LH. J. Clin. Endocrinol. 28:1214–1219.

Alexander, G. 1958. Heat production of newborn lambs in relation to type of birth coat. Proc. Aust. Soc. Anim. Prod. Biennal Conf. 2nd 10:14.

Altman, P. L., and D. S. Dittmer. (eds.). 1966. Environmental biology–Biological handbook. Fed. Am. Soc. Exp. Biol., Bethesda, Md.

Alvarez, M. B. 1968. Relation of environmental temperature to the activity of the adreno-sympathetic system of cattle. Ph.D. Dissertation, University of Missouri.

Alvarez, M. B., L. Hahn, and H. D. Johnson. 1969. Cutaneous moisture loss in the bovine during heat exposure and catecholamine infusion. J. Anim. Sci. 30:95.

Amer, M. 1968. An improved assay of parathyroid hormone. Endocrinology 82: 166.

American Physiological Society. 1959. Handbook of physiology. Neurophysiology, Vol. 1. Washington, D.C.

American Physiological Society. 1960. Handbook of physiology. Neurophysiology, Vol. 2. Washington, D.C.

American Physiological Society. 1960. Handbook of physiology. Neurophysiology, Vol. 3. Washington, D.C.

American Physiological Society. 1962. Handbook of physiology. Circulation, Vol. 1. Washington, D.C.

American Physiological Society. 1963. Handbook of physiology. Circulation, Vol. 2. Washington, D.C.

American Physiological Society. 1963. Handbook of physiology. Circulation, Vol. 3. Washington, D.C.

Andersson, B., and B. Larsson. 1961. Influence of local temperature changes in the preoptic area and rostral hypothalamus of the regulation of the food and water intake. Acta. Physiol. Scand. 52:75–89.

Arai, K., and T. H. Lee. 1967. A double-antibody radioimmunoassay procedure for ovine pituitary prolactin. Endocrinology 81:1041–1043.

Archibald, R. M. 1957. Nitrogen by the Kjeldahl method. Vol. 2, p. 91. Seligson, D. (ed.) Standard methods of clinical chemistry. Academic Press, New York.

Arnaud, C., and T. Littledike. 1966. The measurement of thyrocalcitonin in human and pig plasma by radioimmunologic means. J. Clin. Invest. 45:982.

Arnaud, C., T. Littledike, and T. Kaplan. 1968. Radioimmunoassay of calcitonin–A preliminary report. Mayo Clin. Proc. 43:496–502.

Avery, M. E. 1964. Lung and its disorders in the newborn infant. W. B. Saunders Co., Philadelphia.

Baghtiantz, A., G. V. Foster, A. Edwards, M. A. Kuman, E. Stack, H. W. Soliman, and I. MacIntyre. 1964. Extraction and purification of calcitonin. Nature 203: 1027.

Baird, D. T. 1968. A method for the measurement of estrone and estradiol-17$\beta$ in peripheral human blood and other biological fluids using $^{35}S$ pipsyl chloride. Endocrinol. Metab. 28(2):244–258.

Bangham, D. R., M. V. Mussett, and M. P. Stack-Dune. 1963. The second international standard for prolactin. Bull. World Health Organ. 29:721.

Barker, E. S., and J. K. Clark. 1960. Measurement of renal blood flow by the application of the Fick principle. Vol. 8. *In* H. D. Bruner (ed.) Methods in medical research. Year Book Medical Publ., Chicago.

Barrada, M. S. 1957. Responses of dairy cattle to hot environments with special emphasis on respiratory reactions. Ph. D. thesis. The Johns Hopkins University, Baltimore, Maryland.

Bates, R. W., M. M. Garrison, and J. Cornfield. 1963. An improved bio-assay for prolactin using adult pigeons. Endocrinology 73:217.

Beleslin, D., G. W. Bisset, J. Haldor, and R. L. Polak. 1967. The release of vasopressin without oxytocin in response to haemorrhage. Proc. Roy. Soc. (London) Ser. B. 166:443.

Ben-Daird, M. 1967. A sensitive bioassay based on $H^3$-methylthymidine uptake by the pigeon-crop mucous epithelium. Proc. Soc. Exp. Biol. Med. 125(3):705.

Benjamin, M. 1961. Outline of veterinary clinical pathology. 2nd ed. Iowa State University Press, Ames. pp. 124–135.

Benzinger, T. H., C. Kitzinger, and A. W. Pratt. 1963. The human thermostat in temperature—Its measurement and control in science and industry. Vol. 3, Part III. Reinhold Publishing Co., New York.

Bergman, R. K. 1963. Effects of a prolonged high environmental temperature on the glucocorticoid in the bovine. Ph.D. Dissertation. University of Missouri, Columbia.

Berman, A. 1957. Influence of some factors on the relative evaporation rate from the skin of cattle. Nature 179:1256, 4572.

Berson, S. A., and R. S. Yalow. 1964. Immunoassay of protein hormones. Vol. 4. *In* G. Pincus, K. V. Thimann, and E. B. Astwood (ed.) The hormones. Academic Press, New York.

Berson, S. A., and R. S. Yalow. 1968. Radioimmunoassay of ACTH in plasma. J. Clin. Invest. 47(12):2725.

Bianca, W. 1961. Heat tolerance in cattle. Intern. J. Biometeorol. 5:5–30.

Bianca, W. 1965. Reviews for the progress of dairy science. Section A—Physiology. Cattle in a hot environment. J. Dairy Res. 32:291.

Birge, C. A., G. T. Peake, I. K. Mariz, and W. H. Daughaday. 1967. Radioimmunoassayably growth hormone in the rat pituitary gland: effect of age, sex and hormonal state. Endocrinology 81:195.

Blaxter, K. L. (ed.). 1965. Energy metabolism. *In* Proceedings of third symposium, European association for animal production. Academic Press, New York.

Blaxter, K. L. 1966. The energy metabolism of ruminants. 2nd ed. Hutchinson, London.

Blaxter, K. L., N. McC. Graham, and J. A. F. Rook. 1954. Apparatus for the determination of the energy exchange of calves and of sheep. J. Agr. Sci. 45: 10–18.

Blaxter, K. L. 1966. The energy metabolism of ruminants. 2nd ed. Hutchinson, London.

Blinn, K. A., and W. K. Noell. 1949. The development of a method for continuous recording of alveolar carbon dioxide tension during hyperventilation test in routine EEG work. USAF School Aviat. Proj. 21-02-068.

Bourke, J. R., R. W. Hawker, and S. W. Manley. 1969. *In vitro* bioassay of serum thyrotrophin (TSH) in the normal rat. Endocrinology 84:1292–1295.

Bowers, C. Y., A. V. Schally, G. A. Reynolds, K. L. Lee, and W. D. Hawley. 1967. Some features of the regulation of thyrotropin (TSH) secretion. (Abstr.) Clin. Res. 15:61.

Bradley, S. E. 1963. The hepatic circulation. *In* Handbook of physiology. Circulation, Vol. 2, p. 1396. The American Physiological Society, Washington, D.C.

Brauer, R. W. 1963. Liver circulation and function. Physiol. Rev. 43:115–213.

Brobeck, J. R. 1955. Neural regulation of food intake. Ann. N.Y. Acad. of Sci. 63: 44–55.

Brodie, B. B. 1951. Measurement of total body water. Vol. 4. *In* M. B. Visscher (ed.) Methods in medical research. Year Book Medical Publ., Chicago.

Brodish, A. 1968. A review of neuroendocrinology 1966–67. Yale J. Biol. Med. 41(2):143.

Brody, S. 1945. Bioenergetics and growth. Reinhold Publishing Co., New York. pp. 310 and 314.

Brown, P. S. 1955. The assay of gonadotrophin from urine of non-pregnant human subjects. J. Endocrinol. 13:59.

Brown, J. B., and M. G. Coyle. 1963. Urinary excretion of oestriol during pregnancy. J. Obstet. Gynaec. Brit. Comonw. 70:219.

Brown, M. A., and M. M. Motasem. 1965. Comparisons of two techniques for measuring relative rates of moisture evaporation from limited skin areas of dairy cattle. J. Dairy Sci. 48:1643–1646.

Brozek, J., and A. Henschel. 1961. Techniques for measuring body composition. Nat. Acad. Sci.-Natl. Res. Coun., Washington, D.C. (Now available as Gov't. Research Report AD 286506, Office of Technical Services, U.S. Dept. Commerce.)

Bryant, G. D., and F. C. Greenwood. 1968. Radioimmunoassay for ovine, caprine and bovine prolactin in plasma and tissue extracts. Biochem. J. 109:831.

Buck, W. B. 1964. Functional and biochemical changes in brain tissues and cerebrospinal fluid during induced convulsions of sheep. Ann. N. Y. Acad. Sci. 111: 753–754.

Buck, W. B. 1968. Personal communication.

Buck, W. B., and K. Bond. 1966. A single-stage operation for bilateral adrenalectomy of sheep and goats. Am. J. Vet. Res. 27:155–160.

Buck, W. B., K. Bond, and N. C. Lyon. 1965. A technic for permanent cannulation of the carotid artery of sheep and goats. Cornell Vet. 55:154–163.

Burrows, W. 1939. Artificial insemination of chickens and turkeys. U.S. Dept. Agr. Cir. 525.

Bush, I. E. 1962. Chemical and biological factors in the activity of adrenocortical steroids. Pharm. Rev. 14:317.

Bustad, L. K., and R. O. McClellan. 1966. Swine in biomedical research. Frayn Publishing Company, Seattle, Washington.

Butt, W. R. 1967. Hormone chemistry. D. Van Nostrand, London.

Cain, J. R., U. K. Abbott, and V. L. Ragelo. 1967. Heart rate of the developing chick embryo. Proc. Soc. Exp. Biol. Med. 126:507–510.

Calatayud, J. B., P. A. Gorman, and C. A. Caceres. 1965. Electronic monitoring of physiological phenomena in experimental animals. Vol. II, Chap. 11. *In* W. I. Gay (ed.) Methods in animal experimentation. Academic Press, New York.

Cargill-Thompson, H. E. C., and G. P. Green. 1963. Studies on the effect of hormone administration on bodyweight and tibial epiphysical cartilage width in intact, hypophysectomized and adrenalectomized rats. J. Endocrinol. 25:473.

Carstensen, H. 1967. Steroid hormone analysis. Vol. I. Marcel Dekker, New York.

Casida, L. E. 1969. Research techniques in physiology of reproduction in the female. *In* Techniques and procedures in animal production research. Amer. Soc. Anim. Prod. c/o Q Corp., Albany, New York.

Caspi, E., and F. G. Scrimshaw. 1967. Elementary aspects of infra-red spectroscopy of steroids. Vol. I, p. 55. H. Carstensen (ed.) Steroid hormone analysis. Marcel Dekker, New York.

Chan, L. T., D. Wudde, and M. Saffran. 1969. Comparison of assays for corticotrophin releasing activity. Endocrinology 84(4):967.

Clamans, H. G. 1952. Continuous recording of oxygen carbon dioxide and other bases in sealed cabins. J. Aviat. Med. 23:330.

Clark, P. M., and S. J. Folley. 1953. Some observations on prolactin assays by the pigeon crop weight method. *In* Ciba Found. Colloquia Endocrinol. 5:90. Churchill, London.

Coghlan, J. P., and J. R. Blair-West. 1967. Aldosterone. *In* C. H. Gray and A. L. Bacharach (ed.) Hormones in blood. Academic Press, New York.

Cole, H. H., and P. T. Cupps. 1969. Reproduction in domestic animals. Vol. 2, 2nd ed. Academic Press, New York.

Coleman, G. L., M. L. Crandall, and A. J. Guidry. 1966. Surgical ablation of the thymus gland in neonatal dairy calves. Am. J. Vet. Res. 27:1123–1126.

Collins, W. P., and I. F. Sommerville. 1964. Quantitative determination of progesterone in human plasma by thin layer and gas liquid radiochromatography. Nature (Lond.) 203:836.

Comroe, J. H., Jr., R. E. Forster II, A. B. Dubois, W. A. Briscoe, and E. Carlsen. 1963. The lung–Chemical physiology and pulmonary function tests. 2nd ed. Year Book Medical Publishers Inc., Chicago. 390 pp.

Condliffe, P. G., and J. Robbins. 1967. Pituitary thyroid-stimulating hormone and other thyroid stimulating substances. 2nd ed., Vol. 1. *In* C. H. Gray, and A. L. Bacharach (ed.) Hormones in blood. Academic Press, New York.

Consoliazio, C. R., R. E. Johnson, and E. Marek. (ed.). 1951. Metabolic methods. Mosby Company, St. Louis.

Cooper, C. W., P. F. Hirsch, S. U. Tonerud, and P. O. Munson. 1967. An improved method for the biological assay of thyrocalcitonin. Endocrinology 81:610.

Coppedge, R. L., and A. Segaloff. 1951. Urinary prolactin excretion in man. J. Clin. Endocrinol. 11:465.

Copp, D. H., D. W. Cocheroft, and Y. Kuch. 1967. Calcitonin from ultimobranchial glands of dogfish and chicken. Science 158:924.

Cornelius, C. E., and J. K. Kaneko. (ed.). 1963. Clinical biochemistry of domestic animals. Academic Press, New York. 678 pp.

Crighton, D. B. 1968. Depletion of rat ovarian ascorbic acid by a factor other than leutinizing hormone present in the blood of the pig. J. Reprod. Fertil. 15:457.

Croft, P. G. 1964. An introduction to the anesthesia of laboratory animals. 2nd ed., Universities Federation for Animal Welfare, London.

Cullen, G. E. 1922. Studies of acidosis. XIX. The colorimetric determination of the hydrogen ion concentration of blood plasma. J. Biol. Chem. 52:501.

Damm, H. C., P. K. Besch, and A. J. Goldwyn. 1966. Handbook of biochemistry and biophysics. World, Cleveland.

Danowski, T. S. 1944. Use of thiourea as a measure of change in body water. J. Biol. Chem. 152:207.

Davis, T. R. A., and J. Mayer. 1954. Imperfect homeothermia in the hereditary obese hyperglycemic syndrome of mice. Am. J. Physiol. 177:22.

Daughaday, W. H., G. T. Peake, C. A. Birge, and I. K. Mariz. 1967. The influence of endocrine factors on the concentration of growth hormone in the rat pituitary. Excerpta Med. 142:14–15 (Abstr.).

Deane, N. 1952. Methods of study of body water compartments (space techniques) Vol. 5. *In* A. C. Corcoran (ed.) Methods in medical research. Year Book Medical Publ., Chicago.

Deane, N., G. E. Schreiner, and J. S. Robertson. 1951. The velocity of distribution of sucrose between plasma and interstitial fluid, with reference to the use of sucrose for the measurement of extracellular fluid in man. J. Clin. Invest. 30: 1463.

Deboo, G. J., and R. S. Jenkins, 1965. A technique for recording a noise-free electrocardiogram from a chicken embryo still in its shell. Med. Elec. Biol. Eng. 3: 443–445.

Denenberg, V. H., and E. M. Banks. 1962. Techniques of measuring and evaluation. Chap. 9. *In* E.S.E. Hafez (ed). The behavior of domestic animals. Williams & Wilkins Co., Baltimore.

Dhariwal, A. P. S., R. Naller, M. Batt, and S. M. McCann. 1965. Separation of FSH-releasing factor from LH-releasing factor. Endocrinology 76:290.

Dhariwal, A. P. S., S. L. Krulich, and S. M. McCann. 1969. Purification of a growth hormone inhibiting factor (GIF) from sheep hypothalamus. Neuroendocrinology 4:282.

Dickerman, E., A. Negro-Vilar, and J. Meites. 1969. *In vitro* assay for growth hormone releasing factor (GHRF). Neuroendocrinology 4(2):75.

Dixon, P. F., M. Booth, and J. Butler. 1967. Corticosteroids. *In* C. H. Gray and A. L. Bachavoch (ed.) Hormones in blood. Academic Press, New York.

Dominguez, O. V. 1967. Vol. I., p. 35. *In* H. Carstensen (ed.) Steroid hormone analysis. Marcel Dekker, New York.

Donald, R. A. 1967. Rapid method for extracting corticotrophin from plasma. J. Endocrinol. 39:451–452.

Dorfman, R. I. (ed.). 1962. Methods in hormone research, Vol. 2. Academic Press, New York.

Dorfman, R. I., and A. S. Dorfman. 1954. Estrogen assays using the rat uterus. Endocrinology 55:65.

Dougherty, R. W. 1955. Permanent stomach and Internal fistulas in ruminants: Some modifications and simplifications. Cornell Vet. 45(3):331-357.

Douglas, W. W., and A. M. Poisner. 1964. Stimulus-secretion coupling in a neurosecretory organ. The role of calcium in the release of vasopressin from the neurohypophysis. J. Physiol. Lond. 172:1.

Earle, D. P., and R. W. Berliner. 1946. A simplified clinical procedure for measurement of glomerular filtration rate and renal plasma flow. Proc. Soc. Exp. Biol. Med. 62:262.

Eberlein, W. R., J. Winter, and R. L. Rosenfield. 1967. Androgens. *In* C. H. Gray and A. L. Bacharach (ed.) Hormones in blood. Academic Press, New York.

Eisenstein, A. B. (ed.) 1967. The adrenal cortex. Little Brown, Boston.

Elson, J. 1963. Use of electrocardiograph. Vol. 2, Chap. 110. *In* S. Frankel and S. Reitman (ed.) Clinical laboratory methods and diagnosis. Mosby Company, St. Louis.

Engleman, K., B. Portony, and W. Lovenberg. 1968. A sensitive and specific double isotope derivative method for the determination of catecholomines in biological specimens. Am. J. Med. Sci. 255:259.

Erlanger, J. 1916. Mechanism of oscillatory criteria. Am. J. Physiol. 39:401.

Evans, H. M., L. L. Sparks, and J. S. Dixon. 1966. The physiology and chemistry of adrenocorticotrophin. Vol. 1, p. 317. *In* G. W. Harris and B. T. Donovan (ed.) Pituitary gland. University of California Press, Los Angeles, California.

Findlay, J. D. 1950. The effects of temperature, humidity, air movement, and solar radiation on the behavior and physiology of cattle and other farm animals. Hannah Dairy Res. Inst. Bull. 9. Ayr, Scotland.

Finkelstein, M. 1952. Fluorometric determination of micro amounts of oestrone-oestradiol and oestriol in urine. Acta. Endocrinol. Copenh. 10:149.

Finkelstein, M., E. Forchielli, and R. I. Dorfman. 1961. Estimation of testosterone in human plasma. J. Clin. Endocrinol. Metab. 21:98.

Fishman, J., and J. B. Brown. 1962. Quantitation of urinary estrogens by gas chromatography. J. Chromatogr. 8:21.

Forsyth, I. A. 1967. Prolactin and placental lactogens. Vol. I., p. 234. *In* Hormones in blood. 2nd ed. Academic Press, New York.

Fortier, C. 1962. Adenohypophysis and adrenal cortex. Ann. Rev. Physiol. 24:223.

Fowler, R. C. 1949. A rapid infra-red gas analyzer. Rev. Sci. Instr. 20:175.

Franke, E. K. 1966. Physiologic pressure transducers. Vol. 11, pp. 137–161. *In* Methods in medical research. Year Book Medical Publishers, Chicago.

Franklin, D. L., W. A. Schlegel, and N. W. Watson. 1963. Ultrasonic doppler shift flood flowmeter. Vol. 1, pp. 309–315. *In* E. Alt (ed.) Circuitry and practical consideration in biomedical sciences instrumentation. Plenum Press, New York.

Fraschini, F., M. Motta, and L. Martini. 1966. Methods for evaluation of hypothalmic hypophysiotropic principles. *In* P. Mantegazza and F. Piccinini (ed.) Methods in drug evaluation. North Holland, Amsterdam.

Freiden, E. H., and J. M. Bozer. 1951. Effect of administration of intermedin upon melanin content of the skin of *Rana pipiens*. Proc. Soc. Exp. Biol. Med. 77:35.

Fry, D. L. and J. Ross, Jr. 1966. Survey of flow detection technics. Vol. 11, Sec. 2, p. 50. *In* R. F. Rushmer (ed.) Methods of medical research. Year Book Medical Publishers, Chicago.

Fry, F. E. J. 1949. A simple gas analyzer. Can. J. Res. 27:188.

Futterweit, W., N. L. McNiven, L. Narcus, C. Lantos, M. Prosdowsky, and R. I. Dorfman. 1963. Gas chromatographic determination of testosterone in human urine. Steroids 1:628–642.

Gans, E., and G. P. Van Rees. 1966. Studies on the testicular augmentation assay method for follicle stimulating hormone. Acta. Endocrinol. 52:573.

Garcia, J. F., and I. I. Geschwind. 1968. Investigation of growth hormone secretion in selected mammalian species. p. 267. *In* A. Recile and E. E. Muller (ed.) Growth hormone. Excerpta Med. Intern. Cong. Ser. No. 158.

Gay, W. I. 1965. Methods in animal experimentation. Academic Press, New York.

Geddes, L. A., and L. E. Baker. 1968. Principles of applied biomedical instrumentation. John Wiley & Sons, New York.

Glick, S. M., P. Jumaresan, A. Kagan, and M. Wheeler. 1968. Radioimmunoassay of oxytocin. p. 81. *In* Protein and polypeptide hormones. Excerpta Med. Intern. Cong. Ser. 161.

Goepfert, H., A. W. Von Eiff, and C. Howind. 1953. Quantitative Beziehungen zwischen Energiestoff-Wechsel und reflektorischen Muskeltonus bei der Thermoregulation. Z. Ges. Exp. Med. 120:308.

Gordon, E. E., and C. A. Villee. 1956. An *in vitro* assay for estradiol 17$\beta$ and esterone. Endocrinology 58:150.

Gradwohl, R. B. H. 1956. Clinical laboratory methods and diagnosis, 5th ed., Vol. 1. Mosby Co., St. Louis.

Graham, G. S. 1918. Genzidine as a peroxidase reagent for blood smears and tissue. J. Med. Res. 39:15.

Gray, C. H., and A. L. Bacharach. (ed.). 1967. Hormones in blood. 2nd Ed. Academic Press, New York.

Gray, S. J., and K. Sterling. 1950. Determination of circulating red cell volume by radioactive chromium. Science 112:179.

Greenman, D. L., and M. X. Zarrow. 1961. Steroids and carbohydrate metabolism in the domestic bird. Proc. Soc. Exp. Biol. Med. 106:459.

Greenspan, F. S., C. Li., M. E. Simpson, and H. M. Evans. 1949. Bioassay of hypophyseal growth hormone: The tibia test. Endocrinology 45:455.

Greenwood, A. W., J. S. S. Blyth, and R. K. Callow. 1935. Quantitative studies on the response of the capon's comb to androsterone. Biochem. J. 29:1400.

Gregersen, M. I., J. G. Gibson, and E. A. Stead. 1935. Plasma volume determination with dyes; errors in colorimetry; use of the blue dye T-1824. Am. J. Physiol. 113:54.

Grosvenor, C. E., S. M. McCann, and R. Nallar. 1965. Inhibition of nursing induced and stress induced fall in pituitary prolactin concentration in lactating rats following injection of acid extracts of bovine and rat hypothalamus. Endrocrinology 76:883.

Guidry, A. J., and R. E. McDowell. 1966. Tympanic membrane temperature for indicating rapid changes in body temperature. J. Dairy Sci. 49:74–77.

Guidry, A. J., and H. S. Hofmeyr. 1968. A hygrometric tent system for measuring evaporative losses in sheep. Proc. S. Afr. Soc. Anim. Prod. 7:195–197.

Guillemin, G. 1968. Hypothalamic hormones regulating the secretions of anterior pituitary. *In* G. Josmin (ed.) Endocrine aspects of disease processes. Warren Green Inc., St. Louis, Mo.

Guillemin, R., and A. V. Schally. 1963. Concentration of corticotropin releasing factor (CRF) in acid extracts of sheep hypothalamus by gel filtration and counter distribution. Texas Rep. Biol. Med. 21:541.

Guillemin, R., G. W. Clayton, H. S. Lipscomb, and J. D. Smith. 1959. Fluoremetric measurement of rat plasma and adrenal corticosterone concentration. J. Lab. Clin. Med. 53:830.

Guillemin, R. H., G. W. Clayton, J. D. Smith, and H. S. Lipscomb. 1958. Simultaneous measurements of plasma corticosterone levels and adrenal ascorbic acid levels. Fed. Proc. 17:247.

Gupta, K. K. 1968. Estimating plasma anti-diuretichormone levels. Brit. Med. J. 5624:185.

Hafez, E. S. E. 1959. Reproductive capacity of farm animals in relation to climate and nutrition. J. Am. Vet. Med. Assoc. 135:606–614.

Hafez, E. S. E. 1961. Behavior of farm animals in relation to management and experimental designs. Proc. 8th Intern. Tierzuchtkongress in Hamburg. Verlag Eugen, Ulmer, Stuttgart.

Hafez, E. S. E. 1968. Adaptation of domestic animals. Lea and Febiger, Philadelphia, Pa.

Hahn, P. F., and G. Rouser. 1950. Vol. 1, p. 424. *In* O. Glasser (ed.) Medical physics. Year Book Medical Publishers, Chicago.

Hall, L. W. 1966. Veterinary anaesthesia and analgesia. 6th ed. Williams & Wilkins Co., Baltimore.

Hall, S. R. 1944. Prolactin assay by a comparison of the two crop-sacs of the same pigeon after local injection. Endocrinology 34:14–26.

Hammel, H. T., D. C. Jackson, J. A. J. Stolwijk, J. D. Hardy, and S. B. Stromme. 1963. Temperature regulation by hypothalamic proportional control with an adjustable set point. J. Appl. Physiol. 18:1146–1154.

Harris, G. W. 1960. Central Control of pituitary secretion. Vol. 2, p. 1007. *In* J. Field (ed.) Handbook of physiology. I. Neurophysiology. Am. Physiol. Soc., Washington, D.C.

Harris, J. I., and A. B. Lerner. 1957. Amino acid sequence of the $\alpha$-melanocyte stimulating hormone. Nature 179:1346.

Harris, J. I., and P. Roos. 1956. Amino acid sequence of a melanophore stimulating peptide. Nature 178:90.

Harrison, T. A. 1969. The introduction and maintenance of permanently indwelling catheters in the portal and hepatic veins of sheep. J. Physiol. 200:1–28.

Hartog, M., A. D. Wright, T. R. Fraser, and C. R. C. Heard. 1967. A radioimmunoassay for dog plasma growth hormone. Excerpta Med. 142:29 (Abstr.).

Hays, L., A. Goswity, and B. E. Murphy. 1967. Radioisotopes in medicine *in vitro* studies. USAEC Symp. Ser. 13, U.S. Nat. Techn. Infor. Serv., Springfield, Virginia

Heald, T. J., B. E. Furnival, and K. A. Rookledge. 1967. Changes in the levels of leutinizing hormones in the pituitary of the domestic fowl during ovulatory cycle. J. Endocrinol. 37:73–81.

Heap, R. B. 1964. A fluorescence assay of progesterone. J. Endocrinol. 30:293.

Heatley, N. G., E. P. McElheny, and S. Lepkovsky. 1965. Measurement of the rate of flow of pancreatic secretion in the anesthetized chicken. Comp. Biochem. Physiol. 16:29–36.

Hems, R., B. D. Ross, M. N. Berry, and H. A. Krebs. 1966. Gluconeogenesis in the perfused rat liver. Biochem. J. 101:284–297.

Henderson, L. J., and W. W. Palmer. 1912. The intensity of urinary acidity in normal and pathological conditions. J. Biol. Chem. 13:394.

Heroux, O., J. S. Hart, and F. Depocas. 1956. Metabolism and muscle activity of anesthetized warm and cold acclimated rats on exposure to cold. J. Appl. Physiol. 9:399.

Hertz, R. 1950. Growth in the hypophysectomized rat sustained by pituitary grafts. Endocrinology 65:929.

Hill, K. J., and R. A. Gregory. 1951. The preparation of gastric pouches in the ruminant. Vet. Rec. 63:647–652.

Hill, K. J., and R. A. Gregory. 1957. The preparation of gastric pouches in the ruminant. Vet. Rec. 63:41.

Hiller, A., J. Plazin, and D. D. Van Slyke. 1948. A study of conditions for Kjeldahl determination of nitrogen in protein. Description of methods with mercury as catalyst, and titrimetric and gasometric measurements of ammonia formed. J. Biol. Chem. 176:1401–1420.

Hipsley, E. H., and J. W. McKellar. 1959. The capacity of plasma for binding vasopressin in pregnant and non-pregnant human subjects. J. Endocrinol. 19:345.

Hodges, J. R., M. T. Jones, and M. A. Stockham. 1968. Control of corticotrophin secretion. Vol. 14(3), p. 215. *In* Handbook of experimental pharmacology. Springer–Verlag, New York.

Hooker, C. W., and T. R. Forbes. 1947. Bioassay for minute amounts of progesterone. Endocrinology 41:158.

Howe, C. M. 1960. Tests for liver function in domestic animals. Vet. Rev. Annot. 6:1–26.

Huang, W. Y. 1968. A rapid reliable method of urinary estriol determination in pregnancy. Steroids 11:453.

Hudson, B., J. Coghlan, A. Dulmanis, M. Wintour, and I. Ekkel. 1963. The estimation of testosterone in biological fluids. Aust. J. Exp. Biol. Med. Sci. 41:235.

Hughes, J. P. 1962. A simplified instrument for obtaining liver biopsy in cattle. Am. J. Vet. Res. 23:1111–1113.

Hume, D. M. 1958. A method of hypothalamic regulation of pituitary and adrenal secretion in response to trauma. p. 217. *In* S. B. Curri and L. Martini (ed.) Pathophysiologia diencephalica. Springer-Verlag, Wien, Austria.

Hunter, W. M. 1967. Radioimmunological assay of FSH and LH. p. 92. *In* E. T. Bell and J. A. Loraine (ed.) Recent research on gonadotrophic hormones. Livingstone, Edinburgh.

Ibayashi, H., N. Nakamura, T. Ichikawa, T. Tanioka, and K. Nakao. 1964. The determination of urinary testosterone using thin layer chromatography and gas chromatography. Steroids 3:559.

Igarashi, M., and S. M. McCann. 1964. A new sensitive bio-assay for follicle-stimulating hormone (FSH). Endocrinology 74:440.

Igarashi, M., R. Nallar, and S. M. McCann. 1964. Further studies on the follicle stimulating hormone releasing action of hypothalamic extracts. Endocrinology 75:901.

Ishida, Y., A. Kuroshima, C. Y. Bowers, and A. V. Schally. 1965. *In vivo* depletion of pituitary growth hormone by hypothalamic extracts. Endocrinology 77:759.

Ismail, A. A. A., and R. A. Harkness. 1966. Factors associated with alterations in urinary testosterone levels. Proc. Soc. Endocrinol. 35:xx.

Ittrich, G. 1960. Zeitschrift für physiologische Chemie. Hoppe-Seylers, Z. Physiol. Chem. 320:103.

James, V. H. T., and J. London (ed.). 1968. The investigation of hypothalamic pituitary adrenal function. University Press, Cambridge.

John, E. R. 1966. Recording gross neural potentials from unanaesthetized animals. Vol. 11, Sec. V, p. 251. *In* R. F. Rushmer (ed.) Methods in medical research. Year Book Medical Publishers, Chicago.

Johnson, J. A., J. D. Nash, and F. M. Fusaro. 1963. An enzymatic method for the quantitative determination of glycogen. Anal. Biochem. 5:379–387.

Kanematsu, S., and C. H. Sawyer. 1963. Effects of intrahypothalmic and intrahypophyseal estrogen implants on pituitary prolactin and lactation in the rabbit. Endocrinology 72:243.

Kasten, A. J., and A. V. Schally, 1966. MSH activity in pituitaries of rats treated with hypothalmic extracts. Gen. Comp. Endocrinol. 7:452.

Katz, L. N. and A. Pick. 1956. Clinical electrocardiography. Lea and Febiger, Philadelphia.

Katz, S., L. Krulich, and S. M. McCann. 1966. Effect of insulin induced hypoglycemia on the concentration of growth hormone releasing factor (GHRF) in plasma and hypothalamus of rats. Fed. Proc. 25(2) pt I:Abs. No. 17.

Katz, S. H., A. P. S. Dhariwal, and S. M. McCann. 1967. Effect of hypoglycemia on the content of pituitary growth hormone (GH) and hypothalamic growth hormone releasing factor in the rat. Endocrinology 81:333.

Kay, R. H. 1964. Experimental biology. *In* Measurement and analysis. Reinhold, New York. 416 p.

Kelly, C. F., T. E. Bond, and C. Lorenzen, Jr. 1949. Instrumentation for animal shelter research. Agr. Eng. 30(6):297–304.

Kelly, W. G., S. R. Ranucci, and J. C. Shaver. 1968. A rapid and precise method for the estimation of production rate of cortisol in man. Steroids 11:429.

Kemp, A., and A. J. H. Kits Van Heijningen. 1954. A colorimetric micromethod for the determination of glycogen in tissue. Biochem. J. 56:646–648.

Kibler, H. H. 1960. Energy metabolism and related thermoregulatory reactions in Brown Swiss, Holstein, and Jersey calves during growth at 50° and 80°F temperatures. Mo. Agr. Exp. Sta. Res. Bull. 743.

Kibler, H. H. 1965. Mercury avoidance in collecting and displacing gas samples. J. Dairy Sci. 48:405–406.

Kibler, H. H., and S. Brody. 1950. Influence of temperature, 5° to 95°F, on evaporative cooling from the respiratory and exterior body surfaces of Jersey and Holstein cows. Mo. Agr. Exp. Sta. Bull. 461.

Kibler, H. H., and R. G. Yeck. 1959. Vaporization rates and heat tolerance in growing Shorthorn, Brahman, and Santa Gertrudes calves raised at constant 50° and 80°F temperatures. Mo. Agr. Exp. Sta. Res. Bull. 701.

Kirkham, K. E. 1966. The detection and measurement of thyrotropic substances in health and disease. *In* R. S. Harris, I. G. Wool, and J. A. Loraine (ed.) Vitamins and hormones. Academic Press, New York.

Kleiber, M. 1940. A respiration apparatus for serial work with small animals particularly rats. Univ. Calif. Publ. Physiol. 8:207–220.

Kleiber, M. 1958. Some special features of the California apparatus for respiration trials with large animals. *In* Symposium on energy metabolism. European Association for Animal Production, Copenhagen.

Kleiber, M. 1961. The fire of life. John Wiley & Sons, New York. 454 pp.

Knobil, E. 1966. The pituitary growth hormone—An adventure in physiology. Physiologist 9:25.

Kovacic, N. 1968. Mouse hyperaemic corpora lutea assay of prolactin. J. Reprod. Fertil. 15:259.

Kragt, C. L., and J. Meites. 1965. Stimulation of pigeon pituitary prolactin release by pigeon hypothalamic extract *in vitro*. Endocrinology 76:1169.

Krulich, L., A. P. S. Dhariwal, and S. M. McCann. 1965. Growth hormone-releasing activity of crude ovine hypothalamic extracts. Proc. Soc. Exp. Biol. Med. 120: 180.

Krulich, L., A. P. S. Dhariwal, and S. M. McCann. 1967. Stimulatory and inhibitory hypothalamic factors regulative secretion of growth hormone. p. 87. Proc. 49th Meeting, Endocrinol. Bal Harbour, Florida.

Krulich, L., A. P. S. Dhariwal, and S. M. McCann. 1968. Stimulatory and inhibitory effects of purified hypothalamic extracts on growth hormone release from rat pituitary *in vitro*. Endocrinology 83(4):783.

Kulin, H. W., A. B. Rifkind, and G. T. Gross. 1968. Human luetinizing hormone (LH) activity in processed and unprocessed urine measured by radioimmunoassay and bioassay. J. Clin. Endocrinol. 28:543.

Kwa, H. G., and F. Verhofstad. 1967. Radioimmunoassay of rat prolactin. Biochem. Biophys. Acta. 133:186.

Lamb, E. J., W. J. Dignam, R. J. Pinon, and H. H. Simmer. 1964. Plasma androgens in women. Acta. Endocrinol. Copenh. 45:243.

Landgrebe, F. W., and H. Waring. 1944. Biological assay and standardization of melanophore expanding pituitary hormone. Quart. J. Exp. Physiol. 33:1.

Lane-Petter, W. 1963. Animals for research–Principles of breeding and management. Academic Press, New York.

Lee, D. H. K. 1953. Manual of field studies on the heat tolerance of domestic animals, FAO Develop. paper #38, United Nations, Rome, Italy.

Lee, D. H. K., and R. W. Phillips. 1948. Assessment of the adaptability of livestock to climatic stress. J. Anim. Sci. 7:391.

Liebel, D. S., and G. A. Wrenshall (ed.). 1965. On the nature and treatment of diabetes. Excerpta Med. Intern. Cong. Ser. No. 84.

Lerner, A. B., K. Shigume, and I. Bunding. 1954. The mechanism of endocrine control of melanin pigmentation. J. Clin. Endocrinol. Metab. 14:1463.

Lewis, D., K. J. Hill, and E. F. Annison. 1957. Studies on the portal blood of sheep. Biochem. J. 66:587–592.

Lilly, J. C. 1950. Physical methods of respiratory gas analysis. Vol. 2. *In* J. H. Comre, Jr. (ed.) Methods in medical research Year Book Medical Publishers, Chicago.

Lodge, J. R., R. C. Lewis, and E. P. Reineke. 1957. Estimating thyroid activity of dairy heifers. J. Dairy Sci. 40:209.

Loraine, J. A., and E. T. Bell. 1966. Hormone assays and their clinical application. Williams and Wilkins Co., Baltimore

Lucas, A. M., and C. Jamroz. 1961. Atlas of avian hematology Agr. Monogr. 25. U.S. Dept. Agr., Washington, D.C.

Luft, K. 1943. Über eine neue Methode der registrieren Gas Analyse mit Hilfe der Absorption ultraroter Strahlen ohne spektrale Zeriegung. Ztschr. f. techn. Phys. 24:97.

Lumb, W. V. 1963. Small animal anesthesia. Lea and Febiger, Philadelphia, Pa.

Lurie, A. O., C. A. Villee, and D. E. Reid. 1966. Progesterone in the blood. A quantitative method employing gas liquid chromatography. J. Clin. Endrocrinol. Metab. 26:742.

Machlin, L. J., M. Horino, D. M. Kipnis, S. L. Phillips, and R. S. Gordon. 1967. Stimulation of growth hormone secretion by median eminence extracts in the sheep. Endocrinology 80:205.

Maickel, R. P. 1970. Interaction of drugs with autonomic nervous function and thermoregulation. Fed. Proc. 29:1973.

Manley, S. W., J. R. Bourke, and R. W. Hawker. 1969. An *in vitro* bioassay of thyrotrophin (TSH). Endocrinology 84(5):1286.

Mark, D. D., and A. Zimmer. 1967. Atlas of clinical laboratory procedures. Vol. I. Clinical chemistry. McGraw-Hill, New York.

Markowitz, J., J. Archibald, and H. G. Downie. 1964. Experimental surgery. 5th ed. pp. 191–243. Williams & Wilkins Co., Baltimore

Martin, J. M. 1968. Personal communication.

Martin, L. 1960. The use of 2,3,5-triphenyltetrazolium chloride in the biological assay of oestrogens. J. Endocrinol. 20:187.

Martini, L., F. Fraschini, and M. Motta. 1967. 2nd International congress on hormonal steroids. Excerpta Med. 132:1173.

McCann, S. M., and A. P. S. Dhariwal. 1966. Hypothalamic releasing factors and neurovascular link between the brain and the anterior pituitary. Vol. 1. *In* L. Martini and W. Ganong (ed.) Neuroendocrinology. Academic Press, New York.

McCann, S. M., and J. C. Porter. 1969. Hypothalamic pituitary stimulating and inhibiting hormones. Physiol. Rev. 49(2):241.

McCarthy, J. L., R. C. Corley, and M. X. Zarrow. 1959. Effect of goitrogens on adrenal gland of rat. Am. J. Physiol. 197:693.

McCarthy, J. L., R. C. Corley, and M. X. Zarrow. 1960. Diurnal rhythm in plasma corticosterone and lack of diurnal rhythm in plasma compound F-like material in the rat. Proc. Soc. Exp. Biol. Med. 104:787.

McCullagh, D. R., and W. K. Cuyler. 1936. The response of the capon's comb to androsterone. J. Pharmacol. 66:379.

McDowell, R. E., D. H. K. Lee, and M. H. Fohrman. 1953. The relationship of surface area to heat tolerance in Jerseys and Sindhi-Jersey ($F_1$) crossbred cows. J. Anim. Sci. 12:747–764.

McDowell, R. E., D. H. K. Lee, and M. H. Fohrman. 1954. The measurement of water evaportation from limited area of a normal body surface. J. Anim. Sci. 13:405–416.

McDowell, R. E. 1966. The role of physiology in animal production for tropical and sub-tropical areas. World Rev. Anim. Prod. 1:39–46.

McDowell, R. E., P. C. Underwood, R. H. Lehmann, and M. S. Barrada. 1966. Procedure for exteriorizing the carotid artery in the bovine. J. Dairy Sci. 49: 77–80.

McKenzie, J. M. 1960. Bioassay of thyrotropin in man. Physiol. Rev. 70:398.

McLean, J. A. 1963a. Measurement of cutaneous moisture vaporization from cattle by ventilated capsules. J. Physiol. 167:417–426.

McLean, J. A. 1963b. The partition of insensible losses of body weight and heat from cattle under various climatic conditions. J. Physiol. 167:427–447.

Meites, J., and N. J. Fiel. 1965. Effect of starvation of hypothalamic content of "somatotropin releasing factor" and pituitary growth hormone content. Endocrinology 77:455.

Meites, J., and N. J. Fiel. 1967. Effects of thyroid hormone on hypothalamic content of GHRF. Intern. Symp. Growth. Milan, Italy.

Menini, E. 1965. Gas-liquid chromatography of urinary oestrogens. Biochem. J. 94:15.

Midgley, A. R., Jr. 1966. Radioimmunoassay: A method for human chorionic gonadotropin and human luteinizing hormone. Endocrinology 79:10.

Mitchell, M. L., S. Collins, and J. Byron. 1969. Radioimmunoassay of growth hormone by enzyme partitition. J. Clin. Encocrinol. Metab. 29(2):257.

Moncola, F., F. G. Peron, and R. I. Dorman. 1959. The fluorometric determination of corticosterone in rat adrenal tissue and plasma: Effect of administering ACTH subcutaneously. Endocrinology 65:717.

Monroe, S. E., A. F. Parlow, and A. R. Midgley. 1968. Radioimmunoassay for rat LH. Endocrinology 85(5):1004.

Moodie, E. W., A. I. T. Walker, and P. H. Hutton. 1963. The collection of portal and hepatic venous blood in conscious sheep. J. Exp. Physiol. 48:379–387.

Mount, L. E. 1968. The climatic physiology of the pig. Williams & Wilkinson Co., Baltimore.

Muller, E. E., A. Arimura, S. Dawano, T. Saito, and A. V. Schally. 1967a. Growth hormone-releasing activity in the hypothalamus and plasma of rats subjected to stress. Proc. Soc. Exp. Biol. Med. 125:874.

Muller, E. E., S. Shinji, and A. V. Schally. 1967b. Growth hormone-releasing activity in the hypothalamus of animals of different species. Gen. Comp. Endocrinol. 9:349.

Mullenax C. H., and R. W. Dougherty. 1963. Physiological responses of swine to high concentrations of inhaled carbon dioxide. Am. J. Vet. Res. 24:329–333.

Munson, P. L., and M. C. Sheps. 1958. An improved procedure for the biological assay of androgens by direct application to the combs of baby chicks. Endocrinology 62:173.

Munson, P. L., and W. Toepel. 1958. Detection of minute amounts of adrenocorticotrophic hormone by the effect of adrenal venous ascorbic acid. Endocrinology 63:785.

Murphy, B. E. P. 1967. Some studies of protein-binding of steroids and their application to the routine micro and ultramicro measurements of various steroids in body fluids by competitive protein-binding radioassay. J. Clin. Endocrinol. Metab. 27:973.

Nalbandov, A., and L. E. Card. 1943. Effect of hypophysectomy of growing chicks. J. Exp. Zool. 94:387.

Nalbandov, A. V. (ed.). 1963. Advances in neuroendocrinology. University of Illinois Press, Urbana, Illinois.

Nallar, R., and S. M. McCann. 1965. Luteinizing hormone-releasing activity in plasma of hypophysectomized rats. Endocrinology 76:272.

National Academy of Sciences–National Research Council. 1966. Chickens—Standards and guidelines. NAS Publ. No. 1464. Washington, D.C.

Neher, R. 1964. Steroid chromatography. Elsevier, Amsterdam.

Nelson, D. H., and L. T. Samuels. 1952. A method for the determination of 17-hydroxycorticosteroids in blood: 17-hydroxycorticosterone in peripheral circulation. J. Clin. Endocrinol. Metab. 12:519.

Ney, R. L., N. Shimizu, W. E. Nicholson, D. P. Island, and G. W. Liddle. 1963. Correlation of plasma ACTH concentrations with adrenocortical response in normal human subjects, surgical patients with Cushing's disease. J. Clin. Invest. 42:1669.

Nicoll, C. S. 1965. Neural regulation of adenohypophysial prolactic secretion in tetrapods. Indications from *in vitro* studies. J. Exp. Zool. 158:203.

Niswender, G. D., A. E. Midgley, S. E. Monroe, and L. E. Reicher. 1968. Radioimmunoassay for rat luteinizing hormone with antiovine SH serum and ovine LH $^{131}$I. Proc. Soc. Exp. Biol. Med. 128:807.

Niswender, G. D., L. E. Reichert, Jr., A. R. Midgley, and A. V. Nalbandov. 1969. Radioimmunoassay for bovine and ovine luteinizing hormone. Endocrinology. 84(5):1166.

Odell, W. D., J. F. Wilber, and W. E. Paul. 1965. Radioimmunoassay of thyrotropin in human serum. J. Clin. Endocrinol. Metab. 25:1179.

Odell, W. D., G. T. Ross, and P. L. Rayford. 1967. Radioimmunoassay for luteinizing hormone in human plasma or serum: Physiological studies. J. Clin. Invest. 46:248.

O'Donnell, V. J., and J. R. K. Preedy. 1967. Oestrogens. *In* C. H. Gray and A. L. Bacharach (ed.) Hormones in blood. Academic Press, New York.

Ohara, K. 1964. Differential method for recording the rate of water evaporation in a small scale area. Jap. J. Physiol. 14:468–478.

Olsen, J. D., R. W. Dougherty, and K. Bond. 1967. Permanent catheterization of the aorta in cattle. Cornell Vet. 57:171–177.

Oser, B. L. 1965. Hawk's physiological chemistry. McGraw-Hill, New York.

Pace, N., L. Kline, H. K. Schachman, and M. Harfenist. 1947. Studies on body composition: IV. Use of radioactive hydrogen for measurement *in vivo* of total body water. J. Biol. Chem 168:459.

Paesi, F. J. A., S. E. de Jongh, M. J. Hoogstra, and A. Englebregt. 1955. The FSH content of the hypophysis of the rat as influenced by gonadectomy and oestrogen treatment. Acta. Endocrinol. 19:49.

Painter, E. E. 1940. Total body water in the dog. Am. J. Physiol. 127:744.

Papkoff, Harold and C. H. Li, 1962. Hypophyseal growth hormone. *In* R. I. Dorfman (ed.) Methods in hormone research. Academic Press, New York

Parkhie, M. R., and H. D. Johnson. 1969. Growth hormone releasing activity in the hypothalamus of rats subjected to prolonged heat stress. Proc. Soc. Exp. Biol. Med. 130:843.

Parlow, A. F. 1961. Experience with an assay for determination of luteinizing hormon (LH). p. 300. *In* A. Albert (ed.) Human pituitary gonadotrophins. Charles C Thomas, Springfield, Illinois.

Pearse, A. G. E. 1966. 5-Hydroxytryptophan uptake by dog thyroid "C" cells and its possible significance in polypeptide hormone production. Nature (Lond.) 211:598.

Pecile, A., E. Muller, G. Falconi, and L. Martini. 1965. Growth hormone releasing activity of hypothalamic extracts at different ages. Endocrinology 77:241.

Permutt, M. A., C. W. Parker, and R. D. Utiger. 1966. Immunochemical studies with lysine vasopressin. Endocrinology 78:809.

Peterson, L. H. 1966. Measurement of displacement and strain. Vol. 11, Sec. 1, p. 1. *In* R. F. Rushmer (ed.) Methods of medical research. Year Book Medical Publications, Chicago.

Peterson, R. E., A. Karrer, and S. L. Guerra. 1957. Evaluation of Silber-Porter procedure for determination of plasma hydrocortisone. Anal. Chem. 29:144.

Pfeiffer, C. A., C. W. Hooker, and A. Kirschbaum. 1944. Deposition of pigment in the sparrow's bill in response to direct applications as a specific and quantitative test for androgen. Endocrinology 34:389.

Phillipson, A. T., and J. R. M. Innes. 1939. Permament stomach fistulae in ruminants. Quart. J. Exp. Physiol. 29:33–341.

Pincus, G., T. Nakao, and J. F. Tait. (ed.). 1966. Steroid dynamics. Academic Press New York.

Popovic, V., and P. Popovic. 1960. Permanent cannulation of aorta and vena cava in rats and ground squirrels. J. Appl. Physiol. 15:727–728.

Porter, G., and W. Lane-Petter. 1962. Notes for breeders of common laboratory animals. Academic Press, New York.

Post, T. B., and J. P. Mixner. 1961. Thyroxine turnover methods for determining thyroid secretion rates in dairy cattle. J. Dairy Sci. 44:2265.

Potts, J. T., Jr., L. J. Deftos, R. M. Buckle, L. M. Sherwood, and G. D. Aurbach. 1967. Radioimmunoassay of parathyroid hormone: Studies of the central of secretion of hormone and parathyroid function in clinical disorders. *In* Radioisotopes in medicine, *in vitro* studies. U.S. At. Energy Comm. Symp. Ser. 13. U.S. Nat. Techn. Infor. Serv., Springfield, Va.

Preedy, J. R. K., and E. H. Aitken. 1961. Column partition chromatography of estrone, estradiol-17, and estriol in phenolic extracts of urine: Fluorescence characteristics of interferring material. J. Biol. Chem. 236:1297.

Pullar, J. D. 1958. Direct colorimetry of animals by the gradient layer principle. Proc. Symp. Energy Metab. Publ. No. 8, Copenhagen.

Purves, H. D., and N. E. Sirett. 1968. Liability of endogenous and exogenous corticotrophene in rat plasma as established by bioassay in intact rats. J. Endocrinol. 41:491.

Quin, J. I., and J. G. van der Wath. 1938. Studies on the alimentary tract of merino sheep in South Africa. IV. Description of experimental technique. Onderstepoort J. Vet. Res. 11:341.

Quinlan, J., G. DeKock, and I. P. Marais. 1935. The operation of splenectomy in horses, cattle, sheep, goats, pigs, dogs, and some South African antelopes: A summary of the results of 98 splenecomies. Onderstepoort J. Vet. Sci. Anim. Indus. 5:27.

Raiti, S., and R. M. Blizzard, 1968. Measurement of immunologically reactive follicle stimulating hormone in human urine by radioimmunoassay. J. Clin. Endocrinol. Metab. 28:1719.

Ranck, J. B., Jr. 1966. Electric stimulation of neural tissue. Vol. 11, Sec. V, p. 262. *In* R. F. Rushmer (ed.) Methods in medical research. Year Book Medical Publishers, Chicago.

Randle, P. J., and K. W. Taylor. 1960. Insulin in blood. Brit. Med. Bull. 16:209.

Redding, T. W., C. Y. Bowers, and A. V. Schally. 1966. An *in vivo* assay for thyrotropin releasing factor. Endocrinology 79:229.

Reeve, E. B. 1960. Measurement with $P^{32}$ labeled red cells. Vol. 8. *In* H. D. Brunner (ed.) Methods in medical research. Year Book Medical Publishers, Chicago.

Reichlin, S. 1964. Brain thyroid relationships. *In* M. P. Cameron and M. O'Connor (ed.) Ciba Foundation Study Group No. 18. Little, Brown & Co., Boston.

Reichlin, S. 1966. Control of thyrotropic hormone secretion. *In* L. Martini and W. F. Ganong (ed.) Neuroendocrinology. Academic Press, New York.

Reid, J. T., A. Bensadoun, L. S. Bull, J. H. Burton, P. A. Gleeson, I. K. Han, Y. D. Joo, D. E. Johnson, W. R. McManus, O. L. Paladines, J. W. Stroud, H. F. Tyrell, B. D. H. Van Niekerk, and G. W. Wellington. 1968. Interaction of human and animal research on body composition in body composition in animal and man. Natl. Acad. Sci. Publ. No. 1598, Washington, D.C.

Reineke, E. P., and F. L. Lorscheider. 1967. A quantitative "direct output" method for determination of thyroid secretion rate in the rat. Gen. Comp. Endocrinol. 9:362.

Reinhold, J. G., V. L. Yonan, and E. R. Gershman. 1963. Measurement of total estrified fatty acid and triglyceride concentrations in serum. Vol. 4, p. 85. *In* D. Seligson (ed.) Standard methods of clinical chemistry. Academic Press, New York.

Riddle, O., and R. W. Bates. 1939. The preparation, assay and actions of lactogenic hormone. Chap. 20. *In* E. Allen, C. H. Danforth, and E. A. Doisy (ed.) Sex and internal secretions. Williams & Wilkins Co., Baltimore.

Rieser, P. 1967. Insulin, membranes and metabolism. Williams and Wilkins, Co., Baltimore.

Riondel, A., J. F. Tait, M. Gut, S. A. S. Tait, E. Joachim, and B. Little. 1963. Estimation of testosterone in human peripheral blood using $S^{35}$ thiosemi carbozide. J. Clin. Endocrinol. 23:620.

Robinson, S., and A. A. Robinson. 1954. Measurement of sweating. Vol. 6. *In* J. N. Field (ed.) Methods in medical research. Year Book Medical Publishers, Chicago.

Ryan, K. J. 1959. Biological aromatization of steroids. J. Biol. Chem. 234:268.

Saito, T., A. Arimura, E. Muller, C. Y. Bowers, and A. V. Schally. 1967. *In vivo* release of follicle-stimulating hormone following administration of hypothalamic extracts in normal, castrated and castrated testosterone-treated rats. Endocrinology 80:313.

Sakuma, M., M. Irie, K. Shizume, T. Tusushima, and K. Nakao. 1968. Measurement of urinary human growth hormone by radioimmunoassay. J. Clin. Endocrinol. Metab. 28:103.

Sanz, M. C. 1957. Ultramicro methods and standardization of equipment. Clin. Chem. 3:406–419.

Sawano, A., A. Arimura, C. Y. Bowers, and A. V. Schally. 1967. Effect of CNS-depressants, dermethasone and growth hormone on the response of growth hormone-releasing factor. Endrocinology 81:1410.

Sayers, G. 1967. Adrenocorticotrophin. Vol. 1. *In* C. H. Gray and A. L. Bacharach (ed.) Hormones in blood. 2nd ed. Academic Press, New York.

Sayers, G., M. A. Sayers, H. L. Lewis, and C. N. H. Long. 1944. Effect of adrenotrophic hormone on ascorbic acid and cholesterol content of the adrenal. Proc. Soc. Exp. Biol. Med. 55:238.

Schalk, A. F., and R. S. Amadon. 1928. Physiology of the ruminant stomach. N. Dakota Agr. Exp. Sta. Bull. 216.

Schally, A. F., and C. Y. Bowers. 1964. Purification of luteinizing hormone-releasing factor from bovine hypothalamus. Endocrinology 75:608.

Schally, A. V., C. Y. Bowers, T. W. Reddings, and J. F. Barrett. 1966. Isolation of thyrotropin releasing factor (TRF) from porcine hypothalamus. Biochem. Biophys. Res. Commun. 25:165.

Schally, A. V., A. J. Kastin, W. Locke, and C. Y. Bowers. 1967. Hypothalamic releasing factors Vol. 1. *In* C. H. Gray and A. L. Bacharach (ed.) Hormones in blood. Academic Press, New York.

Schally, A. V., E. E. Muller, A. Arimura, T. Saito, S. Sawano, and C. Y. Bowers. 1968. Growth hormone-releasing factor (GRF): Physiological and biochemical studies with GRF preparations of bovine and procine origin. Ann. N.Y. Acad. Sci. 148 (Art. 2):372.

Schalm, O. W. 1961. Veterinary hematology. 2nd ed. Lea and Febiger, Philadelphia, Pa.

Schreiner, G. E. 1950. Determination of insulin by means of resorcinol. Proc. Soc. Exp. Biol. Med. 74:117.

Schwartz, I. L., D. Schachter, and N. Freinkel. 1949. The measurement of extracellular fluid in man by means of constant infusion techniques. J. Clin. Invest. 28:117.

Seifter, S., S. Dayton, B. Noric, and B. Muntwyler. 1950. The estimation of glycogen with the anthrone reagent. Arch. Biochem. 25:191–200.

Seiverd, C. E. 1964. Hematology for medical technologists. 3rd ed. Lea and Febiger, Philadelphia.

Selenhow, A., S. M. Wool, and S. Refetoff. 1967. Radioimmunoassay of anterior pituitary hormones. Radiolog. Clin. N. Am. 5(2):317.

Selkurt, E. E. 1948. Measurement of renal blood flow. Vol. 1. Sec. 2, p. 191. *In* V. R. Potter (ed.) Methods of medical research. Year Book Medical Publishers, Chicago.

Shannon, J. A. 1939. Renal tublar excretion. Physiol. Rev. 19:63.

Shannon, J. A. and S. Fisher. 1938. The renal tublar reabsorption of glucose in the normal dog. Am. J. Physiol. 122:765

Sherwood, L. M., J. T. Potts, A. D. Care, G. P. Mayer, and G. D. Aurbach. 1966. Evaluation by radioimmunoassay of factors controlling the secretion of parathyroid hormone. Nature (Lond.) 209:52.

Shizume, K., A. B. Lerner, and T. P. Fitzpatrick. 1954. *In vitro* bioassay for the melanocyte stimulating hormone. Endocrinology 54:553.

Short, J. and D. P. Woodknott. 1963. The A.T.A. manual of laboratory practice and techniques. C. C Thomas, Springfield, Illinois.

Short, R. V. 1958. Progesterone in blood. I. The chemical determination of progesterone in peripheral blood. J. Endocrinol. 16:415.

Silber, R. H. 1966. Fluorimetric analysis of corticoids, Vol. 14, p. 63. *In* D. Glick (ed.) Methods in biochemical analysis. Interscience, New York.

Silber, R. H., R. D. Busch, and R. Oslapas. 1958. Practical procedure for estimation of corticosterone or hydrocortisone. Clin. Chem. 4:278.

Sinha, D., and J. Meites. 1965. Effects of thyroidectomy and thyroxine on hypothalamic concentration of thyrotropin releasing factor and pituitary content of thyrotropin in rats. Neuroendocrinology 1:4.

Slater, L. (ed.). 1962. Bio-telemetry (The use of telemetry in animal behavior and physiology in relation to ecological problems). Proceedings of inter-disciplinary conference, New York. MacMillan Company, New York. 372 pp.

Sloper, J. C. 1966. The experimental and cytopathological investigation of neurosecretion in the hypothalamus and pituitary. Vol. 3. *In* G. W. Harris, and B. T. Donovan (ed.) The pituitary gland. Butterworth, London.

Smith, R. E., and D. J. Hoijer. 1962. Metabolism and cellular function in cold acclimation. Physiol. Rev. 42:60-142.

Soberman, R. J., B. B. Brodie, B. B. Levy, J. Axelrod, V. Hollander, and J. M. Steele. 1949. The use of antipyrine in the measurement of total body water in man. J. Biol. Chem. 179:31.

Sourkes, T. L. 1961. Methods of study of pharmacologically active substances and chemical activity of the nervous system. Vol. 9, Sec. 2, p. 121. *In* J. H. Quastel (ed.) Methods in medical research. Year Book Medical Publishers, Chicago.

Spector, N. H., J. R. Brobeck, and C. L. Hamilton. 1968. Feeding and core temperature in albino rats: Changes induced by preoptic heating and cooling. Science 161:286–288.

Spoor, H. J. 1948. Application of the infra-red analyzer to the study of human energy metabolism. J. Appl. Physiol. 1:369.

Staub, A., and O. K. Behrens. 1954. The glucagon content of crystalline insulin preparations. J. Clin. Invest. 33:1629.

Stebbins, W. C. 1966. Behavioral technics. Vol. 11, Sec. V., p. 270. *In* R. F.

Rushmer (ed.) Methods in medical research. Year Book Medical Publishers, Chicago.
Steelman, S. L., and F. M. Pohley. 1953. Assay of the follicle-stimulating hormone based on the augmentation with human chorionic gonadotrophin. Endocrinology 53:604.
Steinke, J., A. Sirek, V. Lauris, F. D. W. Lukens, and A. E. Renald. 1962. Measurement of small quantities of insulin-like activity with rat adipose tissue. III. Persistence of serum insulin-like activity after pancreatectomy. J. Clin. Invest. 42:1322.
Talwalker, P. K., A. Ratner, and J. Meites. 1963. *In vitro* of pituitary prolactin synthesis and release by hypothalamic extract. Am. J. Physiol. 205:213.
Taubenhaus, M., and S. Soskin, 1941. Release of luteinizing hormone from the anterior hypophysis by an acetylcholine-like substance from the hypothalamic region. Endocrinology 29:958.
Taylor, L. W. (ed.). 1949. The fertility and hatchability of chicken and turkey eggs. John Wiley & Sons, New York.
Taylor, R. B. 1960. A method for collection of pancreatic juice in the conscious sheep. Res. Vet. Sci. I:111–116.
Thompson, H. J., R. M. McCrosky, and S. Brody. 1949. Influence of ambient temperature, O° to 105°F., on insensible weight loss and moisture vaporization in Holstein and Jersey cattle. Mo. Agr. Exp. Sta. Res. Bull. 451.
Thompson, H. K., and H. D. McIntosh. 1966. Cannulation and catheterization procedure. Vol. 11, pp. 162–179. *In* R. F. Rushmer (ed.) Methods in medical research. Year Book Medical Publ., Chicago.
Townsend, J., and V. H. T. James. 1968. A semi-automated fluorimetric procedure for the determination of plasm corticosteroids. Steroids 11:497.
Tromp, S. W. 1967. Clinical applications of human biometerology. Abbottempo, Book 4, pp. 8–15. Abbott Laboratories, Chicago.
Tsuda, T. 1964. Absorption of water from the rumen by the use of rumen pouch Method. Tohuku J. Agr. Res. 15:83–85.
Turner, C. D. 1966. General endocrinology. W. B. Saunders Co., Philadelphia.
Unger, R. H., and A. M. Eisentraut. 1967a. Glucagon. Vol. 1. *In* C. H. Gray and A. L. Bacharach (ed.) Hormones in blood. 2nd ed. Academic Press, New York.
Unger, R. H., and A. M. Eisentraut. 1967b. Journess de diabetologie hoteldieu. p. 8. Editions medicales. Flammarion, Paris, France
U.S. Public Health Service. 1968. Guide for laboratory animal facilities and care. U.S. P.H.S. Publ. 1024:1–57.
Utiger, R. D. 1965. Radioimmunoassay of human plasma thyrotropin. J. Clin. Invest. 44:1277.
van der Molen, H. J., and A. Aakvaag. 1967. Progesterone. *In* C. H. Gray and A. L. Bacharach (ed.) Hormones in blood., Academic Press, New York.
Van Dyke, H. B., K. Adamson, Jr., and S. L. Engel. 1955. Aspects of the biochemistry and physiology of the neurohypophyseal hormones. Recent Progr. Hormone Res. 11:1.
Veghte, J. H., and G. Solli. 1962. Determining arctic clothing design by means of infrared radiometry. Mil. Med. 127:242.
Vermeulen, A., and J. C. M. Verplancke. 1963. A simple method for the determination of urinary testosterone excretion in human urine. Steroids 2:453.

Vernikos-Danellis, J., J. E. Anderson, and L. Trigg. 1966. Changes in adrenal corticosterone concentration in rats: Method of bioassay for ACTH. Endocrinology 79:624.

Walker, J. M. 1967. Vasopressin. Vol. 1. *In* C. H. Gray and A. L. Bacharach (ed.) Hormones in the blood. 2nd ed. Academic Press, New York.

Whipp, S. C. 1968. Personal communication.

Whipp, S. C., E. T. Littledike, and S. Wangsness. 1968. A technic for hypophysectomy of calves. Proc. Am. Assoc. Lab. Anim. Sci. Las Vegas, Nevada.

Wiest, W. G., T. Kerenji, and A. I. Csapo. 1966. A double isotope derivative dilution assay for progesterone in biological fluids and tissues. Excerpta Med. Intern. Cong. Ser. III. p. 114.

Wilber, J. F., and R. D. Utiger. 1967. Immunoassay studies of thyrotropin in rat pituitary glands and serum. Endocrinology 81:145.

Winget, C. M., E. G. Avertin, and T. B. Fryer. 1965. Quantitative measurements by telemetry of ovulation in oviposition in the fowl. Am. Med. J. Physiol. 209: 853–858.

Wintrobe, M. M. 1961. Clinical hematology. Lea and Febiger, Philadelphia, Pa.

Witzel, D. A., and C. H. Mullenax. 1964. A simplified approach to splenectomy in the horse. Cornell Vet. 54:628–636.

Woolever, C. A., and A. Goldfien. 1963. A double-isotope derivation method for plasma progesterone assay. Intern. J. Appl. Radiat. Isotop. 14:163.

Worden, A. R., and W. Lane-Petter. 1959. The U.F.A.W. handbook on the care and management of laboratory animals. 2nd Ed. The Universities Federation for Animal Welfare, E.S. Livingstone Ltd., London.

Wotiz, H. H., and S. C. Chattoraj. 1964. Determination of estrogens in low and high titre urines using thin layer and gas liquid chromatography. Anal. Chem. 36:1466.

Wotiz, H. H., and S. J. Clark. 1966. Gas chromatography in analysis of steroid hormones. Plenum Press, New York.

Wright, A. D., and K. W. Taylor. 1967. Immunoassay of hormones. Vol. 1, p. 23. *In* C. H. Gray and A. L. Bacharach (ed.) Hormones in blood. Academic Press, New York.

Yalow, R. S., and S. A. Berson. 1966. Purification of $^{131}I$ parathyroid hormone with microfine granules of precipitated silica. Nature (Lond.) 212:357.

Yalow, R. S., and S. A. Berson. 1967. Radioimmunoassay of human growth hormone in plasma: Principles, practices, and techniques. A. Pecile, and E. E. Muller (ed.) Excerpta Med. Intern. Cong. Ser. 158.

Yeates, N. T. M. 1965. Modern aspects of animal production. Butterworth, London. 372 p.

Yeck, R. G., and H. H. Kibler. 1956. Moisture vaporization by Jersey and Holstein cows during diurnal temperature cycles as measured with a hygrometric tent. Mo. Agr. Exp. Sta. Res. Bull. 600.

Yonaga, T. 1967. A new quantitative bioassay for growth hormone using time markers for the growth rate of hair, teeth and bone. Excerpta Med. Vol. 142:38.

Young, A. C. 1952. $CO_2$ analyzer. USAF School of Aviation Medicine, Proj. No. 22:1301–1302.

Yoshimi, T., and M. B. Lipsett. 1968. The measurement of plasma progesterone. Steroids 11:527.

Yousef, M. K., and H. D. Johnson. 1967. A rapid method for estimation of thyroxine secretion rate of cattle. J. Anim. Sci. 26:1108.

Yousem, H. L. 1964. Simple gas chromatographic method for estimation of urinary estriol in pregnant women. Am. J. Obstet. Gynecol. 88:375.

Zander, J. 1962. Progesterone. Vol. 1, p. 91. *In* R. I. Dorfman (ed.) Methods in hormone research. Academic Press, New York.

Zarrow, M. X., I. M. Yochim, and J. L. McCarthy. 1964. Experimental endocrinology. Academic Press, New York.

Zarrow, M. X., and J. H. Clark. 1969. A modified ovarian cholesterol depletion assay for luteinizing hormone. J. Endocrinol. 43(3):459.

# 4

# ANIMAL CHARACTERISTICS IN RELATION TO ENVIRONMENTAL RESPONSE

## INTRODUCTION

This chapter discusses the factors that are important in describing variations in an animal's response to changes in environmental conditions. It is not meant to be either an exhaustive review of the literature or a discussion on physiology or morphology. The question addressed is how may physioanatomical characteristics be evaluated as we examine differences within and between animals?

Domestic animals use materials undesirable for human consumption and convert them into desirable products. These products are usually in the form of meat, milk, eggs, fiber, and sometimes work. The conversion of material to a usable product has to be economical; i.e., the value of the input material has to be less than the value of the usable product after the cost of conversion has been paid for. Since physiological events are involved in these conversions, it becomes paramount to consider the efficiency of their function under various environmental conditions.

Although the primary research interest is in the function of domestic animals, probing types of experiments are often necessary, using laboratory animals such as mice, rats, or rabbits. Information obtained from such animals, if used in the explanation of how some domestic animals respond to some environmental changes, must be utilized with caution.

There are many anatomical and physiological differences between animals which affect their function in a changing environment. In fact, no two animals are exactly the same nor do two animals have exactly the same environment. Differences in response between animals to given stimuli are usually classified under the vague term "biological variation." This simply means that a portion of the variation in response among animals cannot be logically explained on the basis of known facts. Therefore, it is the problem of biological variation that makes the study of a bioclimatic system so difficult but also very interesting.

To make the situation even more confusing and difficult to study, there exists an interaction between animals and their environment. Simply stated, this means that the same degree of physiological change may not necessarily occur in two animals subjected to a standard change in environment.

Factors that may vary in an animal and influence function can be placed in broad categories for purposes of discussion. Consideration of these factors is absolutely essential when one selects the experimental animals for specific types of investigation.

## GENETICS

One of the first, and perhaps major, sources of variation among animals is that due to differences in their genes. A well-known example is the differences observed between a "native" animal and one that is the result of selection for some production trait. This phenomenon has been recently reviewed for milk production (Branton *et al.*, 1966) and sheep production (Miss Agr. Exp. Sta., 1966). The genotype of animals native to an environment is the result of natural selection and produces an animal that can best live and reproduce in that environment. This selection involves the concept of "survival of the fittest." A population of animals so selected usually does not have the production characteristics desired by man. The concept also suggests an interaction between the genotype of an animal and the environment surrounding it, as measured by that animal's ability to function.

One common method for removing the effects of genetic differences between animals in an investigation is by the use of identical twins. However, a sufficient number of twins is difficult to obtain. In any event, animals with similar genotypes should be selected for investigations on differences in function because of different environmental conditions. Differences in genotype (as measured by a physiological function) can be studied equally well by subjecting two animals with distinctly different genes to a common environment (see Chapter 6). For example, one may question why an animal native to a tropical environment can tolerate high ambient temperatures better than an animal native to a more temperate environment. The question can be phrased: Which physiological event is influenced by differences in genotype, and what is the nature of this influence when an animal is subjected to tropical conditions?

The major concern inherent in these situations should be the danger encountered when reasoning from observations made on animals with one set of characteristics to predicting how other animals with different sets of characteristics may respond to changes in environment.

## TIME OF OBSERVATION

### Physical Condition

Another major source of variation among animals in their response to environment is their physical condition at the time of observation. An animal at any given time is the result of all the stresses to which he has been exposed during his lifetime. An animal that has had a very serious respiratory disease, for instance, may not function as he would have had he not been exposed to the disease. An easily observable difference among animals is body size. It is a factor that can influence many physiological functions, such as growth rate. An animal, therefore, is what he is because of all of the events in his history.

### Age

One of the more significant factors, one to which all animals are subjected, is that of accumulated time since birth. This is commonly called the aging process. This process can be measured and discussed in two different ways.

#### CHRONOLOGICAL AGE

This is merely the number of days since birth and certainly is the most common way of measuring the degree of aging. However, there is considerable variation in the rate of maturing among animals within a species. This simply means that not all animals of a given species are in the same state of maturity by 1 year after birth. Practically any environmental factor is a potential source of variation on the rate of maturity.

#### PHYSIOLOGICAL AGE

A much more precise measure of maturity is to use those physiological changes associated with aging. This is sometimes called biological age. Changes in body function and form occur in an orderly sequence from the time of conception through the time of senility and to death from "old age." Simple examples of aging are the changes that occur in teeth. The first permanent pair of teeth appear in a sheep at about 1 year of age. At approximately each year thereafter, another pair of teeth appears until 4 permanent pairs are visible. Another example of a measurement is the amount of nitrogen in the lenses of the eye in swine (Kauffman *et al.*, 1967). Puberty, the time when ability to reproduce has been reached, is another common benchmark in the development of an animal. Further, the natural rate of aging is influenced by factors in the environment (Brues and Sacher, 1965). Radiation is a common example.

Many other physiological events change with aging, such as increase in blood pressure, decrease in heart rate, and calcification of bone. These factors all become important when investigations are made into the effects of environment on body function. Animals of different physiological ages may not respond similarily to a given environmental stress.

### Cyclic Variation

Body functions of animals go through cyclic changes. One of the more common is the estrous cycle of the female (discussed below). Other, less obvious cycles must be considered in environmental investigations.

#### DAILY

Many changes in body function follow the 24-hr light–dark cycle of a day, often called the circadian (diurnal) rhythm. Metabolic rate, for example, is lowest in the morning, but body temperature is highest in the evening. When the physiological function being measured is influenced by time of day, that variation in measurement must be taken into account when the results of different environmental stresses are interpreted.

### SEASONAL

Changes in animal function and behavior because of yearly changes in environment are commonplace and are well known. Examples of such changes range from hibernation to migration. One common observation is that of anestrus, the cessation of the reproductive process during certain periods of the year. Part of the seasonal variation is obviously the result of seasonal food supply, prevalence of parasites, transmission of disease, and those other factors that are intermediary steps between the climate and an animal's well-being. Also, because animals tend to adjust to external conditions, an animal's immediate past environmental conditions may influence how it will respond. Again, the important point seems to be that an animal may not respond to a given environment in the same manner at all seasons of the year because of past conditions. Consequently, comparing results made at one time of year with those made at a different time of year may be confusing.

## FUNCTIONAL RESPONSES

A number of excellent texts are available that describe the details of physiological and anatomical characteristics, as well as many additional, more specialized volumes (Dukes, 1955; Best and Taylor, 1961; Guyton, 1966; Ruch and Patton, 1966).

The specific examples discussed are intended to serve only as illustrations and to support the concept that an animal's response to a given environment may be better assessed with a comprehension of the complex relationships and interactions of animal and environment.

### Degree of Acclimatization*

Because of an animal's ability to alter its physiological function, adjustments occur in these functions to produce a set of stable conditions within limits of normalcy. There are two important points that need to be considered in planning studies on environmental effects on animals. First, not all animals will adjust to environment, i.e., become ac-

*See also Chapter 8.

climatized, at the same rate. Obviously, animals in different phases of this adjustment may respond differently to a given environmental stress. Therefore, it is important in sound experimental procedure to have a preconditioning period that is common for all animals going into the experiment (Shaw, 1967).

Second, although it is never possible to remove all differences among animals due to differences in physiological state, it is absolutely imperative that these differences be recognized when one interprets results of an experiment that involved the action of environment on a physiological function.

## Homeostasis

Homeostasis may be defined as the tendency to uniformity or stability of normal body states of the individual. Such a concept applies to each component of each body system and results from the establishment of a balance between stimulation and inhibition of a given element. Examples that might be cited include circulating levels of hormones, blood glucose levels, oxygen levels in the tissues, and homeothermy.

### HOMEOTHERMY

Homeothermy refers to the ability of the body to maintain body temperature within relatively narrow limits. Therefore, this homeostatic mechanism becomes of particular interest in environmental studies. The body temperature of homeotherms is species specific and is relatively constant when compared with poikilotherms. However, body temperature, as such, is difficult to establish because of differences between areas of the body. The temperature of each area can be influenced by a wide variety of circumstances, such as muscular activity, feeding, water consumption, and many other factors including diurnal and other rhythmic cycles. Body temperature is dependent on a balance between heat production and heat loss. In the absence of such a balance, body temperature is either increased (hyperthermia) or decreased (hypothermia).

Upper and lower critical temperatures have been defined as the temperatures beyond which the homeotherm cannot maintain its body temperature at the basal metabolic rate (Folk, 1966). Therefore, the range between these upper and lower temperatures has been referred to as the zone of thermal neutrality. Within this zone, thermal regulation is accomplished by the physical methods of heat transfer (Hafez,

1964) with a minimum of physiological activity. These methods of heat transfer are discussed in Chapters 2 and 8. At temperatures above or below the zone of thermal neutrality, physiological mechanisms interact with the physical methods to provide homeothermic conditions. Hardy (1961) reviewed the physiology of temperature regulation, and Hafez (1964) has reviewed behavioral thermoregulation.

Homeothermic ability depends on the temperature-regulating system, which basically consists of receptors for detection of temperature within and surrounding the body and effectors for controlling heat production and heat loss. In addition, there must be a thermoregulatory center for coordination of regulatory functions. An oversimplified analogy might therefore be that the overall function is comparable to maintenance of room temperature by a system composed of a thermometer (receptor) for detection of temperature, an air conditioning system (for heat removal), a heating system (for heat production), and a thermostat (the thermoregulatory control) for initiating heat production or cooling. The room temperature would be maintained without heating or cooling at the preset point for a relatively narrow temperature range (zone of thermal neutrality). Outside this zone the heating system (heat production) or the cooling system (heat removal) must operate to maintain the preset temperature. When climatic conditions are more severe than the upper and lower capabilities of the equipment, the room temperature will either increase or decrease as when such conditions exist for an animal. When an increased (hyperthermy) or decreased (hypothermy) temperature exists, the zone of homeothermy has been exceeded. The upper and lower body temperatures compatible with life establish what is sometimes referred to as the zone of survival.

The above analogy is oversimplified when the actual biological system is considered. For example, there are thermoreceptors located in various areas of the body and heat production–heat loss mechanisms are not restricted to any one method or unit as is the case with a furnace or an air conditioning system.

## HEAT PRODUCTION

The amount of heat produced by the animal during metabolic activity is additive to heat that may be gained by physical means. The quantity of heat produced within the body varies greatly in the different organs and with the activity of each organ under its particular conditions at the time of determination.

Within the digestive system, for example, there is an increase in metabolism subsequent to food intake. This so-called specific dynamic action may amount to as much as 30 percent of the caloric value of the protein ingestion. This transient increase becomes apparent within approximately 1 hr of food consumption in a fasting subject.

In ruminants, the activity of the rumen provides an additional source of heat production and rumen temperature is normally greater than rectal temperature. However, this temperature is markedly affected by the temperature of ingested materials. Bailey *et al.* (1962) and others have reported a sharp temperature drop of the rumen and a corresponding but smaller rectal temperature change following consumption of water.

The total quantity of ingested material also alters the amount of heat produced. At high temperatures the animal consumes less food. The consumed food substrate results in decreased heat production or energy utilization and thus aids in heat balance. In general, the opposite response occurs at low temperatures. The relationship of environmental temperature and feed regulatory mechanisms has been discussed (Wilson, 1967; Johnson, 1967). Such responses require caution in determination of such parameters as body temperature or any other value that might be influenced by food consumption. It is usually necessary to standardize time of determination with respect to food and water intake as well as other activities of the animal that occur as the result of management practices.

Both muscular activity and form are an important source of heat. The relatively small muscular activity involved with standing results in 10 percent greater heat production in cattle and sheep than in reclining animals (Bianca, 1968). Therefore, it must be expected that an animal undergoing muscular activity such as for draft purposes or grazing would respond to a given environment in a different manner than an animal not undergoing voluntary muscular activity. However, feathers trap heat and small movements will cause this heat to escape.

Pregnancy and lactation also contribute to the total heat production and should be considered when metabolic determinations are made in relation to environment. A 15 percent decrease of energy metabolism was noted in lactating dairy cows exposed to a 29°C (84°F) environment after being in a 65°F (18.4°C) environment for 4 to 6 weeks (Kibler *et al.*, 1965). Two to 4 weeks were required for values to reach their lowest point and establish a degree of equilibrium.

The Van't Hoff-Arrhenius law also results in increased temperature upon chemical activity. The ratio of the rate at which a chemical reaction proceeds at a given temperature ($T_1$) to that at which the reaction proceeds at $T_1 - 10°C$ is called the $Q_{10}$ of the reaction. The $Q_{10}$ of chemical reactions is typically 2.0–3.0. Thus, a two- to threefold increase in heat production via chemical reactions would be expected if the thermoregulatory mechanism permitted a 10°C change in tissue temperature, and other factors (such as enzyme activity) were not altered by the change of temperature.

One additional regulatory mechanism should be mentioned as a further example of heat production. The endocrine system regulates the overall metabolic rate through control of the rate of reactions as well as such mechanisms as nonshivering thermogenesis. These mechanisms are discussed below.

#### HEAT LOSS

Equations given elsewhere (Chapter 2) describe heat transfer by radiation, conduction, convection, and evaporation and the factors that influence such transfer in loss of heat from animals where the body temperature exceeds that of the environment. These heat loss methods are exclusively under physical control except that homeothermic mechanisms, such as body conformation, hair coat, or heat production, may alter such heat loss.

### Other Stabilizing Functions

One particularly significant physiological mechanism interacting with physical factors is *respiration* (see also Chapter 3). Alteration of respiration rate is thought to arise from stimulation of central thermoreceptors and/or peripheral receptors. Waites (1962) has demonstrated the effectiveness of such peripheral receptors in the sheep and goat.

Differences exist between species with regard to importance of respiratory evaporation as a percent of the total evaporative water loss. Macfarlane (1968) has summarized the maximum sweat secretion rates for sheep and Brahman cattle. In sheep, 75 percent of the total evaporative losses occur via the respiratory system. In contrast, only 12 percent of the total loss was ascribable to this mechanism in Brahman cattle. In Shorthorn cattle, a more heat-susceptible breed, 24 percent of evaporative loss is by respiration. Thus, the significance of thermoregulation via respiratory evaporation varies considerably between species as well as physical factors of the environment, particularly moisture content of the air.

Bianca (1968) states that the increased ventilation is generally accomplished by an increased rate with a decreased depth. Thus, exchange within the lungs is minimized as most evaporation occurs from the upper portions of the respiratory tract. However, under severe heat stress the rapid shallow breathing becomes slower and deeper. It has been demonstrated that such respiration results in respiratory alkalosis through excessive loss of carbon dioxide (Bianca and Findlay, 1962). Blood pH can reach values as great as 7.8 even though renal compensation is initiated. In cats, Frankel and Ferrante (1966) have observed hypocapnic alkalosis below 42°C and metabolic acidosis above 42°C.

Forrest *et al.* (1968) describe responses of "stress-susceptible" and "stress-resistant" strains of hogs to elevated temperatures. Important differences in heart and respiration rates were observed during exposure to 42°C. In addition, the stress-susceptible animals showed significant increases in levels of carbon dioxide and concomitant decreases in levels of oxygen, whereas stress-resistant animals did not experience significant alterations. Conversely, the amount of oxygen in stress-resistant animals increased slightly and carbon dioxide decreased significantly. An awareness of the possibility of such strain difference and an intimate knowledge of the animal employed are thus essential for accurate evaluation of environmental responses.

In addition to respiration losses, *water loss* occurs via the skin, through the urine and the feces. The colon of the alimentary tract is capable of removing water from fecal material as a conserving mechanism (Macfarlane, 1968). Yet, cattle normally absorb only 20 percent of the water present in feces. Some types, such as *Bos indicus*, can reduce fecal water output 10 percent more than European breeds, so that as much as 35 percent of the water can be reabsorbed under conditions of water deprivation. Sheep are capable of similar conservation by reducing water content of fecal pellets from 60 percent water to 45 percent water.

Species differences also exist with respect to urinary water loss. Antidiuretic hormone (ADH) produced by the posterior pituitary regulates the amount of reabsorption occurring in the kidney tubules. ADH acts on the tubules to increase the rate of reabsorption. Therefore, a reduction in circulating levels of ADH results in an increased urinary water loss.

Reduction of osmotic pressure results in decreased ADH levels through decreased stimulation of osmoreceptors of the hypothalamus. Aldosterone, produced by the adrenal cortex, increases the rate of reabsorption of sodium from the tubules resulting in decreased amounts of sodium excreted in the urine (Guyton, 1966).

These regulatory mechanisms are therefore operative in maintenance of homeostasis where evaporation has altered interstitial and plasma volumes. Macfarlane (1968) has discussed species and breed differences with respect to water balance.

It should be reemphasized that the above are only examples of homeostatic mechanisms. In actuality, each physiological function proceeds within definite limits and thus maintains conditions within certain limits that are compatible with life and preservation of the species. An awareness and understanding of the physiological mechanisms by which homeostasis occurs becomes absolutely essential for the evaluation of within and between animal response to bioclimatic conditions.

## Production of Utility to Man

Bioclimatic conditions have been demonstrated to alter the ability of the domestic animal to convert energy to a more desirable form for human utilization. In general, the adverse effects of elevated temperatures have received particular attention. This is due, at least in part, to the fact that thermoregulatory mechanisms are less affected by elevated temperatures than by low temperatures usually encountered in areas of domestic animal production (Bianca, 1968).

### GROWTH AND BODY COMPOSITION

Both prenatal and postnatal growth rates are examples of characteristics affected by environment (Hafez, 1968). Species, breed, and strain differences exist. For example, Shorthorn cattle reared at 27°C weigh less than those raised at 10°C. However, Brahman cattle grew at a similar rate at the two temperatures. But again, the numbers studied are small; therefore, interpretations may need to be reevaluated. Factors that result in such differences require consideration, both in studies specifically directed toward environmental influences and in studies where such factors are not controlled and influence some other factor such as feed quality. Judge (1969) has reviewed the influences of bioclimatic conditions upon meat quality. Bovine characteristics have been reviewed by Hedrick (1968).

Early work at Missouri (Moulton *et al.*, 1922) showed that the percentage of fat, bone, and lean in the carcasses of Hereford cattle was influenced by level of nutrition, with full-fed animals having a higher percentage of fat, especially from 30 to 48 months of age. The chemical composition of the fat-free animal body is practically constant in normal mature animals (Hankins *et al.*, 1959). The development of bone, muscle, and fat occurs at different rates in the postnatal animal. Bone growth occurs first, followed by muscle and fat (Hammond, 1933).

### EGGS

Hafez (1968) has also reviewed the role of bioclimatic conditions upon egg production. Quantity, size, composition, fertility, and hatchability have been shown to fluctuate. Again, breed and strain differences are evident. The relatively rapid progress in genetic selection of poultry provides an excellent example of the caution that should be exercised in interpretation of responses to a given set of environmental conditions. Such genetic progress requires that previous observations be re-examined as improved strains become available.

### LACTATION

Recent discussions include (Kamal *et al.*, 1961; Johnson, 1967; Johnson *et al.*, 1962, 1966; Hafez, 1968) the influence of environment upon milk production and composition. Production was more affected by high (27°C) temperature than by low (0°C). Johnson *et al.* (1962) have pointed out that, in general, high producers are more susceptible to heat than low producers.

Breed differences do exist, with the Brahman being most heat tolerant and the Holstein most cold tolerant. Milk production decreases approximately 4 lb per °F increase in rectal temperature. Relative humidity of approximately 65 percent becomes a critical factor in maintenance of milk production if the ambient temperature reaches 80°F, as this is the point at which heat balance is disturbed resulting in decreased food consumption.

Milk composition is also altered with high environmental temperatures, causing an increase of nonprotein nitrogen, and palmitic and stearic acids. The same conditions cause decreases in total solids, solids-not-fat, total nitrogen, lactose, oleic acid, and short-chain fatty acids. The percent of fat decreases at ambient temperatures of 21°–27°C and increases beyond 27°C. Low (0°) environmental temperatures also cause increases of fat by as much as 10–35 percent in Jersey cattle. However, changes occurring in Holsteins are less.

### Endocrine Balance

The endocrine system of the animal is subject to alteration of function because of a variety of external forces. Therefore, it is involved in the interaction of an animal and its bioclimatic environment.

The thyroid gland has received considerable attention by virtue of the relationship between its activity and metabolic activity. Measures of production of thyroid hormone include protein-bound iodine (PBI), iodine uptake, and thyroxin reaction and degradation rates. All tend to confirm the generalization that elevated temperatures depress production of thyroid hormones and thereby decrease heat production through a reduced metabolic rate. The opposite response occurs upon exposure to low temperatures. Cattle studies do not support the concept that the decreased food consumption associated with high temperature is the primary factor limiting thyroid function (Yousef and Johnson, 1966). It is probable that afferent nerve impulses from thermoreceptors impinge upon the hypothalamus and reduce thyrotrophin releasing factor, thus reducing thyroid activity.

The adrenal gland has also received attention in relation to environment. Secretions of the medulla and the cortex have been implicated in a variety of physiological responses. For example, in the chicken exposed to 0°C temperatures, circulating levels of the catecholamines reached a maximum in 4 weeks, production being 60 percent greater than in control birds after 20 weeks (Linn and Sturkie, 1968). Such hormones are thought to be involved with nonshivering thermogenesis. Exposure to high temperatures (31°C) had no effect on plasma levels of norepinephrine or epinephrine. However, hormone levels within the adrenal were increased, indicating the possibility of inhibition of hormone release from the gland. In contrast, hyperthermia of the bovine does result in increased plasma levels of catecholamines (Robertshaw and Whittow, 1966). The greatest increase occurred during the initial phase of hyperthermia.

The production and function of adrenal cortical hormones have been described in various texts (Eisenstein, 1967). The results of Judge *et al.* (1966) and Forrest *et al.* (1968) indicate that adrenocortical insufficiency is associated with the response of stress-susceptible swine to elevated temperature. Such an insufficiency results in inferior carcass quality. In contrast, hyperactivity of the adrenal cortex of thermally stressed sheep is a factor contributing to impaired female reproductive performance (Howarth and Hawk, 1968).

Potentially every endocrine gland can be involved with the response of an animal to its environment. Therefore, the above should serve only as examples rather than a complete listing of endocrine–bioclimatic interactions.

## Reproduction

The available evidence tends to indicate that reproduction may be one of the most sensitive physiological mechanisms when considering measurement of changes in bioclimatic conditions.

Ample evidence is available to substantiate the generally accepted fact that environmental conditions affect reproductive performance. These influences include ecological characteristics (Hafez, 1964), male and female fertility (Ulberg and Burfening, 1967), and influences on birth weight and vitality (Shelton and Huston, 1968). Physiological mechanisms have not been established for all instances of lowered performance, perhaps due at least in part to the number of steps in the reproductive process that can, and apparently are, influenced. In the absence of relatively precise control of conditions, the physiological mechanism may vary. The environment affects both the male and female, and temperatures above the critical temperature for a species appears to be most detrimental.

Within the male this effect may be on various stages of spermatogenesis. Therefore, evaluation of the experimental conditions must take into consideration the interval of time required for completion of spermatogenesis. This interval differs between species. In addition, the available evidence indicates that the male gamete can be influenced by temperatures encountered within the female's reproductive tract (Howarth *et al.*, 1965; Burfening and Ulberg, 1968). The female gamete is also directly affected by increased maternal temperature (Alliston *et al.*, 1965).

Within the female, the evidence indicates that environmental factors may also exert an effect at various stages of the reproductive process. Included are impairment of ovarian function of the rat (Chang and Fernandez-Cano, 1959), failure of fertilization and early morphological abnormality in sheep (Dutt *et al.*, 1959), low birth weight of lambs (Shelton and Huston, 1968), and effects on the early sheep zygote that cause embryonic death relatively late in gestation (Alliston and Ulberg,

1961; Alliston *et al.*, 1965). Plasse *et al.* (1968) examined breed differences in onset of puberty and ovulation frequency in a subtropical environment. The degree of diversity requires careful evaluation for accurate interpretation of results. Although complete evidence is not available, it seems probable that the nature of the response varies between species, nutritional states, strains of animals, duration and intensity of the thermal stress, as well as degree of acclimation of the animal. In addition, pregnant animals should be expected to respond differently from nonpregnant animals; and seasonally polyestrous animals, such as sheep, might respond differentially.

#### SEX

Consideration must also be given to the sex of the animal when evaluating the effect of environment upon growth and/or composition. Hedrick (1968) has recently completed an extensive review of bovine growth and composition. Characteristics influenced by sex include daily gain, dressing percent, carcass grade, and finish. Similar differences between sexes exist in other species. Such differences are, of course, associated with the androgenic and estrogenic hormones of the male and female, respectively.

These and other steroid hormones have marked effects upon the activity and behavior of animals. A most striking example of their influence is associated with mating behavior, at which time activity of both sexes is at a maximum. Species differences are marked. For example, sheep are seasonally polyestrous in contrast to cattle that are typically polyestrous throughout the year.

#### GESTATION

In the female, gestation is an additional source of variation. Gestation imposes an additional requirement upon the female. This requirement results in increased nutritional requirements and acts as an additional stress in any environmental condition.

## BODY SHAPE AND EXTERIOR

### Shape and Form

Some authors (Bonsoma, 1967) hold that under extreme climatic conditions the body conformation of wild species such as the polar bear in the cold and the camel on the desert aids response to a specific climate. However, it should be emphasized that external traits, such as conformation, hair coat, and skin traits, may influence the animal's response to bioclimatic conditions. It is the responsibility of the physiologist to determine the interrelationship of these factors in various bioclimatic systems.

Recent work (McDowell, 1967) with the zebu has given evidence that the body appendages may not be as important as was formerly thought. These workers removed the hump and dewlap and showed that these structures were not critical in the regulation of body temperature. There may be an optimum external body conformation for certain environments, but the less obvious internal physiology and anatomy, discussed previously, may be either as important or more important than the readily observed external traits. Also, caution should be exercized in interpreting the correlations of certain external traits and response to environment, or to productive traits. Correlations do not mean an automatic cause and effect relationship. McDowell (1967) has indicated that in milk production the "physiological adaptability" and "productive adaptability" in some cases showed a negative association.

### Hair Coat

There is great variation in hair color in the various breeds and types of animals, but probably man has selected coat color to satisfy his needs rather than the needs of the animals. Under native conditions the wild animals have evolved certain colors that offer advantages for survival under certain environments. Such an example is the advantage of the coat color in the zebra, which affords a camouflage pattern as protection from predators. There is great variation in the colors and color patterns of the criollo cattle in Africa and Latin America (Phillips, 1963), which would indicate that color *per se* has little selective advantage. Dowling (1961) indicates coat color is of minor importance in determining heat tolerance of cattle in Australia. However, (Bonsma, 1967) cattle with pigmented skin such as the zebu, Brown Swiss, and Africander can withstand hazards of radiation best. It should be noted that skin pigmentation is associated with certain hair color and skin thickness, so it is not possible to delineate the precise effects of these specific external factors.

Coat type, which is a subjective description of the hair and includes a measure of length, diameter, curliness, and weight per unit area, showed significant breed and season differences, with a repeatibility of 0.6 (Turner and Schleger, 1960). A significant correlation (–0.503) was shown between coat score and postweaning gain in British breed calves. However, there was no such relationship in the British-zebu crossbred calves. Also, there was a significant correlation of –0.232 of coat score of dam and growth of their calves for the British breed calves but none for the crossbred zebu calves. The greater gain of calves with the sleek, short coats indicates some physiological advantage under grazing conditions in Northern Australia. Coat score was slightly better correlated with preceding gain than subsequent gain, indicating that coat score is probably an index of the internal animal physiology rather than the reverse relationship. However, the exact cause and effect relationship of hair coat traits and animal response is difficult to assess. Schleger and Turner (1960) found that depth of hair coat and hair diameter were the best individual characteristics influencing response to environment.

Bonsma (1967) has indicated that woolly-coated condition can be recognized at 4 days after birth and is related to mature weights of Herefords in South Africa. A 197-lb difference was shown in mature weights of Herefords at the Mara Station between sleek and woolly coats, whereas the difference in similar cattle in the Temperate Zone was only 41 lb. The sleek-coated Herefords were the heaviest in both locations, but the difference was much greater where heat stress was a problem.

Differences in hair coat between Sahwial zebu and Jersey cattle were studied in Australia (Pan, 1964). Hair length was 8.8 mm in Jersey vs. 4.8 mm in zebu cattle, while hair diameter was 60.9 $\mu$m in Jersey and 67.9 $\mu$m in zebu, a nonsignificant difference in diameter. The zebu had the greatest variability overall in hair coat. The diameter of the medulla of the hair was 39.42 $\mu$m in the zebu compared to 26.7 $\mu$m in Jerseys, while the percent medullated hairs, diameter of medullated hairs, and diameter of nonmedullated hairs showed no significant difference between the two breeds.

Seasonal changes influence hair coat in both quantity and quality, with lowest coat scores in the summer and highest in the winter. The ability of the animal to shed the winter hair coat is associated with animal response to hot climate, as shown by movement of cattle from Montana to Florida. There is a steady decline in hair coat score from calves to 3 years of age and a further decline in mature cows (Turner and Schleger, 1960). Thus the hair coat becomes more sleek with increasing age; but the physiological explanation for this phenomenon is not known.

The effect of sex on coat score was studied; bulls had the lowest score (sleeker) followed by heifers and steers (Turner and Schleger, 1960). The exact effect of the endocrine system on hair coat and its subsequent influence on adaptation needs further study. Pregnancy and lactation had no effect on coat score, whereas animals on a higher plane of nutrition had sleeker coats than animals on pasture where nutrition was a limiting factor at certain seasons of the year. This again emphasizes the necessity to carefully describe the total environment as well as the characteristics of the animal and his previous environment in studies of the physiology of the animal response.

## Skin

### THICKNESS

There are significant differences in skin thickness between the zebu and Shorthorn and the various crosses of the two breeds, with the Shorthorn skin thinnest (0.1904 in.) and the zebu thickest (0.2333 in.) (Carpenter, 1959). A thicker hide may be helpful in combating ravages of insects in tropical environments. Also, animals with a well-developed panniculus muscle, such as the horse and Africander cattle, can displace insects by movement of this muscle (Bonsma, 1967). Another index of hide thickness is shown by a hide weight that was 10.4 percent of the live body weight in the zebu compared to 7.7 percent hide weight in the 1/4 zebu–3/4 Shorthorn steers (Carpenter *et al.*, 1964). The specific advantage of hide thickness in maintaining body temperature and resisting radiated heat effects under tropical conditions is still open for discussion and research. There were significant differences in the skin thickness for the five European breeds of dairy cattle in Australia (Nay and Hayman, 1963). Skin thickness was greater during the winter (July) than in the summer (January) in all breeds except the Friesian.

### SWEAT GLANDS

Sweat glands are important in the heat-regulation mechanism of the horse and man; in cattle and sheep they are not easily measured. Also, the importance of the sweating mechanism is limited under a climate of high humidity. Australian workers (Nay and Hayman, 1956) studied sweating in cattle. They indicated that each hair follicle is accompanied by a sweat gland (Table 4.1). There are significant differences in average number of sweat glands per square centimeter among the different

**TABLE 4.1 Number of Sweat Glands for Cattle, by Breed**

| | Dewlap | | Side | |
|---|---|---|---|---|
| Breed | Average | Range | Average | Range |
| Sindhi | 1117 | 1017–1285 | 1509 | 1246–1778 |
| Schiwal | 1066 | 952–1253 | 1507 | 1262–1738 |
| Jersey | – | – | 1005 | 878–1111 |
| Friesan | – | – | 996 | 963–1014 |
| Red Poll | – | – | 981 | 934–1029 |

breeds. Dairy cattle have 5 to 25 percent more sweat glands than beef cattle. The average size of the sweat gland is significantly larger in the zebu than in the Jersey, Friesan, or Red Poll. However, there are significant differences between animals in physiological and adaptive mechanisms under stress conditions. This should be encouraging to the geneticist who is selecting for productivity of the productive European breeds under unfavorable climate conditions.

The sweat glands of the zebu are closer to the epidermis, compared to the European breeds. The sweat glands of the zebu are saclike with few convolutions. Ferguson and Dowling (1955) measured the sweat production by a special absorption technique and showed that zebu × Jersey had 620 $g/m^2/hr$ vs. 140 in the Ayrshire. The sweating rates of Brahman × Shorthorn $F_2$ yearling steers were lower than Hereford × Shorthorn under mild conditions but higher under temperature stress conditions. Also, the Brahman crossbreds increased their sweating rate at lower skin and rectal temperatures compared to the Hereford × Shorthorns. Animals with sleek coats showed higher sweating capacity despite lower thermal stimulus to the skin. Maximum sweating rates occurred in the summer season (Schleger and Turner, 1965).

Here again, the relationship of sweating rates to over-all response and to various productive traits needs further study.

### Subcutaneous Fat

The subcutaneous fat layer in the polar bear serves as protection from extreme cold (Bonsma, 1967). However, the insulative subcutaneous layer of fat may not be as important in maintaining body temperature in swine (Irving, 1956). Also, recent work in England (Fuller, 1965) showed there was no difference in the percent of body fat deposited subcutaneously at temperatures from 10° to 30°C in growing pigs during a 56-day experiment. The maintenance of homeothermy in swine is achieved by the subcutaneous fat, with a change in blood flow to periphery and changes in heat production (Williams, 1967). Again, the fact that subcutaneous fat thickness in pigs has been greatly reduced by the geneticist, without an effect on response to a variety of climate, raises doubts about the importance of subcutaneous fat in response to bioclimatic factors.

## BEHAVIORAL RESPONSES

The psychological well-being of an animal will certainly influence its performance under any environmental condition, such as failure of milk let-down in an animal that is frightened. The emotional state of the animal thus can become very important in the conduct of investigations into environmental effects. It is appropriate to reemphasize here once again that it is practically impossible to make any measurement of a physiological function without the risk of having made some change in the animal. When the change influences the function under study, misleading data result. For example, making a simple observation on a free-ranging animal may cause that animal to act differently, especially if he is aware of the observer. Practically anything that is done to the animal has the potential of interfering with the outcome of a study. The rigid use of control animals is the usual way of trying to reduce the effect of such factors to a minimum (see Chapter 6).

### Freedom of Movement

Ordinarily, domestic animals prefer to have complete freedom of movement. A wild animal placed in an area restricted in size will behave in a very abnormal manner. Obviously, abnormal behavior influences many body functions.

#### RESTRICTED MOVEMENT

The amount of restriction as well as the type of the restriction can influence the well-being of the animal. Degrees of restriction range from complete immobility by caging or squeeze chutes with very little or no opportunity for any movement, to a large paddock with practically no limit on movement. Therefore, two general factors must be considered: the amount of limited movement, i.e., the amount of immobilization through cage size; and the manner in which the animal is restricted. Bond (1967) has reported that cattle in corrals with wire fences, as compared to those in corrals made of wooden fences, were exposed to greater air movement and lower air temperatures, and consequently gained faster.

### FREE MOVEMENT

Even with unlimited movement, problems in handling animals can develop. An animal will search out the most desirable environment, when permitted to do so. Should a great variety of environments be available to a group of animals, it must be assumed that the ability to find the most desirable environment will vary among animals. Obviously, this can cause variations in physiological responses.

#### Behavior under Stress

Perhaps one of the least studied characteristics of an animal, and consequently the least understood, is his behavior under stress. The animal's behavior under stress is generally an attempt to adjust to the stress. For example, a rabbit in high air temperature may place his chin in his water bowl to help keep body temperature low. Many other examples of unusual animal behavior could be cited, but the important point to the researcher is to realize that animals learn how to react to stress conditions and will act in an unorthodox manner. These actions can completely bias experimental data.

#### Response to Training

An animal's attitude toward his total environment can often be altered by training. A wild animal is not going to respond to a given environment in the same way that a docile animal will. A training period may be necessary in order that the animal will become familiar with the activities associated with the collection of data. It may even be necessary to eliminate an animal from consideration, if it lacks sufficient docility during the training period.

#### Human Influence in Taking Behavior Data

It is nearly impossible to completely remove the human element from the collection of experimental data. The reaction between the animal and the way it is handled will often influence the outcome of a study.

A few simple examples follow. Loud and unusual noises made by the caretaker during feeding or cleaning animal quarters will disrupt the daily routine. Rough handling of large animals while moving them, putting them in a restrained position, or even during feeding can cause a change in the animal's behavior. A partial solution is to have the same set of caretakers for the duration of the study. Use those persons that know the animals. Animals respond adversely to strange people and strange things.

## SUMMARY

In the final analysis it must be recognized that practically any external factor can alter one or more physiological mechanisms in an animal. Some changes are great enough so that they can be detected during the experiment; others are insidious so that their presence is not detected. The latter situations are dangerous because they may interfere with the proper interpretation of data or mask real differences in experimental response. In fact, it is nearly impossible to make a measure on the animal without the risk of making a change in the animal's physiology.

In setting up an experiment, it must be remembered that no two animals are exactly alike and therefore will not always respond to various stimuli in exactly the same manner. It is also nearly impossible to put two animals in exactly the same environment, even if the most sophisticated environment control chambers are used. Another point that must be remembered is that two animals may not respond to a change in the environment in the same manner (interaction between environmental change and animal response).

Although all these points are important, in that they make proper interpretation of experimental results more difficult, they do not prohibit sound investigations into environment–animal relationships. It does mean, however, that these factors must be recognized by the researcher in order that he may place the proper limitations on the value of his own experimental results as well as to determine the real significance of statements that are found in the literature.

## REFERENCES

Alliston, C. W., B. Howarth, Jr., and L. C. Ulberg. 1965. Embryonic mortality following culture *in vitro* of one- and two-cell rabbit eggs at elevated temperatures. J. Reprod. Fertil. 9:337–341.

Alliston, C. W., and L. C. Ulberg. 1961. Early pregnancy loss in sheep at ambient temperatures of 70° and 90°F as determined by embryo transfer. J. Anim. Sci. 20:608–613.

Bailey, C. B., R. Hironka, and S. B. Slen, 1962. Effect of the temperature on the environment and the drinking water on the body temperature and water consumption of sheep. Can. J. Anim. Sci. 42:1–8.

Best, C. H., and N. B. Taylor. 1961. The physiological basis of medical practice. A text in applied physiology. 7th ed. Williams & Wilkins Co., Baltimore.

Bianca, W. J. 1968. Thermoregulation in adaptation of domestic animals. E. S. E. Hafez (ed.) Lea & Febiger, Philadelphia.

Bianca, W. J., and J. D. Findlay. 1962. The effect of thermally induced hypernea on the acid–base status of the blood of calves. Res. Vet. Sci. 3:38–49.

Bond, T. E. 1967. Microclimate and livestock performance in hot climates. p. 207–220. *In* R. H. Shaw (ed.) Ground level climatology, AAAS Pub. No. 86, Washington, D.C.

Bonsma, J. 1967. Climatology and breeding for adaption. p. 146–175. *In* T. J. Cunha, A. C. Warnick, and M. Koger (ed.) Factors affecting calf crop. University of Florida Press, Gainesville.

Branton, C., R. E. McDowell, and M. A. Brown. 1966. Zebu–European crossbreeding as a basis of dairy cattle improvement in the U.S.A. South. Coop. Ser. Bull. 114.

Brues, A. M., and G. A. Sacher. 1965. Aging and levels of biological organization. The University of Chicago Press, Chicago.

Burfening, P. J., and L. C. Ulberg. 1968. Embryonic survival subsequent to culture of rabbit spermatozoa at 38° and 40°C. J. Reprod. Fertil. 15:87–92.

Carpenter, J. W. 1959. Slaughter and carcass characteristics of Brahman, shorthorn and Brahman–shorthorn crossbred steers. Ph.D. dissertation, University of Florida, Gainesville.

Carpenter, J. W., A. Z. Palmer, W. G. Kirk, F. M. Peacock, and M. Koger. 1964. Slaughter and carcass characteristics of Brahman and Brahman–shorthorn steers. Fla. Agr. Exp. Sta. Tech. Bull. No. 680.

Chang, M. C., and L. Fernandez-Cano. 1959. Effects of short changes of environmental temperature and low atmospheric pressure on the ovulation of rats. Am. J. Physiol. 196:653–655.

Dowling, D. F. 1961. The significance of the coat in heat tolerance of cattle. II. Effect of solar radiation on body temperatures. Aust. J. Agr. Res. 11:871–874.

Dukes, H. H. 1955. Physiology of domestic animals. 7th ed. Comstock Publishing Associates, Ithaca, New York.

Dutt, R. H., E. F. Ellington, and W. W. Carlton. 1959. Fertilization rate and early embryo survival in sheared and unsheared ewes following exposure to elevated air temperature. J. Anim. Sci. 18:1308–1318.

Eisenstein, A. B. 1967. The adrenal cortex. Little, Brown & Co., Boston.

Ferguson, K. A., and D. F. Dowling. 1955. The function of cattle sweat glands. Aust. J. Agr. Res. 6:640–644.

Folk, G. E., Jr. 1966. Introduction to environmental physiology. Lea & Febiger, Philadelphia.

Forrest, J. C., J. A. Will, G. R. Schmidt, M. D. Judge, and E. J. Briskey. 1968. Homeostasis in animals (*Sus domesticus*) during exposure to a warm environment. J. Appl. Physiol. 24:33–39.

Frankel, H. M., and F. L. Ferrante. 1966. Effect of arterial $PCO_2$ on appearance of increased lactate during hyperthermia. Am. J. Physiol. 210:1269–1272.

Fuller, M. F. 1965. The effect of environmental temperature on the nitrogen metabolism and growth of the young pig. Brit. J. Nutrit. 19:531–546.

Guyton, A. C. 1966. Textbook of medical physiology. 3rd ed. W. B. Saunders Co., Philadelphia.

Hafez, E. S. E. 1964. Behavioral thermoregulation in mammals and birds. Intern. J. Biometerol. 7:231–240.

Hafez, E. S. E. 1968. Adaptation of domestic animals. Lea & Febiger, Philadelphia.

Hammond, J. 1933. How science can help improve the nations food supply. Soc. Chem. Indus. J. 52:637.

Hankins, O. G., A. M. Gaddis, and W. L. Sulzbacher. 1959. Meat research techniques pertinent to animal production research. pp. 195–228. *In* Techniques and procedures in animal production research. Am. Soc. Anim. Prod. Beltsville, Md.

Hardy, J. D. 1961. Physiology of temperature regulation. Physiol. Rev. 41:521–606.

Hedrick, H. B. 1968. Bovine growth and composition. Mo. Agr. Exp. Sta. Res. Bull. No. 928.

Howarth, B., Jr., C. W. Alliston, and L. C. Ulberg. 1965. Importance of uterine environment on rabbit sperm prior to fertilization. J. Anim. Sci. 24:1027–1032.

Howarth, B., Jr., and H. W. Hawk. 1968. Effect of hydrocortisone on embryonic survival in sheep. J. Anim. Sci. 27:117–121.

Irving, L. 1956. Physiological insulation of swine as bareskinned mammals. J. Appl. Physiol. 9:414–420.

Johnson, H. D. 1967. Climatic effects on physiology and productivity of cattle. pp. 189–206. *In* R. H. Shaw (ed.) Ground level climatology. AAAS Pub. No. 86, Washington, D.C.

Johnson, H. D., A. C. Ragsdale, I. L. Berry, and M. D. Shanklin. 1962. Environmental physiology and shelter engineering. LXII. Effect of various temperature-humidity combinations on milk production of Holstein cattle. Mo. Agr. Exp. Sta. Res. Bull. No. 791.

Johnson, H. D., H. H. Kibler, I. L. Berry, O. Wayman, and C. P. Merilan. 1966. Environmental physiology and shelter engineering with special reference to domestic animals. LXX. Temperature and controlled feeding effects on lactation and related physiological reactions of cattle. Mo. Agr. Exp. Sta. Bull. No. 902.

Judge, M. D. 1969. Environmental stress and meat quality. J. Anim. Sci. 28:755–760.

Judge, M. D., E. J. Briskey, and R. K. Meyer. 1966. Endocrine related post-mortem changes in porcine muscle. Nature 212:287–289.

Kamal, T. H., H. D. Johnson, and A. C. Ragsdale. 1961. Influence of the stage of lactation and environmental temperatures on the salt balance of milk. J. Dairy Sci. 46:1655–1657.

Kauffman, R. G., H. W. Norton, B. G. Harmon, and B. C. Breidenstein. 1967. Growth of the porcine eye lens as an index to chronological age. J. Anim. Sci. 26:31–35.

Kibler, H. H., H. D. Johnson, M. D. Shanklin, and L. Hahn. 1965. Environmental physiology and shelter engineering with special reference to domestic animals. LXIX. Acclimation of Holstein cattle to 84°F (29°C) temperatures: Changes in heat producing and heat dissipating functions. Mo. Agr. Exp. Sta. Res. Bull. No. 893.

Lin, Y.-C., and P. D. Sturkie. 1968. Effect of environmental temperature on the catecholamines of chickens. Am. J. Physiol. 214:237–240.

Macfarlane, W. V. 1968. Adaptation of ruminants to tropics and deserts. *In* E. S. E. Hafez (ed.) Adaptation of domestic animals. Lea & Febinger, Philadelphia.

McDowell, R. E. 1967. Factors in reducing the adverse effects of climate on animal performance. pp. 277–291. *In* R. H. Shaw (ed.) Ground level climatology. AAAS Pub. No. 86, Washington, D.C.

Mississippi Agricultural Experimental Station. 1966. Sheep research in the southern region. South. Coop. Ser. 119.

Moulton, C. R., P. F. Trowbridge, and L. D. Haigh. 1922. Studies in animal nutrition. II. Changes in proportions of carcass and offal on different planes of nutrition. Mo. Agr. Exp. Sta. Res. Bull. No. 54.

Nay, T., and R. H. Hayman. 1956. Sweat glands in zebu (*Bos indicus* L.) and European (*Bos taurus* L.) cattle. I. Size of individual glands, the denseness of their population, and their depth below the skin surface. Aust. J. Agr. Res. 7:482–494.

Nay, T., and R. H. Hayman. 1963. Some skin characters in five breeds of European (*Bos taurus* L.) dairy cattle. Aust. J. Agr. Res. 14:294–302.

Pan, Y. S. 1964. Variation in hair characters over the body in Sahiwal zebu and Jersey cattle. Aust. J. Agr. Res. 15:346–356.

Phillips, R. W. 1963. Beef cattle in various areas of the world. *In* T. J. Cunha, M. Koger, and A. C. Warnick (ed.) Crossbreeding beef cattle. University of Florida Press, Gainesville.

Plasse, D., A. C. Warnick, and M. Koger. 1968. Reproductive behavior of *Bos indicus* females in a subtropical environment. I. Puberty and ovulation frequency in Brahman and Brahman X British heifers. J. Anim. Sci. 27:94–100.

Robertshaw, D., and G. C. Whittow. 1966. The effect of hyperthermia and localized heating of the anterior hypothalamus on the sympathoadrenal system of the ox (*Bos taurus*). J. Physiol. 187:351–360.

Ruch, T. C., and H. D. Patton. 1965. Physiology and biophysics. W. B. Saunders Co., Philadelphia.

Schleger, A. V., and H. G. Turner. 1960. Analysis of coat characters of cattle. Aust. J. Agr. Res. 11:875–885.

Schleger, A. V., and H. G. Turner. 1965. Sweating ranges of cattle in the field and their reaction to diurnal and seasonal changes. Aust. J. Agr. Res. 16:92–106.

Shaw, R. H. 1967. Ground level climatology. AAAS Pub. 86, Washington, D.C.

Shelton, M., and J. E. Huston. 1968. Effects of high temperature stress during gestation on certain aspects of reproduction in the ewe. J. Anim. Sci. 27:153–158.

Turner, H. G., and A. V. Schleger. 1960. The significance of coat type in cattle. Aust. J. Agr. Res. 11:645–663.

Ulberg, L. C., and P. J. Burfening. 1967. Embryo death resulting from adverse environment on spermatozoa or ova. J. Anim. Sci. 26:571–577.

Waites, G. M. H. 1962. The effect of heating the scrotum of the ram on respiration and body temperature. Quart. J. Exp. Physiol. 47:314–323.

Williams, C. M. 1967. Livestock production in cold climates. pp. 221–232. *In* R. H. Shaw (ed.) Ground level climatology. AAAS Pub. No. 86, Washington, D.C.

Wilson, W. O. 1967. Environmental temperature and feed regulating mechanisms. pp. 247–264. *In* R. H. Shaw (ed.) Ground level climatology. AAAS Pub. No. 86, Washington, D.C.

Yousef, M. K., and H. D. Johnson. 1966. Blood thyroxine degradation rate in cattle as influenced by temperature and feed intake. Life Sci. 5:1349–1363.

# 5

# ENVIRONMENT AND PHYSIOPATHOLOGY

## INTRODUCTION

The average annual economic loss due to infectious and noninfectious livestock diseases from 1951 to 1960 was estimated to have been at least $1,422,020,000 (U.S. Dept. Agr., 1965), a rather significant figure. Of the 78 important infectious diseases of domestic livestock described by Merchant and Barner (1964), at least 53 diseases are affected by seasonal factors.

Although seasonal incidence of disease might be due to, among other things, the number of susceptible animals and husbandry practices favoring transmission of disease, it must be recognized that climatic factors may have a direct effect on the host, intermediate host, or disease agent and may indirectly affect the host by affecting the quality or quantity of available foodstuffs. In some circumstances environmental conditions may favor concentration of toxic elements in feedstuffs.

A search of the literature reveals a few known instances of climatic factors influencing disease susceptibility. Although specific data are rarely encountered, statements often are made concerning the effects of environmental conditions on the health and susceptibility of many diseases. It must be understood that in most cases the expressions, "it is believed . . . ," "it is probable . . . ," and others indicate the lack of definite experimental evidence. These expressions of the effects of environment on the incidence and susceptibility to diseases are the results of observations of highly trained professional people. They cannot be discounted, but neither can they be taken at face value. Controlled experimental evidence is needed before the importance of environmental factors is fully appreciated.

Before studies can be properly evaluated to determine the effects of environmental factors on increasing (or decreasing) the susceptibility of domestic animals to specific diseases, it will be necessary to establish normal physiological values of the species with respect to sex and age under controlled environmental conditions. Such studies can be most effectively made in environmental facilities where temperature, humidity, light, wind, and altitude can be controlled and programmed. These investigations will become much more informative as better methods are developed for continuously sampling the physiological function of undisturbed animals in their natural environment. Without doubt, better methods of biotelemetry will be required. Special considerations should also be made concerning adequate measurement of environmental factors (see Chapter 2).

For many specific effects of environment on animals not included in this section, the reader is referred to *Environmental Biology* (Attman and Dittmer, 1966), *Adaptation* (Dill *et al.*, 1964), and *Medical Biometeorology* (Tromp, 1963).

## DIRECT EFFECTS OF THE ENVIRONMENT

### Temperature and Humidity

#### LOW TEMPERATURE

A study of the provocation of swine influenza by exposure to adverse weather indicated that some feature of the cold weather to which the animals were exposed was responsible for provoking masked (latent) influenza virus to infectivity. The exact constituent of the uncontrolled environment could not be determined from the data obtained (Shope, 1955).

Environmental conditions have been reported to be a definite factor in the severity of virus pneumonia of swine (Eernstman, 1962a, b, 1963).

A report has been made on the pathogenesis of diarrhea in the newborn calf, with special reference to physiological functions and environmental conditions. In the northwestern United States the most serious beef cattle losses have occurred in "early calving" under harsh, varied climatic conditions on crowded, low-lying, snow- and ice-covered calving areas with relatively high daytime temperatures and very low night temperatures. The interaction of several factors warrants consideration. Experimental data (Brandley and McClurkin, 1964; Moll, 1957, 1965) indicate that the susceptibility of colostrum-deprived calves varies considerably during seasons of the year, with those born in late fall, winter, or early spring being more likely to develop disease than calves born during the summer. Many lambs, born to ewes that had experienced long periods of undernourishment and climatic stress during pregnancy, and which died at ages up to 10 days, had no $\gamma$-globulin in their serum in spite of the fact that some of these lambs had consumed colostrum (Halliday, 1965).

A dysentery of newborn lambs due to *Escherichia* spp. and resulting from cold, wet weather and unsanitary conditions of corrals and lambing sheds has been reported (Marsh and Tunnicliff, 1938).

A study has been made of *Pasteurella multocida* infection in chickens exposed to cold (Juszkiewicz, 1967). A group of 15-week-old chickens was placed, after inoculation, in a temperature-controlled chamber at 1°–2°C below zero. In a second trial, chickens were placed for 4 hr in a heat chamber at 36°C, inoculated, and left for an additional 2 hr. Thereafter, they were transferred to a temperature of 1°–2°C below zero. A corresponding *Pasteurella*-inoculated control group was maintained at 18°C for each trial. No effect of cold could be demonstrated on mortality rate in the first trial, though the development of *Pasteurella* infection was lower in the control group. In chickens exposed to alternated temperature, increased mortality was associated with enhanced pasteurellosis and increased secondary infection.

Neonatal hypoglycemia is a metabolic disease of baby pigs and is caused by restriction of food intake. Studies have indicated that gluconeogenesis is poorly developed in baby pigs until the seventh day after birth (Elneil and McCance, 1965). The blood glucose level is then unstable and largely dependent upon dietary source. Blood glucose levels below 50 mg/100 ml will produce symptoms. A study was made of blood glucose concentration in newborn piglets kept without food or water for 24 hr at 14°, 35°, and 40°C (McCance and Widdowson, 1959). Blood sugar was normal at 40° and 35°C, but very reduced at 14°C (body temperatures dropped precipitously after 12 hr at 14°C in one litter). Their results also indicated that the newborn pig loses liver glycogen at high as well as at low environmental temperatures.

Muscle glycogen (especially skeletal, diaphragm, and, to lesser degree, heart) was slightly lower at 40°C than at 35°C but was greatly reduced at 14°C. Therefore, the low blood sugar may not be simply an exhaustion of liver glycogen as previously reported (Goodwin, 1955).

A study of factors affecting death loss in 810 litters of baby pigs farrowed in autumn 1962, and 492 litters in spring 1963 indicated that deaths during the first 21 days were related to farrowing house temperature at the time of farrowing (Bauman *et al.*, 1966). Deaths from chilling were greater among spring than autumn litters, but the reverse was true of deaths from enteric causes.

Postparturient hemoglobinuria is a disease of high-producing dairy cows. Hypophosphatemia is a predisposing cause of the disease (Parkinson and Sutherland, 1954; Penny, 1956). The ingestion of cold water or exposure to extremely cold weather has been reported to precipitate attacks of the disease (Tarr, 1947).

Hypomagnesemic tetany is a highly fatal disease of ruminants. There is much confusion in relation to the etiology and pathogenesis of this disease; however, the incidence varies from year to year depending largely on climatic conditions and management practices. The disease occurs in beef or dry dairy cattle at pasture during winter, usually with inadequate nutrition and shelter in periods of changeable weather (Allcroft, 1947; Allcroft and Green, 1938; McBarron, 1952). It has been found that reduced serum magnesium levels occur in adult cattle and sheep exposed to cold, wet, windy weather.

Prolonged exposure to severe cold, especially of unacclimatized animals, can, of course, cause freezing of the extremities such as ears, tails, teats, and occasionally feet. The injured tissue is a potential site for secondary bacterial infection.

## HIGH TEMPERATURE

Polish workers (Kita, 1966; Wojtatowicz, 1966) have studied the effect of relatively high environmental temperatures on pigs experimentally infected with *Erysipelothrix insidiosa*. Pigs kept at a temperature of 30°C were found to be more susceptible to infection by *E. insidiosa* than those kept at 13°–15°C, as judged by febrile response and clinical signs of the disease. The following changes were found in pigs exposed to 30°C: (1) a decrease in the amount of complement in blood and a marked decrease in the phagocytic index; (2) decrease in the concentration of ascorbic acid in the blood and an increase in the concentration of magnesium; (3) decrease in the number of acidophilic leukocytes in blood; and (4) reduced concentration of sodium and an increase in potassium excreted in the urine. Histopathological examination of skin samples from pigs inoculated intradermally with *E. insidiosa* and held at 30°C demonstrated more abundant cellular infiltration, mainly in the cutis and deeper necrotic lesions than infected pigs at 13°–15°C.

Prolonged exposure to high environmental temperature, especially when accompanied by severe muscular exertion and with high humidity, will cause hyperthermia. If the body temperature exceeds a critical point, obvious signs of distress such as panting, increased heart rate, salivation, and sweating may be noted. If the body temperature is not lowered, respiration becomes shallow, the animals collapse, showing convulsions and terminal coma. Death occurs in most species after a prolonged body temperature of 107°–109°F.

Anydrosis is a disease characterized by absence of sweating, resulting in hyperthermia or heat exhaustion. It occurs mainly in horses (Evans and Smith, 1954; Evans *et al.*, 1957) and less commonly in cattle (Stewart, 1943, 1956).

## Light

### BOVINE PINK EYE

This area of study is notable for its dearth of material; however, recent studies have demonstrated a new approach to the etiology of a specific disease. Bovine keratoconjunctivitis (pink eye) is an acute infectious disease of cattle, occurring principally in the summer months. *Moraxella bovis*, the one organism consistently associated with the disease, may or may not reproduce the disease under experimental conditions. Starting with the hypothesis that sunlight might be a contributing factor, it has been found that exposure to light from a mercury sunlamp enhanced the effect of *Moraxella bovis* infection of the bovine eye. Exposure of the eyes to ultraviolet light (2800–3200Å) plus *Moraxella bovis* increased the severity of the keratoconjunctivitis (Hughes *et al.*, 1965). In recent unpublished work the same authors have demonstrated that exposure of the eyes to ultraviolet light increased the severity of the disease and shortened the incubation period. Sunlight (Allen, 1919), dust (Anthony, 1957), wind and flies (Baldwin, 1945), and nutritional deficiencies (Jackson, 1953) have all been mentioned as contributing factors.

## CANCER EYE

A field survey of the effects of sunlight on the incidence of bovine ocular squamous carcinoma (cancer eye) has been reported (Anderson and Skinner, 1961). Records from nine herds (4960 animals) whose ages ranged from 2½ to 18 years were used. The incidence of "cancer eye" (after adjusting for age effect) increased significantly with: (1) increase in average hours of sunshine; (2) decrease in average latitude; and (3) increase in average altitude. These criteria were thought to reflect differences in the ultraviolet component of sunlight. Other evidence of an association between sunlight and cancer in bovine and laboratory animals and man was also mentioned.

## SUNBURN

Sunburn can occur when young pigs with tender skin, or white pigs of any age, are exposed to bright sunlight without acclimatization.

## PHOTOSENSITIZATION

Photosensitization is a disease state elicited by the action of direct sunlight. Porphyrinemia (porphyrins in the blood) may develop into photosensitization. In domestic animals, chiefly ruminants, porphyrinemia leads to cutaneous hypersensitivity to light, especially in areas of low pigmentation. The areas of low pigmentation, when exposed to direct sunlight, develop severe erythema, inflammatory edema, serous exudation, and necrosis, which may result in gangrenous sloughing of patches of skin. Porphyrinemia may be due to hepatic injury or to congenital anomalies in which hemoglobin destruction leads to excessive amounts of porphyrin in the blood.

An inherited susceptibility to photosensitization in Southdown sheep has been reported in New Zealand (Hancock, 1950). An inherited susceptibility to photosensitization also occurs in cattle (Fourie, 1953) and pigs (Claire and Stephens, 1944).

Photosensitization can also occur from ingesting other sensitizing agents, for example, phenothiazine sulfoxide (Smith and Jones, 1966) and aphids (McClymont, 1955). Some plants (e.g., *Fagopyrum esculentum* and *Hypericum* spp.) liberate a red fluorescent pigment that causes photosensitization without liver injury. The effective wavelengths of light are from the middle of the visible spectrum, differing in this respect from photosensitization of hepatogenous origin (Smith and Jones, 1961). The great majority of photosensitization in animals is not due to porphyrins of hemoglobin origin. Usually the porphyrins are derived from the chlorophyll in the herbivorous animal's food. In most cases, liver injury is demonstrable. Hepatotoxins, associated with ingested food, injure the liver, which then is no longer able to remove the porphyrin (phyloerythrin) from the portal blood. Claire (1952) lists 16 disorders accompanied by hepatosis and icterus and also 55 plants suspected or known to cause photosensitization.

In the case of "plant-produced" photosensitization, the indirect effects of climatic factors are very important.

#### VITAMIN D AND ULTRAVIOLET DEFICIENCY

Vitamin D deficiency may result from insufficient exposure of animals to ultraviolet light irradiating animals or from insufficient irradiation of their roughage. When both the direct and indirect effects of too little solar ultraviolet irradiation are combined, clinical signs of vitamin D deficiency may be noted, especially in young animals. Husbandry practices such as controlled environments for rearing animals and rapid curing methods of hay-making could significantly reduce vitamin D levels and, without supplementation, could result in a deficiency state.

Vitamin D deficiency in sheep has been studied (Ewer, 1950, 1951a, b, 1953; Ewer and Bartrum, 1958). Marked differences in blood levels of vitamin D between sheep with long and short fleece have been observed, especially during periods of maximum sunlight (Quarterman *et al.*, 1961). Rickets in calves under farm conditions has also been reported (Hibbs *et al.*, 1945).

### Altitude

#### BRISKET DISEASE

Symptoms or signs of high mountain or brisket disease of cattle are characterized by severe pulmonary hypertension (high blood pressure of lungs) and concomitant right heart failure (Thorne *et al.*, 1965; Velasquez, 1947). Histopathological study of the lungs from these cattle demonstrated medial muscular hypertrophy (increase in size of smooth muscle cells) in the small arteries and arterioles proportional to the degree of hypertension. Other species are reported to be affected with varying susceptibility. The disease appears to be directly related to chronic hypoxia (low oxygen tension). Cattle do not show a sustained hyperventilatory response at high altitude. They do not develop a significant erythropoetic response under the stimulus of lowered atmospheric $pO_2$ (partial pressure of oxygen). Hemoglobin and packed cell volume (Hct) values show only slight increases (Alexander and Will, 1967).

#### ACUTE PULMONARY EMPHYSEMA

Acute pulmonary emphysema was first reported in Britain. It was called by some "fog" or "feg" fever (old Anglo-Saxon term referring to pasture aftermath) and was eventually associated with animals grazing on meadow aftermath (plant growth after last hay removal) (Leslie, 1949). It was first reported in the United States (1940) in Montana (Farquharson and Butler, 1944). Because it was first noticed in the United States in mountainous country, i.e., when cattle were brought from fairly high altitudes to graze on the aftermath of relatively low-lying meadows after removal of the last hay crop, it was thought to be due to changes in altitude. Almost identical lesions have been found in cattle that received similar changes in pasture management, i.e., from dry to lush green pastures, but with no change in altitude. This raises considerable doubt about the part played by abrupt altitude changes.

### Air

#### POLLUTION

The subject of air pollution is not new, but it is attracting wide-spread attention at the present time, probably because air pollutants are increasing at an alarming rate.* Although a review of the literature reveals little experimental work showing the effects of air pollution on animal health, some work has been done in regard to its effect on human health.

Environmental facilities may be useful for air pollution studies. Pollutants can be "fed" into the system at known rates and concentrations under controlled conditions of temperature, humidity, and air movement.

Some of our large industrial cities have been plagued with atmospheric contamination with smog. "Smog" is a term used to denote the presence in the atmosphere of a mixture of smoke and fog.

*Methods of measuring air composition are described in Chapter 2.

Three smog disasters have been reported in the last century (Tromp, 1963). In December 1930, in the Mense Valley near Liège in Belgium a fog laden with gaseous impurities persisted for 5 days, causing the deaths of 63 people and serious illness to many more. The second reported disaster occurred at Donora near Pittsburgh, Pennsylvania, in November 1948, and resulted in 19 deaths. The greatest smog disaster occurred in London in 1952. This smog lasted 4–5 days and caused at least 4000 deaths and much additional illness. It has been estimated that this fog covered an area of 450 square miles, contained 380 tons of smoke, and 800 tons of sulfuric acid. It would be unusual if domestic animals were not affected by the air pollution conditions.

The harmful effects of smogs may be due to different chemical contaminants such as those produced by action of ozone on hydrocarbons, which are released to the air, and to other contaminants.

## FLUORIDE INTOXICATION

Occasional cases arise where it is suspected that cattle have been damaged through ingestion of abnormal amounts of fluoride that have been deposited on or assimilated by forage or as a result of emission of fluorine compounds to the air by industrial or other operations (Shupe and Alther, 1966).

Emissions into the air of fluorides from industrial operations have been carried by the winds to surrounding land, contaminating the forage and thus causing fluorosis in cattle grazing on these lands. The symptoms vary from a slight mottling of the teeth to severe bone and joint involvement and decreased weight and milk production; in severe cases, death may occur. Normal bones may contain as much as 600 ppm of fluorine (Mitchel, 1942). Bones from animals that have been damaged by fluorosis will contain 3000 to 4000 ppm of fluorine.

The long-term effects of fluoride on the hemopoietic system, liver, and thyroid have been studied (Hoogstratten *et al.*, 1965).

Excessive and damaging fluorine contamination has been found near a chemical plant in Iowa (Buck, 1968). As fertilizer and chemical plants increase, this problem may be amplified.

## MISCELLANEOUS AIR AND WATER POLLUTANTS

Solid-state rocket fuels give off materials toxic to plants and animals. Two toxic materials known to be released are beryllium and fluorine (Shupe and Alther, 1966).

Sulfur dioxide given off in the fumes from certain industrial plants is toxic to vegetation. In range country this indirectly affects animal health. Sulfur dioxide in concentrations of less than 1 ppm are believed to be injurious to plant foliage. Amounts of 400–500 ppm are immediately dangerous to animal life, and 6–12 ppm will cause immediate irritation to the nose and throat. It is dangerous to the eyes in 20 ppm concentrations. It chiefly affects the respiratory tract and bronchi and may cause edema of the lungs (Sax, 1963; Schmidt and Rand, 1952).

There was much concern for the suffering of wildlife from oil pollution following the wreck of the SS *Torrey Canyon* off the English coast. Either ingestion of oil or its absorption through the skin results in poisoning.

Recently, large-scale leakage from a Santa Barbara channel oil well created a serious oil pollution problem, causing severe death losses in fish and other wildlife along the adjacent California coast.

### Diseases Due to Exposure to Radionuclides

The results of radionuclide studies show that the lower large intestine usually receives the highest radiation dose following ingestion of radioactive materials. The contents of both the omasum and lower large intestine have a relatively long transient time and are quite concentrated, so that if radiated material has been ingested, the radiation dose in this area is high (Bustad *et al.*, 1965).

The deposition of fallout on the skin of animals can cause lesions often referred to as "beta burns" since they are due to the $\beta$-component of $\beta$–$\gamma$-emitting radionuclides. The most noticeable reaction to beta burns is a localized change in wool growth and in pigmentation. The area of skin affected is principally the back region, where dust settles easily.

Bone-seeking radionuclides may cause osteogenic sarcomas (bone tumors) or, in very large doses, death of some animals.

The present state of knowledge indicates that the whole-body dose from $\gamma$-emitting radionuclides will be the limiting exposure from the standpoint of survival.

The threshold dose for epilation (removal of hair by the roots) of the sheared sheep skin lies between 2500 and 5000 rep (Luschbaugh and Spalding, 1957). The threshold dose for epilation of the normally wooled skin of the face lies between 5000 and 10,000 rep. The threshold dose for epilation of a thickly wooled sheep (staple length of 33 mm) is probably between 10,000 and 15,000 rep.

Unshorn sheep are naturally well protected from radiation. The amount of $\beta$-irradiation that would have been required to produce lesions as extensive as those caused by the infectious pustular dermatitis is estimated to be between 50,000 and 150,000 rep.

A study of the secretion of $^{131}I$ in the milk of dairy cows showed that maximum levels of $^{131}I$ were found at the seventh day in milk and blood and after the tenth day in the thyroids (Lengeman and Swanson, 1957). From 5 to 10 percent of a daily dose appeared in the milk each day during continuous $^{131}I$ ingestion. Massive daily doses of sodium iodide reduced the radioactivity of the milk by one half in 2 days and removed two thirds of the isotope from the thyroid in 10 days.

In experiments carried on after termination of a Russian nuclear test series (Kahn *et al.*, 1962), certain cows were placed under shelter and provided with water and hay that had been protected from recent fallout. Under these sheltered conditions, the $^{131}I$ in milk remained at or below 20 pc (picocuries) per liter, whereas levels in other milk during the same period were as high as 270 pc per liter. The milk of the cows on pasture contained $^{131}I$, $^{140}Ba$, and $^{137}Cs$. With a few possible exceptions, the sheltered cows, eating stored feed, gave milk containing no detectable $^{131}I$ or $^{140}Ba$.

It has been suggested (Noyes *et al.*, 1963) that exposure to a nuclear explosion imposes conditions that lead to the invasion of the blood by microorganisms normally held in restraint (largely by phagocytic cellular elements), but that these conditions have no primary influence on pathogens controlled largely by humoral antibody mechanisms.

## Diseases Due to Stresses Imposed by Climatic Changes

### RESPIRATORY DISEASES

During the process of shipping, cattle are exposed to unusual factors, such as cold, fatigue, and change of feed and water. These factors, while thought to be important in the development of shipping fever, infectious bovine rhinotracheitis, and possibly calf diphtheria, have not been evaluated singly by controlled experimentation (Jensen and Mackey, 1965).

The hypothesis that a latent virus is activated by the "stress" of shipment to make a favorable environment for secondary bacteria finds support in experiments (Carter and McSherry, 1955; Hoerlein and Marsh, 1957; Reisinger *et al.*, 1959). In another series of experiments, the number of sick animals was always higher when inadequate shelter was provided for recently "shipped-in" animals.

The incidence of shipping fever is greatest during wet, cold periods, especially during changeable spring and fall weather. The disease usually develops before the cattle reach the stockyards. It may be due to the exposure en route (Farley, 1932). Shipping fever generally occurs in late fall and early winter when cattle are shipped through stockyards. The disease occurs repeatedly under or following conditions causing stress. Other pasteurelloses, including fowl cholera, are usually initiated by unfavorable environmental conditions (Carter, 1954).

Besides the factors producing stress that have been referred to previously, it should be realized that many of the animals from western Canada have clinically inapparent chronic pneumonias. These latent infections no doubt become acute in some instances and may facilitate entry of other pathogens (Carter and McSherry, 1955).

Infectious bovine rhinotracheitis (IBR) is thought to be influenced by sudden changes in temperature and by cold, wet weather. The incidence is greatest during fall and winter.

The incidence of calf diphtheria is seasonal, being more common from November through February. During attempts to reproduce the disease, transmission failures led to the postulation of predisposing causes such as inclement weather and widely fluctuating temperatures (David, 1952; Mohler and Morse, 1905; Scrivner and Lee, 1934).

Contagious pleuropneumonia of horses spreads in poorly ventilated stables, and the greatest extension of the disease is frequently associated with northern and eastern winds during cold, damp periods of the year.

Equine rhinopneumonitis epizootics in young horses occur from October to December with occasional outbreaks in winter and early spring. This virus disease in horses resembles somewhat the "common cold" in man.

## MASTITIS

Although many statements are made regarding the effect of weather on the incidence of mastitis, the following statement is made by the National Mastitis Council (1963): "There is no conclusive evidence to indicate that the season *per se* influences the incidence of udder infections and mastitis." So many other factors are involved that it will take many controlled experiments in accurately controlled environmental facilities to demonstrate that environment does or does not play a role in establishing the incidence of mastitis.

#### INFECTIOUS DIARRHEA

Infectious diarrhea affects all cattle types and breeds. It occurs during all seasons but is more prevalent during the winter months when cattle are closely confined in stables or corrals. In herds where the disease is enzootic (occurring endemically in animals), it may reappear each autumn following a disease-free summer.

#### NONINFECTIOUS DISEASES

Of the noninfectious diseases, the literature on bloat is a good example of the subjective type of writing. Several statements taken from a text that was standard for many years (Hutyra and Marek, 1926) follow: (1) "It is also claimed that bloating (in cattle) is easily brought about when animals swallow a lot of air while pasturing against the wind." (2) The incidence of bloat increased "in years when the spring has been cold and when the plants, which had been backward for some time, grow more luxuriantly when warm weather sets in." (3) "The same circumstances cause bloating after pasturing on short stubble fields where weeds have been exposed to the sun a short time only and where dropped-off grains have germinated."

Almost all of such hypotheses are untested. They do not have critical supporting evidence; neither have they been proven wrong. The same statement can be made concerning many other diseases.

### Influence of Intensive Methods in Animal Husbandry

Intensive methods in husbandry of poultry and pig units have increased production efficiency and have also presented new health problems. Concentrations of ammonia as high as 50–100 ppm in the air of integrated poultry units have been reported during the winter months when ventilation was poorest (Anderson *et al.*, 1964b). It has been reported (Bullis *et al.*, 1950) that keratoconjunctivitis in chickens might be caused by exposure to ammonia in the integrated poultry units.

Male guinea pigs have been exposed to atmospheric concentrations of 170 ppm of ammonia for periods of 6 hr a day, 5 days a week, for 12 weeks and no evidence was found of chronic intoxication (Weatherby, 1952). However, after 18 weeks there were significant changes in the spleen (congestion and increased amounts of hemosiderin, a dark, yellow pigment containing iron), tubular casts in the kidneys, degeneration of the suprarenal glands, and liver congestion. Guinea pigs, mice, chickens and turkeys exposed to 20 ppm of ammonia have gross and microscopic changes in the lungs (Anderson *et al.*, 1964a). In this series of experiments, chickens exposed for 72-hr periods at 20 ppm of ammonia had significantly increased susceptibility to aerosol exposure of Newcastle disease virus. This is one of the few contemporary controlled experiments testing the hypothesis that certain environmental conditions affect susceptibility to a specific disease.

Since the beginning of 1960, manure in liquid form has been utilized in Sweden. As early as 1963, several acute poisonings occurred, caused by poisonous gases formed in liquid manure. As a rule the acute poisonings were connected with heavy agitation of the liquid manure inside the cowshed. The development of the acute poisonings is very fast, with dyspnea and convulsions, and the cause is supposed to be hydrogen sulfide. High concentrations of hydrogen sulfide have often been measured in agitated liquid manure.

In a considerable number of cattle sheds handling liquid manure, chronic disease has been observed, mostly in dairy cows. It has almost always been noted in sheds where the liquid manure is repumped or agitated. This chronic disease seems to cause a loss of condition and reduction in milk production, accelerated breathing and heart rate, severe pain in the feet, often with a high degree of softening of the hoofs, intensified pulsation in the digital artery, and standing with arched back and often crossed feet. Subcutaneous hematomas are often seen. In some cases important blood changes are seen, with prolonged bleeding time, lowered albumin globulin quotient, and anemia. Hydrogen sulfide in combination with other poisonous products from the liquid manure is again thought to be the cause. Further investigations of the etiology of these cases are planned (Holtenius and Hogsved, 1968).

## ENVIRONMENTAL FACTORS INDIRECTLY AFFECTING ANIMALS

### Direct Effect of Environmental Factors on Disease Agents

#### AIRBORNE INFECTIONS

Environmental factors influencing airborne infection have been extensively studied (Dimick, 1963; Hatch and Gross, 1964; Lepper and

Wolfe, 1966; Riley and O'Grady, 1961; Rosebury, 1947; Wells, 1955). Most studies have been directed to organisms causing disease in humans or diseases transmissible from animals to humans. Because of the abundance of technical information available (see the reference list), no attempt will be made to include detailed information here.

Airborne infections are, in general, perpetuated by the formation of droplet particles in the air. Organism-bearing particles are liberated into the air by two principal means (Riley and O'Grady, 1961): primarily from the respiratory tract and from movements that shed bacteria-bearing particles from the skin, hair, or wounds; and secondarily from the redistribution of organism-bearing particles that have accumulated in dust.

True atomization of liquids may be regarded as complete when particle size approaches 10 $\mu$ and most droplets expelled by violent expiratory processes do not greatly exceed 10 $\mu$ (Wells, 1955). Atomization intensifies evaporation and condensation, and most respiratory droplets evaporate in midair. When they evaporate, residues of nutrient substances and organisms may remain. "For all intents and purposes... droplet nuclei do not settle—they are as much a part of the atmosphere as gases themselves. Indoors they remain suspended until they are breathed or vented."

Environmental factors (especially humidity) determine how rapidly droplets evaporate, the manner and degree of dispersion of droplets, and survival of organisms present in the droplet (Gregory, 1961; Songer, 1967; Wells, 1955; Wright *et al.*, 1968).

It is apparent that good animal health depends on adequate ventilation, provided this ventilation is produced in a suitable manner.

The physiological response to a cold environment may also be conducive to establishing airborne infections. A slow respiratory rate (Olsen, 1969), possibly accompanied by an increased tidal volume, could be expected to carry droplet nuclei deep into the respiratory system (Hatch and Gross, 1964).

### SOILBORNE INFECTIONS

The incidence of several clostridial infections (e.g., black leg and bacillary hemoglobinuria) is influenced by environmental factors; it is highest during periods of wet weather. It is not clear whether this is a factor of transmission or of conditions suitable for propagation of the organism—possibly both.

Leptospirosis is reported to be most common in areas or seasons when the climate is warm, soils are alkaline, and the surface water is abundant (Blood and Henderson, 1963).

Epidemiological evidence (Van Ness, 1961) supports the hypothesis that soilborne infections of anthrax are especially likely in warm weather when the environmental temperature is above 60°F. Periods of high rainfall and overflowing streams may also spread the disease (Merchant and Barner, 1964).

#### PARASITES

The eggs of many parasites undergo development outside the host, especially in pastures (U.S. Dept. Agr., 1941). Moisture, temperature, and rainfall play an important part in the development necessary for reinfestation. After desiccation, the eggs of the human and swine ascaris did not develop embryos in atmospheres less than 80 percent saturated with moisture. All swine ascaris eggs perished after 9 days of desiccation (Swales *et al.*, 1942).

Some hookworm larvae have a rather narrow temperature zone of survival, between 60° and 90°F. Temperature controls the speed of development of eggs and larvae of many parasites. Both low and high temperatures may injure or kill certain parasites. Direct sunlight, under certain conditions, will kill parasite eggs and larvae.

Climate also may act as a deterrent to massive infestations; i.e., seasonal climatic changes may affect the life and distribution of intermediate forms and may also influence management practices, which in turn may limit parasitic invasions.

Temperature and humidity are extremely important to parasites living outside or away from their hosts. Heat injury of the parasites is definite but still poorly understood. It may be due to the inactivation of enzymes, to asphyxiation, to an accumulation of waste products, or to a breaking up of the mitochondria in certain tissues. Cold injury is also poorly understood, but it may be due to an accumulation of toxic products ordinarily eliminated or to the inability of parasites to utilize certain food materials (Steinhaus, 1940).

### Effects of Climate on Intermediate Hosts and Insect Vectors

Climate has pronounced effects on parasitism through its influence on the distribution of intermediate hosts. The eye worm of poultry, *Oxyspirura mansoni*, has as its intermediate host the cockroach,

*Pycnoscelus surinamensis*. Temperature has a definite effect on the distribution of the cockroach.

Seasonal incidence of infectious necrotic hepatitis may result from climatic factors affecting either liver fluke or snail populations (Gee, 1958; Herbert and Hughes, 1956; Ross, 1967a, b; Sinclair, 1956).

Insect vectors are very important in transmission of many virus diseases. The incidence of vesicular stomatitis, blue tongue, and contagious ecthema is much reduced after the first frost.

## Diseases Caused by Climatic Factors Influencing Plants

The metabolic state of the plant is thought to be of great importance in plant-produced photosensitization. The metabolic state of the plant is, of course, affected by environmental conditions.

### FACIAL ECZEMA

Facial eczema of sheep is a photosensitizing disease occurring periodically in New Zealand and eastern Victoria, Australia. The disease is caused by a toxin, sporodesmin, produced by the fungus *Pithomyces chartarum*, which grows on dead plant material in predominantly rye grass pastures. Rapid proliferation of the fungus occurs best under conditions of high temperature and humidity when these climatic conditions follow hot, dry weather. The height of the pasture grass also influences the amount of fungus ingested.

The disease is characterized by edema and eczema of the face and ears, and varying degrees of icterus and phylloerythrin in the circulating blood. The blood serum shows increased transaminase, cholesterol, lipid phosphorus, and bilirubin. The principal and most damaging lesions are in the liver, which is usually enlarged. The bile ducts are edematous and fibrotic and may be plugged with inspissated bile (Smith and Jones, 1966).

### PHALARIS POISONING

A number of outbreaks of phalaris poisoning have been described. They have been associated with drought conditions when generally the only green feed available was phalaris a few inches high. A correlation may occur between the mortality during outbreak and the degree of starvation of the sheep prior to feeding on the plant (Farleigh, 1966).

### OTHER DISEASES

Many other plant poisons may be indirectly caused by climatic conditions. For instance, drought may limit the growth of safe nutritive plants so that grazing animals will eat harmful plants that would not be touched under ordinary grazing conditions. Halogeton poisoning is a good example. Severe drought may also result in deficiency diseases (Peirce, 1945).

Under certain environmental conditions plants that are normally valuable for grazing become toxic by accumulating excess quantities of toxic materials. Hydrocyanic acid substances may be released within plants following freezing, wilting, or crushing, but rumen microflora can also initiate their enzymic release. Young plants and sprouts contain more toxic glycosides than mature ones, and plants growing in the shade contain more glycoside than those in the sun (Kingsbury, 1964; Schmutz *et al.*, 1968).

Nitrate poisoning following ingestion of certain plants is frequently associated with drought, heavy application of nitrate fertilizers, and soils characteristically high in nitrogen. Hot, humid weather following a drought may stimulate rapid growth of cereal and root crops growing on ground highly fertilized with nitrogen, resulting in toxic levels of nitrates (Kingsbury, 1964).

Wet, autumn weather allowing heavy mold growth on corn ears can result in leucoencephalomalacia of horses (Biester and Schwarte, 1940).

Rye grass staggers affects sheep, cattle, and horses and occurs most commonly in the autumn. The disease disappears following rainy periods (Clegg and Watson, 1960; Cunningham and Hartley, 1959; Hopkirk, 1935).

## Effect of Environmental Factors on Food and Water Intake

High environmental temperatures can indirectly result in salt toxicity by increasing intake of water containing 1.2 percent NaCl. The effects of 1.2 percent salt water during summer are more severe than those of 1.25 percent NaCl during winter (Weeth and Haverland, 1961).

The decreased intake of water during the winter time has been suggested as a factor in the development of urinary calculi (Blood and Henderson, 1963).

Warm temperatures stimulating massive "blooms" and concentration of algae near the shore by wind may result in poisoning of livestock drinking the water (Brandenburg and Shigley, 1947; Fitch *et al.*, 1934; O'Donoghue and Wilton, 1951; Olsen, 1953; Steyn, 1943).

## MISCELLANEOUS SPECIAL CONSIDERATIONS FOR EXPERIMENTATION

There is an urgent need for facilities where the effects of extreme and rapidly changing environmental conditions as contributing causes of many infectious and noninfectious diseases, and the susceptibility of animals to these diseases, can be studied. This statement is based on opinions, observations, hypotheses, and some facts presented in scientific and lay communications media by well-trained people in the biomedical and research fields.

Some of the statements and quotations listed below, many of which are based on astute observations, emphasize the need to establish facilities that will make it possible for the scientist to develop facts about the relationship of environmental factors and diseases.

### Environmental Conditions Where Control is Needed for Useful Study

#### TEMPERATURE

In a country as large as the United States, extremes of temperature can be expected and do occur. These extremes occur in a number of states where ranching is an important industry and where severe livestock losses are experienced as a result of these dramatic changes in temperature.

Following are some official Weather Bureau recordings of temperature changes that have been experienced in certain parts of the country:

January 23, 1916—there was a temperature drop in Montana of 100°F, from 44° to −56° during a 24-hr period

December 24, 1924—there was a 77°F drop in 12 hr, 56° at noon to −21° at midnight at Airfield, Montana

January 12, 1911—a 62°F drop, from 49° at 6:00 a.m. to −13°F at 8:00 a.m. at Rapid City, South Dakota

January 2, 1943—a 58°F drop in 27 min, from 54° at 9:00 a.m. to −4°F at 9:27 a.m. at Spearfish, South Dakota

**TABLE 5.1 Wind-Chill Chart[a]**

| Estimated Wind Speed (mph) | ACTUAL THERMOMETER READING °F | | | | | | | | | | | |
|---|---|---|---|---|---|---|---|---|---|---|---|---|
| | 50 | 40 | 30 | 20 | 10 | 0 | −10 | −20 | −30 | −40 | −50 | −60 |
| | EQUIVALENT TEMPERATURE °F | | | | | | | | | | | |
| Calm | 50 | 40 | 30 | 20 | 10 | 0 | −10 | −20 | −30 | −40 | −50 | −60 |
| 5 | 48 | 37 | 27 | 16 | 6 | −5 | −15 | −26 | −36 | −47 | −57 | −68 |
| 10 | 40 | 28 | 16 | 4 | −9 | −21 | −33 | −46 | −58 | −70 | −83 | −95 |
| 15 | 36 | 22 | 9 | −5 | −18 | −36 | −45 | −58 | −72 | −85 | −99 | −112 |
| 20 | 32 | 18 | 4 | −10 | −25 | −39 | −53 | −67 | −82 | −96 | −110 | −124 |
| 25 | 30 | 16 | 0 | −15 | −29 | −44 | −59 | −74 | −88 | −104 | −118 | −133 |
| 30 | 28 | 13 | −2 | −18 | −33 | −48 | −63 | −79 | −94 | −109 | −125 | −140 |
| 35 | 27 | 11 | −4 | −20 | −35 | −49 | −67 | −82 | −98 | −113 | −129 | −145 |
| 40 | 26 | 10 | −6 | −21 | −37 | −53 | −69 | −85 | −100 | −116 | −132 | −148 |
| Wind speeds greater than 40 mph have little additional effect | Little danger for properly clothed person | | | | Increasing danger | | | Great danger | | | | |
| | | | | | DANGER FROM FREEZING EXPOSED FLESH | | | | | | | |

[a]To use the chart, find the estimated or actual wind speed in the left-hand column and the actual temperature in °F in the top row. The equivalent temperature is found where these two intersect. For example, with a wind speed of 10 mph and a temperature of −10°F, the equivalent temperature is −33°F. This lies within the zone of increasing danger of frostbite, and protective measures should be taken.

These are unusual temperature changes and although it might be unreasonably expensive to duplicate them in an experimental environmental facility, such equipment should be capable of 60° F changes in 2 hr. This is a minimum requirement as far as temperature changes are concerned, but is absolutely necessary if comprehensive studies are to be made of the effects of abrupt temperature changes on the normal physiology and susceptibility to disease of farm animals. In other countries, more drastic changes may occur.

#### WIND OR AIR MOVEMENT

Wind or air flow potentials should be an integral part of environmental facilities. Stress due to air flow may occur when animals are shipped relatively long distances in ventilated trucks and in stock cars by rail under various levels of temperature and humidity. Table 5.1 is self-explanatory. Although developed by the U.S. Army for man, it should be applicable to animals with appropriate modifications. Wind or air movement up to 40 mph can be obtained in environmental facilities at reasonable costs of construction and operation. Adequate rate of change of room air is essential to control the concentration of certain gases and odors.

#### HUMIDITY

Humidity is an important factor in the study of the effects of environment on the production and susceptibility of domestic animals to disease and parasitism. Humidity as well as air flow must be measured and controlled along with temperature.

#### ALTITUDE (BAROMETRIC PRESSURE)

Pressure chambers are essential in disease studies where the effects of altitude are important considerations. This is especially true in the study of at least two important cattle diseases: brisket disease and pulmonary emphysema.

### Physical Facilities for Disease Studies

#### SPACE

This is one of the difficult, and important, factors to be considered in the design of an environmental facility. As with all other environmental studies, the space required depends on the kinds of experiments, the physiological recording equipment used, whether physiological data are recorded by direct wire or by radiotelemetry, the temperament of the animals, and the availability of animal caretakers.

*Ceiling* height depends on the physiological recording equipment, observation requirements, and lighting methods used. The minimum should be no less than 9 feet, preferably higher.

If large domestic animals are to be used, adequate *space* should be provided. It requires a minimum of 48 $ft^2$ of floor space for an average size cow. Cattle can be stanchioned individually in 48 $ft^2$ of floor space. Under loose conditions more floor space might be needed. The figures given here must be considered the bare minimum. Two to four sheep can be kept in the same space and about the same number of swine. This, of course, depends on the size, breed, and age of the animals.

In some experiments it is necessary to keep the animals completely *separated*. This is especially true in pigs that have chronic surgical preparations since pigs have a decided tendency to "improve" on the surgical technique by chewing on each others' exposed catheters (this refers to catheters chronically implanted in the cardiovascular system).

## CONSTRUCTION AND EQUIPMENT

Detailed construction is discussed in Chapter 7; however, a few requisites will be included here from the operator's viewpoint.

1. In situations where studies of the effects of environmental factors on specific diseases is concerned, it is necessary to have absolute filters, a competent air balance system, and walls and internal equipment that can be thoroughly disinfected (which eliminates wooden construction).

2. Rapid change of environmental factors should be possible, not only as previously suggested to simulate some of the conditions occurring in nature, but also to allow for quick changes in environmental requirements from one experiment to another. These facilities are expensive to build and to operate and should accommodate the maximum number of experiments.

3. The walls, floor, and ceiling should have adequate numbers of attachment devices for stanchions, pens, and special equipment. These devices should be strong and yet should not interfere with cleaning; in other words, it should be possible to expose them for attachment or seal them off when not in use.

4. The walls and floor should have special heating and cooling facilities so that they could follow the rapid air-temperature changes in the room, or be independently controlled at temperatures differing from that of the air.

5. Special heating elements should be installed in manure troughs and drains to facilitate cleaning during periods of extreme freezing temperatures.

6. The manure troughs and drains should also have cooling coils incorporated in their structure to minimize odors when room temperatures are normal or high.

7. The structure of the rooms should be versatile so that different species could be accommodated with minimum changes.

8. A visual monitoring system should be installed for visualizing animal behavior as well as for the safety of personnel working in the rooms. Closed-circuit television systems are adequate and not too unreasonable in price. A programming system is necessary for the operator's efficiency and convenience. All controllable environmental changes should be capable of being programmed.

9. Equipment controls and all recording and visual monitoring systems should be in one place for convenience and efficiency of operation. This also prevents spread of contaminating microorganisms by the operators.

10. Construction and accessibility of environmental facilities should be such that personnel operating the equipment should be able to do so in one area, and this area should be free of bacterial or viral contamination. Maintenance of much of the equipment should be sufficiently accessible so that there would be little or no danger to personnel involved.

## SAFETY

If altitude (barometric pressure) facilities are required, a number of safety devices must be included:

(1) Structure capable of withstanding the pressure changes to which it is subjected
(2) Special structures in the drainage system
(3) A small compressing–decompressing chamber so that personnel could enter or leave the room without endangering their health
(4) Safety (warning) devices to be installed for the protection of personnel working on altitude conditions

## Selection of Animals

Because of the initial cost and operational expense involved, and the fact that numbers of animals may be limited in environmental facility studies, the animals used must be carefully selected for the particular study.

The term "specific pathogen-free" animals is a much abused, misused term. An appended statement should declare from what pathogen(s) they are "free." In many cases, such animals may have considerable immunity against the test microorganism in spite of no apparent exposure to the agent in question.

In addition, the animals must be as healthy and as near normal as possible and should be parasite free. A series of physiological tests should be made on animals during the preexperimentation period (see Chapter 3 on methods of recording physiological reactions). These physiological values could then be followed through changes imposed by environmental conditions, induced diseases, or both.

## Restrictions

Environmental facilities impose restrictions on methods of recording physiological information. Restraint is necessary when physiological recordings are made by direct wire telemetry. Restraint alone has been shown to change blood pressure, body temperature, cardiac rhythm, blood glucose, and blood calcium. It is possible that restraint may change other physiological values. It should be possible to conduct experiments where little or no restraint is imposed on animals. This means that radiotelemetry should be used as much as possible in recording biophysical data and that animals should have surgically implanted cardiovascular (or other) catheters for blood sampling for biochemical tests. Under these conditions, samples for analysis can be taken with little disturbance to the animal.

## Pitfalls in the Development and Use of Experimental Facilities

Probably the first, and most important, problem in developing adequate facilities for environmental studies is incompetent *supervision*. The services of an engineer who is well trained and experienced in this field is mandatory. Environmental facilities should be under the direct and competent supervision of experienced professional personnel, who, in turn, should be supported by above-average technical people, not only those involved in the experiments but also technicians and engineers who are given the task of seeing that the equipment operates accurately and with little or no unexpected outage.

*Design* responsibilities should never be left to a contractor or an architect familiar only with ordinary air-conditioning problems. Environmental facilities require special construction and equipment, and only a few people have experience in this field. This is especially true if wide ranges of temperature and humidity are required and if these changes must be made in short intervals of time. Lack of competent supervision during construction can be serious. Unless careful supervision is demanded, the original design and the completed facilities may show little correlation.

Another mistake that has been made frequently is the failure of those concerned to appreciate the *cost* of construction and maintenance of environmental facilities. It is not unusual for an institution to venture into this field with little or no appreciation of the costs involved. Probably only a few laboratories can afford elaborate facilities.

In cattle experiments the use of *carefully selected* identical twins might be substituted for large numbers of animals. The construction and maintenance costs of the equipment precludes the use of poorly selected animals. For a further discussion of animal selection, see Chapter 6.

Experiments must be unusually *well planned* because of the operational costs and because of the large number of highly trained technical people involved in even the simplest experiments. This does not mean that "pilot" experiments are impractical, but it does suggest that experiments involving considerable time, help, and numbers of animals should be planned with unusual care.

Usual *methods* of recording physiological events may be completely inadequate under certain environmental conditions. An example is the use of exposed catheters and transducers in measuring blood pressures directly from chronically implanted cardiovascular catheters. The exposed catheters and transducers will freeze at very low temperatures unless protected.

If diseases are to be studied, the *air* handling system and filters must be adequate under all conditions.

## REFERENCES

Alexander, A. F., and D. H. Will. 1967. Stress of high altitude environment on cattle. *In* Biometeorology, Vol. 2. Permagon Press, New York.

Allcroft, W. M. 1947. The fate of calcium and magnesium in the blood stream of the ruminant. Vet. J. 103:157.

Allcroft, W. M., and H. H. Green. 1938. Seasonal hypomagnasaemia of the bovine, without clinical symptoms. J. Comp. Path. 51:176.

Allen, J. A. 1919. Preliminary note on infectious keratitis. J. Am. Vet. Med. Assoc. 54:306.

Altman, P. L., and D. S. Dittmer. 1966. Environmental biology. Fed. Am. Soc. Exp. Bio., Bethesda, Maryland.

Anderson, D. E., and P. E. Skinner. 1961. Studies on bovine ocular squamous carcinoma ("cancer eye"). XI. Effects of sunlight. J. Anim. Sci. 20:474.

Anderson, D. P., C. W. Beard, and R. P. Hanson. 1964a. The adverse effects of ammonia on chickens including resistance to infection with Newcastle disease virus. Avian Dis. 8:369.

Anderson, D. P., F. L. Cherms, and R. P. Hanson. 1964b. Studies on measuring the environment of turkeys raised in confinement. Poultry Sci 43:305.

Anthony, H. D. 1957. Bacteriological and pathological studies of bovine keratitis. M.S. Thesis, Kansas State College of Agriculture and Applied Science, Manhattan.

Baldwin, E. M. 1945. A study of bovine infectious keratitis. Am. J. Vet. Res. 6:180.

Bauman, R. H., J. E. Kadlec, and P. A. Powlen. Some factors affecting death loss in baby pigs. p. 9. Res. Bull. Purdue Univ. Agr. Exp. Sta. No. 810.

Biester, H. E., and L. H. Schwarte. 1940. Further studies on moldy corn poisoning (leucoencephalomalacia) in horses. Vet. Med. 35:636.

Blood, D. C., and J. A. Henderson, 1963. Veterinary medicine, 2nd ed. Williams and Wilkins Co., Baltimore.

Brandenburg, T. O., and F. M. Shigley. 1947. "Water bloom" as a cause of poisoning in livestock in North Dakota. J. Am. Vet. Med. Assoc. 110:384.

Brandly, C. A., and A. W. McClurkin. 1954. Epidemic diarrheal disease of viral origin of new-born calves. Ann. N.Y. Acad. Sci. 66:181.

Buck, W. B. 1968. Personal communication.

Bullis, K. L., G. H. Snoeyenbos, and H. Van Roekel. 1950. A keratoconjunctivitis in chickens. Poultry Sci. 29:386.

Bustad, L. K., R. O. McClellan, and R. J. Garner. 1965. The significance of radionuclide contamination in ruminants. p. 131. *In* R. W. Dougherty (ed.) Physiology of digestion in the ruminant. Butterworth, Washington, D.C.

Carter, G. R. 1954. Observations on the pathology and bacteriology of shipping fever in Canada. Can. J. Comp. Med. Vet. Sci. 18:359.

Carter, G. R., and B. J. McSherry. 1955. Further observations on shipping fever in Canada. Can. J. Comp. Med. Vet. Sci. 19:177.

Claire, N. T. 1952. Photosensitivity in diseases of domestic animals. Commonwealth Bur. Anim. Health Rev., Ser. 3.

Claire, N. T., and E. H. Stephens. 1944. Congenital porphyria in pigs. Nature (London) 153:252.

Clegg, F. G., and W. A. Watson. 1960. Ryegrass staggers in sheep. Vet. Rec. 72: 731.

Cunningham, I. J., and W. J. Hartley. 1959. Ryegrass staggers. New Zealand Vet. J. 7:1.

David, R. W. 1952. Pathology of spontaneous and experimental necrotic laryngitis in the bovine. M.S. Thesis, Colorado State University, Fort Collins.

Dill, D. B., E. F. Adolph, and C. G. Wilbur. 1964. Handbook of physiology. Sec. 4: Adaptation to the environment. Am. Physiol. Soc., Williams and Wilkins Co., Baltimore.

Dimick, R. L. (ed.) 1963. Proceedings of the 1st international symposium on aerobiology. University of California Press, Berkeley.

Eernstman, T. 1962a. The influence of the micro- and the macro-climate and of isolation measures on the occurrence of enzootic or virus pneumonia of pigs. A critical review of the literature. I. Tijd. Diergeneeskund. 87:965.

Eernstman, T. 1962b. The influence of the micro- and the macro-climate and of isolation measures on the incidence of enzootic or virus pneumonia of pigs. II. Tijd. Diergeneeskund. 87:1188.

Eernstman, T. 1963. The influence of the micro- and the macro-climate and of isolation measures on the incidence of enzootic or virus pneumonia of pigs. A critical review of the literature. III. Tijd. Diergeneeskund. 88:1344.

Elneil, H., and R. A. McCance. 1965. The effect of environmental temperature on the composition and carbohydrate metabolism of the new-born pig. J. Physiol. 179:278.

Evans, C. L., and D. F. G. Smith. 1954. Physiological factors in the condition of "dry coat" in horses. Vet. Rec. 69:1.

Evans, C. L., A. M. Nisbet, and K. A. Ross. 1957. A histological study of the sweat glands of normal and dry-coated horses. J. Comp. Path. 67:397.

Ewer, T. K. 1950. Rickets in sheep: Field trials in East Anglia. Vet. Rec. 62:603.

Ewer, T. K. 1951a. Rickets in sheep. I. The experimental production of rickets in young sheep. Brit. J. Nutr. 5:287.

Ewer, T. K. 1951b. Rickets in sheep. II. Measurement of phosphorus absorption. Brit. J. Nutr. 5:300.

Ewer, T. K. 1953. Vitamin D requirements of sheep. Aust. Vet. J. 29:310.

Ewer, T. K., and P. Bartrum. 1958. Rickets in sheep. Aust. Vet. J. 24:73.

Farleigh, E. A. 1966. Acute *Phalaris tuberosa* poisoning of sheep. Vet. Insp. N.S.W. 30:71.

Farley, H. 1932. An epizoological study of shipping fever in Kansas. J. Am. Vet. Med. Assoc. 80:165.

Farquharson, J., and W. J. Butler. 1944. Discussion on pulmonary emphysema. Proc. 47th Ann. Mtg. U.S. Livestock Sanitary Assoc., p. 224. Waverly Press, Baltimore, Md.

Fitch, C. P., L. M. Bishop, W. L. Boyd, R. A. Gortner, C. F. Rogers, and J. E. Tilden. 1934. "Water bloom" as a cause of poisoning in domestic animals. Cornell Vet. 24:30.

Fourie, P. J. J. 1953. Does bovine congenital porphyrinuria (pink tooth) produce clinical disturbances in an animal which is protected against the sun? Onderstepoort J. Vet. Res. 26:231.

Gee, R. W. 1958. Disease problems of dairy cattle in irrigation areas. Aust. Vet. J. 34:352.

Goodwin, R. F. W. 1955. Some common factors in the pathology of the new-born pig. Brit. Vet. J. 111:361.

Gregory, P. H. 1961. The microbiology of the atmosphere. p. 192. *In* N. Polunin (ed.) Plant science monographs. Interscience, New York.

Halliday, R. 1965. Failure of some hill lambs to absorb maternal gamma-globulin. Nature 205:614.

Hancock, J. 1950. Congenital photosensitivity in Southdown sheep, a new sublethal factor in sheep. New Zealand J. Sci., Tech. Sec. A 32:16.

Hatch, T. F., and P. Gross. 1964. Pulmonary deposition and retention of inhaled aerosols. p. 45. Academic Press, New York and London.

Herbert, T. G. E., and L. E. Hughes. 1956. "Black disease" (infectious necrotic hepatitis) in a heifer. Vet. Rec. 68:223.

Hibbs, J. W., W. E. Krauss, C. F. Monroe, and W. D. Pounden. 1945. A report on the occurrence of rickets in calves under farm conditions. J. Dairy Sci. 28:525.

Hoerlein, A. B., and C. L. Marsh. 1957. Studies on the epizotiology of shipping fever in calves. J. Am. Vet. Med. Assoc. 131:123.

Holtenius, P., and O. Hogsved. 1968. Liquid manure gas poisoning in cattle. Reports and Contributions of the 5th Intern. Mtg. Diseases of Cattle. Opatija, Yugoslavia. p. 160.

Hoogstratten, B., N. C. Leone, J. L. Shupe, D. A. Greenwood, and J. Lieberman. 1965. Effects of fluorides on hemopoietic system, liver and thyroid gland in cattle. J. Am. Vet. Med. Assoc. 192:26.

Hopkirk, C. S. M. 1935. Staggers in livestock. New Zealand J. Agr. 51:18.

Hughes, D. E., G. W. Pugh, and T. J. McDonald. 1965. Ultraviolet radiation and *Moraxella bovis* in the etiology of bovine infectious keratoconjuctivitis. Am. J. Vet. Res. 26:1331.

Hutyra, F., and J. Marek. 1926. Pathology and therapeutics of the diseases of domestic animals. Vol. 2, p. 87. Alexander Eger, Chicago.

Jackson, F. C. 1953. Infectious keratoconjunctivitis of cattle. Am. J. Vet. Res. 14:19.

Jensen, R., and D. R. Mackey. 1965. Diseases of feedlot cattle. Lea & Febiger, Philadelphia.

Juszkiewicz, T. 1967. Experimental *Pasteurella multocida* infection in chickens exposed to cold: Biochemical and bacteriological investigations. Polskie Arch. Weter. 10–4:615.

Kahn, F., C. B. Straub, and I. R. Jones. 1962. Radioiodine in milk of cows consuming stored feed and of cows on pasture. Science 138:1334.

Kingsbury, J. M. 1964. Poisonous plants of the United States and Canada. Prentice-Hall, Englewood Cliffs, New Jersey.

Kita, J. 1966. The effect of thermal stimuli on the opsonophagocytic reaction and complement system and the role of these stimuli in the pathogenesis of erysipelas. Polskie Arch. Weter. 10:1.

Lengeman, F. W., and E. W. Swanson. 1957. A study of the secretion of iodine in milk of dairy cows using daily oral doses of $I^{131}$. J. Dairy Sci. 40:216.

Lepper, M. H., and E. K. Wolfe. 1966. Second international conference on aerobiology (airborne infection). Bact. Rev. 30:485.

Leslie, F. J. S. 1949. Fog fever. Vet. Rec. 61:228.

Luschbaugh, C. C., and J. F. Spalding. 1957. The natural protection of sheep from external beta radiation. Am. J. Vet. Res. 18:345.

Marsh, H., and E. A. Tunnicliff. 1938. Dysentery of new-born lambs. Mont. Agr. Exp. Sta. Bull. No. 361:1.

McBarron, E. 1952. Metabolic diseases in dairy cattle. Aust. Vet. J. 28:36.

McCance, R. A., and E. M. Widdowson. 1959. The effect of lowering the ambient temperature on the metabolism of the new-born pig. J. Physiol. 147:124.

McClymont, G. L. 1955. Possibility of photosensitization due to ingestion of aphids. Aust. Vet. J. 31:112.

McEwan, A. D. 1950. The resistance of the young calf to disease. Vet. Rec. 62:83.

Merchant, I. A., and R. D. Barner. 1964. An outline of the infectious diseases of domestic animals. 3rd ed. Iowa State University Press, Ames, Iowa.

Mitchel, H. M. 1942. The fluorine problem in livestock feeding. NRC Comm. Anim. Nutr., NAS-NRC, Washington, D.C.

Mohler, J. R., and G. B. Morse. 1905. *Bacillus necrophorus* and its economic importance. p. 76. *In* U.S. bureau of animal industry, 21st annual report, 1904. Govt. Printing Office, Washington, D.C.

Moll, T. 1957. The etiology and pathogenesis of pneumonia-enteritis in non-colostrum-fed and colostrum-fed calves. Vet. Med. 52:115.

Moll, T. 1965. The pathogenesis of diarrhea in the newborn calf, with special reference to physiologic functions and environmental conditions. J. Am. Vet. Med. Assoc. 147:1364.

National Mastitis Council. 1963. Current concepts of bovine mastitis. p. 7. Washington, D.C.

Noyes, H. E., J. R. Evans, and H. J. Baker. 1963. Effects of nuclear detonation on swine—Bacteriologic studies. Ann. N.Y. Acad. Sci. 105:651.

O'Donoghue, J. G., and G. S. Wilton. 1951. Algal poisoning in Alberta. Can. J. Comp. Med. 15:193.

Olsen, J. D. 1969. Physiologic response of cattle to controlled cold exposure: Aortic blood pressure, heart rate, and respiratory rate. Am. J. Vet. Res. 30:47.

Olsen, T. A. 1953. Minnesota research report. U.S. Army Chem. Corp., Biological Laboratories, Fort Detrick, Maryland.

Parkinson, B., and A. K. Sutherland. 1954. Post-parturient haemoglobinuria of dairy cows. Aust. Vet. J. 30:232.

Peirce, A. W. 1945. The effect of intake of carotene on the general health and on the concentration of carotene and of vitamin A in the blood and liver of sheep. Aust. J. Exp. Biol. 23:295.

Penny, R. H. C. 1956. Post-parturient haemoglobinuria (haemoglobinaemia) in cattle. Vet. Res. 68:233.

Quarterman, J., A. C. Dalgarno, and I. McDonald. 1961. The natural source of vitamin D for sheep. Proc. Nutr. Soc. 20:xxviii.

Reisinger, R. C., K. L. Heddleston, and C. A. Manthei. 1959. A myxovirus (SF-4) associated with shipping fever of cattle. J. Am. Vet. Med. Assoc. 135:147.

Riley, R. L., and F. O'Grady. 1961. Airborne infection: Transmission and control. p. 27. Macmillan Co., New York.

Rosebury, T. 1947. Experimental air-borne infection. Williams & Wilkins Co., Baltimore.

Ross, J. G. 1967a. An epidemiological study of fascioliasis in sheep. Vet. Rec. 80: 214.

Ross, J. G. 1967b. A further season of epidemiological studies of *Fasciola hepatica* infections in sheep. Vet. Rec. 80:368.

Sax, N. I. 1963. Dangerous properties of industrial materials. 2nd ed. Reinhold Publishing Co., New York.

Schmidt, H. J., and W. E. Rand. 1952. A critical study of the literature on fluoride toxicology with respect to cattle damage. Am. J. Vet. Res. 13:38.

Schmutz, E. M., B. N. Freeman, and R. E. Reed. 1968. Livestock-poisoning plants of Arizona. The University of Arizona Press, Tucson, Arizona.

Scrivner, L. H., and A. M. Lee. 1934. The morphology, culture, isolation, and immunity studies of *Actinomyces necrophorus* in calf diphtheria. J. Am. Vet. Med. Assoc. 83:360.

Shope, R. E. 1955. The swine lungworm as a reservoir and intermediate host for swine influenza virus. V. Provocation of swine influenza by exposure of prepared swine to adverse weather. J. Exp. Med. 102:567.

Shupe, J. L., and E. W. Alther. 1966. The effects of fluorides on livestock, with particular reference to cattle. p. 307. *In* Pharmacology of fluorides, Part I, handbook of experimental pharmacology. Springer and Verlag, New York.

Sinclair, K. B. 1956. Black disease: A review. Brit. Vet. J. 112:196.

Smith, H. A., and T. C. Jones. 1961. Veterinary pathology. 2nd ed. Lea & Febiger, Philadelphia. p. 71.

Smith, H. A., and T. C. Jones. 1966. Veterinary pathology. 3rd ed. Lea & Febiger, Philadelphia. p. 82.

Songer, J. R. 1967. Influence of relative humidity on the survival of some airborne viruses. Appl. Microbiol. 15:35.

Steinhaus, E. A. 1940. Principles of insect pathology. 1st ed., 2nd impression. McGraw-Hill, New York. p. 22.

Stewart, C. M. 1943. A condition amongst military dairy farm cattle in India, with imported blood, resembling "dry coat" or "non-sweating" in horses. (Abst.) Vet. Bull. 13:434.

Stewart, C. M. 1956. Dry coat or non-sweating in horses in India. Irish Vet. J. 10: 189.

Steyn, D. G. 1943. Poisoning of animals by algae on dams and pans. Farming S. Afr. 18:489–510.

Swales, W. E., P. E. Sylvestre, and S. B. Williams. 1942. Field trials of control measures for parasitic diseases of sheep. Conf. Res. Sec. D Zool. Sci. 20:115.

Tarr, A. 1947. A note on the occurrence of puerperal haemoglobinaemia of cattle in South Africa. J. S. Afr. Vet. Med. Assoc. 18:167.

Thorne, J. L., H. Kuida, and H. H. Hecht. 1965. Altitude maladjustment in the bovine. Ann. N.Y. Acad. Sci. 127:646.

Tromp, S. W. 1963. Smog, its origin and prevention. p. 142. *In* Medical biometerology. Elsvier Publishing Co., Amsterdam.

U.S. Dept. Agr. 1941. Climate and man. p. 517. Yearbook of agriculture. U.S. Govt. Printing Office, Washington, D.C.

U.S. Dept. Agr. 1965. Losses in agriculture. Agriculture handbook No. 291. U.S. Govt. Printing Office, Washington, D.C.

Van Ness, G. B. 1961. Anthrax in the plains states. Southwestern Vet. 14:290.

Velasquez, Q. J. 1947. Enfermedad de las alturas. Rev. Med. Vet. (Bogota) 16:53.

Weatherby, J. H. 1952. Chronic toxicity of ammonia fumes by inhalation. Proc. Soc. Exp. Biol. Med. 81:300.

Weeth, H. J., and L. H. Haverland. 1961. Tolerance of growing cattle for drinking water containing sodium chloride. J. Anim. Sci. 20:518.

Wells, W. F. 1955. Airborne contagion and air hygiene: An ecological study of droplet infections. Harvard University Press, Cambridge, Mass. p. 42.

Wojtatowicz, Z. 1966. Effect of increased environmental temperature on the cellular protective mechanism in the skin of an experimental pig infected with *Erysipelothrix insidiosa*. Polskie Arch. Weter. 10:2.

Wright, D. N., G. D. Bailey, and M. T. Hatch. 1968. Role of relative humidity in the survival of airborne *Mycoplasma pneumoniae*. J. Bact. 96:970.

# 6

# DESIGN AND EXECUTION OF EXPERIMENTS USING DOMESTIC ANIMALS

## INTRODUCTION

The planning, design, and execution of experiments for evaluating physiological responses and obtaining shelter-engineering data require careful consideration in environmental studies. The objectives of this chapter are to identify acceptable experimental designs and procedures for evaluating measurements of physiological and/or production functions. Consideration is also given to limitations and advantages of environmental laboratories, the execution of experiments, and research trends in environmental physiology and shelter-engineering. Special attention is given to the use of laboratory facilities; Chapter 8 deals with field research.

## DEFINING THE PROBLEM

Before designing any experiment the question(s) or objective(s) need to be clearly established. Some experiments require a number of observations of many animal functions; others require observations of only a limited number of functions—one function may suffice. Some measurements and evaluations may be simple, others complex. It is extremely important for the investigator to attempt not to make the experiment any more complex than necessary. Questions that can be answered by the investigator are limited by the capacity and training of supporting personnel and the availability of facilities, funds, and experimental animals.

Experiments using domestic animals in environmental laboratories are designed to serve five major purposes: (1) to identify and differentiate the effects of specific environmental elements on animals; (2) to determine and quantify the effects of specific environmental factors or their interaction on the animal's thermal balance, water balance, and other concomitant physiological functions; (3) to develop techniques for animal selection to fit natural environments; (4) to develop engineering technology for shelter design; and (5) to provide data for simulation modeling.

## EXPERIMENTAL DESIGN

The experimental design will be largely determined by the facilities available. The number of simultaneous treatments is usually limited. This factor becomes increasingly critical as the investigator attempts to increase the number of environmental test conditions.

The major objective in any experimental design is to provide the maximum degree of confidence in treatment effects. A single chamber has greater limitations than a multiple chamber facility. It is most useful when a response can be observed in the same animal under successively different conditions of environmental treatment. This is commonly called "the individual serving as its own control." It is appropriate for many types of studies, but it possesses the disadvantage that the experimental treatment is confounded with time.

Units that will hold more than one animal, or multichamber facilities, reduce the influence of time trends, provide more data, and permit more treatments. Treatments can be imposed in such a way that some animals may serve as a control group (held at some predetermined standard conditions) while the others are subjected to the experimental conditions. Treatment effects can then be expressed as a deviation from observations on the control group. The usual concepts for statistical analysis normally require that a new group of animals be randomly assigned to both control and experimental treatments for each experiment.

Residual experimental effects on control animals may preclude their legitimate use in subsequent experiments. It is frequently not practical to provide new animals for each treatment especially when large, extremely valuable, or scarce animals are used.

Just as the increase from one to two controlled chambers greatly increases the amount of useful information, the availability of four units further enhances the efficiency of experimentation. Factorial treatment arrangements are very efficient experimentally. The use of four chambers permits, for example, simultaneous conduction of a 2 × 2 factorially designed experiment. In this way one can achieve greater precision in determining the importance of interaction effects.

Thus a challenging problem confronting investigators in which facilities and animal numbers are limited is the selection of an appropriate experimental design. Examples of an experimental design for each of several situations follow.

With multiple-animal chambers, a choice of experimental design is available: (1) designs appropriate for crossover trials; and (2) designs appropriate for continuous trials. Lucas (1959) describes these designs as follows:

> Continuous trials are those in which an animal, once placed on an experimental treatment, remains on that treatment for the duration of the trial. Cross-over trials are those in which an animal receives a time-sequence of two or more treatments during the course of the trial.
>
> The method of taking advantage of the correlations between productions in different portions of the same lactation for cows, or in different lactations, is ordinarily different in the two types of trials. Also special provision to remove inherent time trends in behavior and the effects of secular trends in non-treatment environment must be made in designing cross-over trials. The method of removing the main effects of items such as starting time and location from experimental error is essentially the same in the two types of trials.
>
> A feature of experimental design ordinarily found in both types of trials—a feature necessary to take advantage of the correlations between production rates at different times—is the use of at least two periods of observation on each cow. In continuous trials there are basically two periods of observation, but in cross-over trials there may be several periods.

Crossover trials (also called switchback or reversal) are used primarily in milk production or growth-rate studies to correct for time trends in lactation or growth rate. Such a design provides treatment comparisons since differences between experimental units can be eliminated from experimental error. However, carryover effects from period to period must be small because of the risk that performance in a given period may reflect not only the direct effect of the current treatment but also residual effects from preceding treatments (Patterson and Lucas, 1962).

### SINGLE CHAMBER–ONE COW CAPACITY

OBJECTIVE Determine thermal-neutral zone for unacclimated* cows.

DESIGN REQUIREMENTS (1) Keep exposure period short to minimize acclimation. (2) Between exposures maintain the cow in an environment that may be considered a standard or base. (3) Use a consistent feeding and watering regime prior to and during tests. (4) Eliminate variations due to excitement or changes in feed quality. (5) Feed the animals at a fixed time prior to exposure in the experimental environment—at least 3 hr.

PROCEDURE (1) Prepare 3 animals for 2 weeks in advance of exposure to environment at test conditions. During this time place the animals daily in the chamber to train them to the physical handling procedures. (2) At 10:00 a.m. on Day One of the experimental period expose Animal A to the first test condition (for example, –10°C) and make measurements as needed. Remove at 6:00 p.m. Day Two: expose Animal B to the same climatic conditions, and on Day Three, continue with Animal C. Replicate as required (at least 3 times in this example). (3) Repeat the schedule for each scheduled climatic condition (i.e., –10°, 0°, 10°, 20°, 30°, and 40°C).

ANALYSIS The collected data can be summarized graphically, showing both the mean values and the degree of variation, as illustrated in Figure 6.1. The graph shows the trend for the group under the different environmental conditions. This may be followed by standard analysis of variance procedures to determine the probable significance of temperature effects on animal response.

If the single chamber has multiple animal capacity, as illustrated in Figure 6.2, the responses of six animals can be determined simultaneously. An example of an approach is outlined in Table 6.1.

*Acclimation is used as the adaptation to environmental test conditions. For further discussion see Folk (1966).

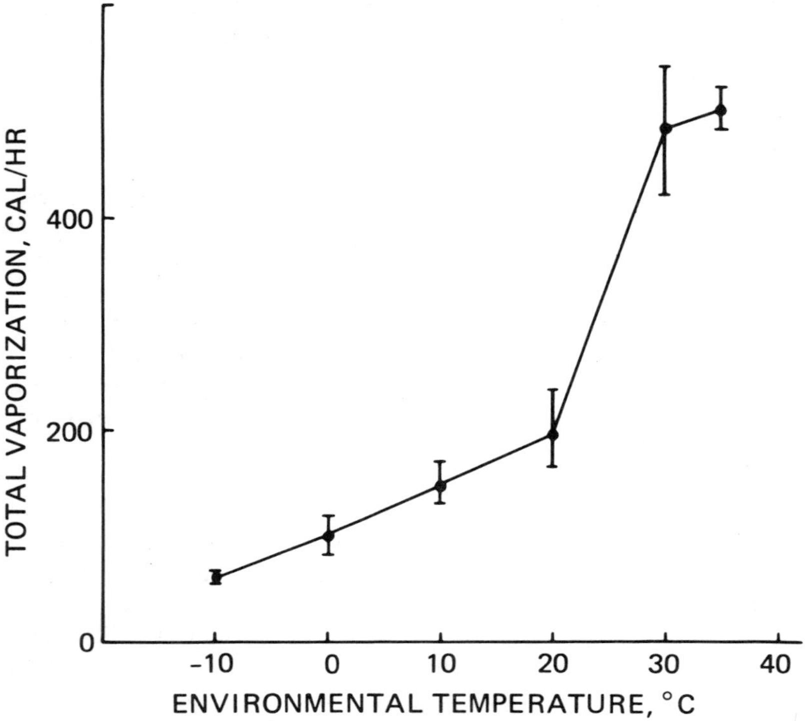


**FIGURE 6.1** Illustration of a simple graphic presentation of the variance in vaporization rate in a group of three cows at six environmental temperatures. The line connects the mean value, and the vertical bar indicates the range of variation (standard deviation).

### SINGLE CHAMBER–SIX COW CAPACITY

OBJECTIVE Determine effect of temperature on energy metabolism.

DESIGN REQUIREMENTS The requirements are much the same as in the preceding example.

PROCEDURES (1) Establish a test schedule such as that illustrated in Table 6.1. The time response to various temperatures is not established for many reactions of the various species of animals. Many gross changes are accomplished within 2 weeks, but longer periods are preferable for many of the slower responding reactions such as growth.

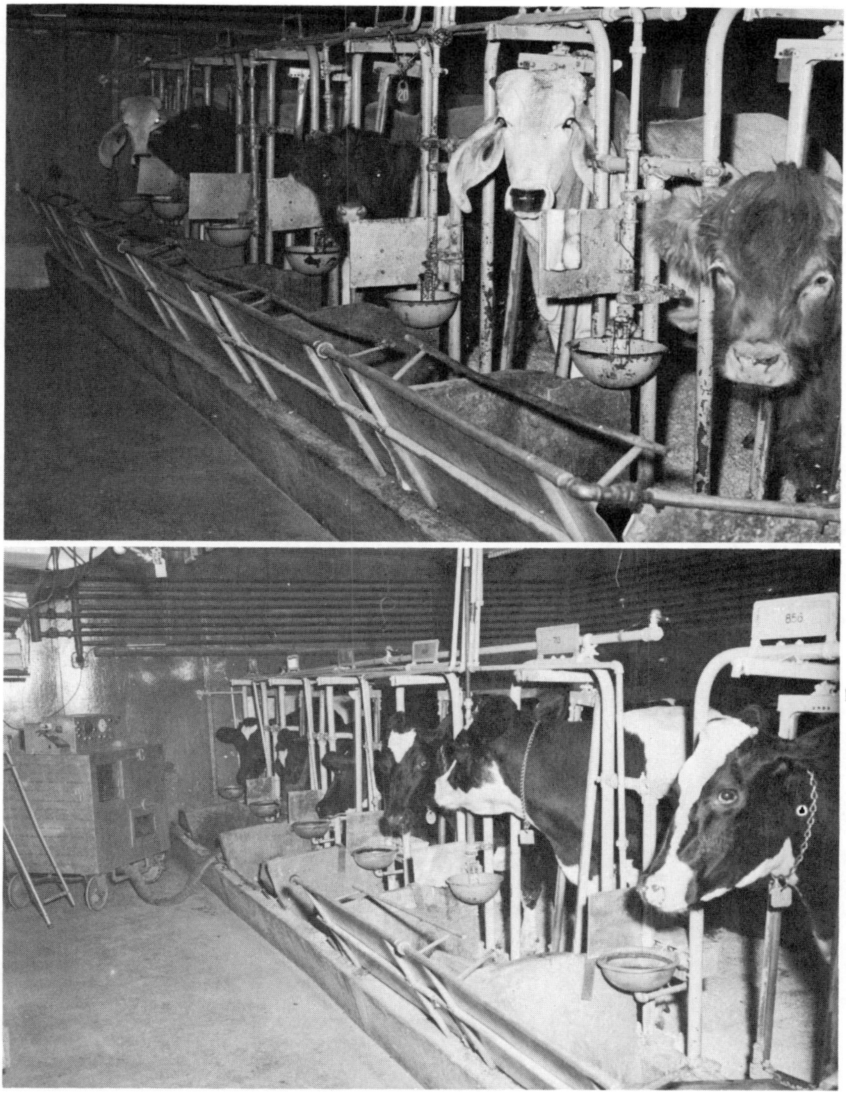

FIGURE 6.2 View of each chamber of the Missouri Climatic Laboratory. Figure 3.7 is a diagram of the over-all laboratory.

**TABLE 6.1 Example of an Acceptable Procedure Using Energy Metabolism (cal/m²/day)**

| Cow number | Energy Metabolism 18°C[a] (Thermoneutral) Period 1[b] Mar. 2–Mar. 22 | −10°C Period 2 Mar. 23–Apr. 12 | 18°C Period 3 Apr. 13–May 3 | 0°C Period 4 May 4–May 24 | 18°C Period 5 May 25–June 15 |
|---|---|---|---|---|---|
| 1 | 3500 | 4500 | 3600 | | |
| 2 | 3400 | 4400 | — | | |
| 3 | 3700 | 4700 | — | | |
| 4 | 3300 | — | — | (continue treatments as necessary) | |
| 5 | 3600 | — | — | | |
| 6 | 3350 | — | — | | |

[a]Random assignments of temperatures are worthy of consideration; however, extensive replication would be necessary.
[b]Measurements for each treatment are made during the last week of each period to minimize residual effects of preceding experimental periods.

Two weeks is considered minimum for observing changes in milk production. Three weeks is preferred for energy metabolism. (2) Use 6 cows, and let each animal provide its own base condition—serve as its own control. This procedure minimizes the gradual changes within the limited periods of time, but progressive changes, such as the normal decline in milk production, are complicating factors. Six test subjects are preferred, but three of one breed are considered minimum for metabolism studies for this example. If one group of base or standard animals were to be compared to a test group, about 25 animals in each group would probably be needed. The actual requirement would depend on the degree of variation among the data within a test condition and the degree of significance required.

ANALYSIS (1) Plot a scatter diagram according to time, with the test conditions blocked in by vertical lines. This will provide a visual indication of misleading data that might be caused, for example, by a rare change in an animal's reaction, deviations from scheduled test conditions, or a measurement error. It will also help in deciding on the type of prediction equation. Although data should not be deleted solely on the basis that it "looks out of place," other records may provide sufficient evidence to verify errors and permit deletion of the data that is affected. In addition to prediction equations, statistical methods may be applied to the establishment of mean values and the deviations within and among breeds under various test conditions.

### TWO CHAMBERS—CAPACITY OF SIX LACTATING COWS EACH

An example of a crossover or reversal design using two chambers is presented in Table 6.2. It is based on an experiment conducted in the Missouri Climatic Laboratory several years ago (Johnson *et al.*, 1963; see also Figure 6.1).

OBJECTIVE Determine temperature and humidity effects on lactation and related functions.

DESIGN REQUIREMENTS Two groups (A and B) of six animals were used. Two animals for each grouping represented early, middle, and late stages of lactation. One group was exposed to a standard base temperature and humidity environment, experimental conditions, and back to base conditions in that order. The second group was exposed to the

**TABLE 6.2 Experimental Design for Temperature and Humidity Studies on Lactating Animals**

| | Chamber 1 (Treatment) | | | Chamber 2 (Control) | | |
|---|---|---|---|---|---|---|
| Period[a] | Temp. | RH (%) | Group | Temp. | RH (%) | Group |
| 1 | 27°C | 50 | B | 18°C | 50 | A |
| 2 | 27°C | 50 | A | 18°C | 50 | B |
| 3 | 27°C | 50 | B | 18°C | 50 | A |
| 4 | 32°C | 50 | A | 18°C | 50 | B |
| 5 | 32°C | 50 | B | 18°C | 50 | A |
| 6[b] | 32°C | 50 | A | 18°C | 50 | B |
| 7 | 32°C | 75 | B | 18°C | 50 | A |
| (To be continued at various humidity levels) | | | | | | |

[a] Acceptable periods for lactation trials are 3 weeks long.
[b] Following period 6, four early lactation animals are introduced and the late-lactation animals are removed.

experimental conditions followed by the base conditions and then the test environment. The animals were exposed to each condition for 2 weeks. Six weeks were required to complete the test for one experimental condition.

PROCEDURE After tests were completed for two experimental conditions, the late-stage-of-lactation animals were replaced by four new animals in early lactation and the animals originally in early and middle stages of lactation were designated as middle- and late-lactating animals, respectively.

The experimental design is diagrammed in Table 6.2.

ANALYSIS Treatment differences were tested as outlined by Brandt (1938) and Lucas (1956), comparing milk production during each treatment period to that under 18°C and 50 percent relative humidity conditions. This method proved somewhat more sensitive than the standard "t-test" for discerning environmental effects and reduced the variation due to normal time trends with advancing lactation from the error and the treatment variance values.

### TWO CHAMBERS—12 PIGS EACH

OBJECTIVES Determine the main effects and interactions of one environmental factor (E) and two dietary factors (A and B) on growth rate of swine.

DESIGN REQUIREMENTS When two chambers are available, two environmental treatments can be compared concomitantly. If each of the chambers has the capacity for 12 animals, a number of factorial treatment arrangements is feasible: 2 × 2; 2 × 3; 2 × 4; 2 × 6; 2 × 2 × 2; 2 × 2 × 3; 2 × 2 × 6; and 2 × 3 × 4. One of the two-level factors in each case is some environmental parameter, and the other variously leveled factors could be, diet and/or hormone treatments. In the 2 × 2 × 6 and the 2 × 3 × 4 cases, replication in time would be required to have residual degrees of freedom in the final analysis of variance.

An example is a 2 × 2 × 2 factorial arrangement of treatments. Two levels—control and experimental—of each of the three factors (E = control environment; E′ = experimental environment) are chosen with regard to prior information of hypothesis on expected effects of the factors. There are eight treatment groups: (EAB); (E′AB); (EA′B); (E′A′B); (EAB′); (E′AB′); (EA′B′); and (E′A′B′).

PROCEDURE For such treatments, it is desirable that the pigs be of the same sex, be as nearly similar in age as possible (1 week) and have comparable growth rates as evidenced by similar body weights for age. The pigs should be allotted randomly to the treatments, three pigs per treatment. Also it is desirable to allow the pigs to become accustomed to experimental surroundings and the routine prior to beginning observations. Adjustment may be judged by behavior, feed intake, and growth rate. During this period all pigs are treated as absolute controls (that is, EAB). The experimental chamber should be maintained at a standard level during pretreatment, and at the same time the pigs are given pre-experimental adjustments to dietary treatments.

Weigh all pigs. Set the experimental chamber for the selected environmental conditions and conduct experiment as required.

ANALYSIS The estimated variance due to treatments may be analyzed as follows: Where E represents the environment and A and B are dietary factors, the sources of variation are E, A, B and concomitant interactions of (E × A), (E × B), (A × B), and (E × A × B). Each of these sources has 1° of freedom. The residual sum of squares constitutes the error term, which in this case has 23 – 7 = 16° of freedom.

If all effects are considered fixed, then each main effect and interaction effect is tested for statistical significance by the ratio of its respective mean square to the error mean square.

### FOUR CHAMBERS: SIX COWS EACH

OBJECTIVE Determine main effects and interactions of two environmental elements and 1 dietary factor on milk yield of lactating cows.

DESIGN REQUIREMENTS With four control chambers, four environmental treatments can be imposed simultaneously. A 2 × 2 factorial of environmental treatments would permit determination of the interaction effects between environmental factors. If each chamber has the capacity for six animals, it is possible to design experiments either as 2 × 2 × 2 or 2 × 2 × 3 factorial arrangement of treatments. Two of the two-level factors in each case could be environmental parameters and the other two- or three-level factors diet and/or hormone treatment, for example.

An example of a 2 × 2 × 2 factorial arrangement of treatments would be to have two levels of environment and two dietary regimes. Such an arrangement gives eight treatment groups: ($TH$D); ($T'H$D); ($TH'$D); ($T'H'$D); ($TH$D'); ($T'H$D'); ($TH'$D') and ($T'H'$D'), where $T$ = temperature, $H$ = relative humidity, and D = diet.

PROCEDURE The 24 cows should be drawn from as large a population as possible and allotted to treatment groups (three cows per group), with balance insofar as possible for age, breed, stage of lactation, indication of previous producing ability, body size, and available information on susceptibility to altered ambient environment.

During pretreatment each cow should be allowed to become accustomed to the experimental procedures including diet in a pretreatment period. All chambers must be maintained at the base or control temperature and humidity conditions during the pretreatment period.

During the experiment, conditions in the chambers and observations on the animals follow the predetermined schedule.

ANALYSIS The experimental data may be analyzed as follows: The sources of variation are $T$, $H$, $D$ and concomitant interactions of ($T \times H$), ($T \times$ D), ($H \times$ D), and ($T \times H \times$ D); each has 1° of freedom.

The residual sum of squares constitutes the error term, which has 16° of freedom.

If all effects are considered fixed, then each main effect and interaction is tested for statistical significance by the ratio of its respective mean square to the error mean square.

Effects of concomitant variables, such as those listed in "Procedure," often can be removed from the experimental data by means of covariance analysis.

## PLANNING TYPICAL LABORATORY EXPERIMENTS WITH LARGE ANIMALS

Several years of experience with climatic or environmental laboratory studies have clearly demonstrated that to conduct a collaborative laboratory experiment using large animals requires close scheduling and planning by investigators of related disciplines. The following discussion will indicate many of the items that must receive attention regardless of the number of chambers available. These procedures are listed in chronological order.

1. State clearly the objectives and/or questions that are to be answered; this statement is to be reviewed by all collaborators.
2. Reassess the availability of trained personnel, animal facilities, funds, and equipment for the current experiment.
3. Outline the environmental conditions, number of animals, treatments, and animal measurements that are to be used for the current study.
4. Outline the experimental design with the cooperation or assistance of a statistician. The aid of the consulting statistician is invaluable in designing efficient experiments within the available assets of the investigator. The sample size required to establish animal differences in relation to variance among animals in very important if the experiment is to make efficient use of the investigator's resources. Gill (1969) provides an excellent discussion and data regarding sample size needed for milk yield studies.

To derive an estimate of the number of animals needed on a study the investigator must supply an upper limit ($L$) to the amount of error that he can tolerate, the desired probability that the estimate will lie within the limit of error, and a guess at $\sigma$, the standard deviation. One may assume a probability of 95 percent or 99 percent and assume that

$\bar{x}$, the sample mean, lies between the limits $\pm L$ of $\mu - 1.96\sigma/\sqrt{n}$ and $\mu + 1.96\sigma/\sqrt{n}$ for 95 percent, where $\mu$ is the universe mean. This gives the equation $1.96\sigma/\sqrt{n} = L$. The formula may be conveniently approximated (Snedecor and Cochran, 1967) as $n = 4\sigma^2/L^2$. The $\sigma$ must be known or estimated from prior data on similar populations. The formula for 99 percent probability is $n = 6.6\sigma^2/L^2$.

Fundamentals of experimental design are discussed in detail by several statisticians (Cochran and Cox, 1957; Kempthorne, 1952; Steel and Torrie, 1960). Generally the number of animals used, the precision of the method, and the increased precision of the measured function as offered by replications or error limit ($L$) will determine the extremes of the environmental treatment necessary to impose on the animal to obtain significant treatment differences.

5. Selection of animals is of prime importance. In selecting animals for laboratory studies, uniformity among individuals is generally considered to be of greater importance than representativeness of the population so that the environmental effects or mechanisms of response may be clearly discerned. The observed difference between means of two groups of animals is due to many causes. The experimenter may wish to reduce that portion due to genetic causes by the use of identical twins in some species or by the use of full siblings or littermates. Also, the researcher may use animals of the same age, sex, and previous treatment or, by the use of a stratified sample, is able to remove this variance through the analysis of variance. Care must be exercised, however, in applying results obtained from animals selected for uniformity to the large herd. The recommended procedure is to test a hypothesis in the laboratory using uniform animals; if significant results are obtained, then extend the study to larger numbers of animals that are representative of the population, using random selection (preferably in a field trial) to bridge the gap that may exist between laboratory and field trial results.

6. Preconditioning to feed and laboratory conditions in a holding area, such as barn or sheltered lot, is extremely important for reliable data. Without preconditioning, the first treatments of an experiment may be confounded by feed, management, or acclimation effects. The preconditioning time depends on physiological parameters and the characteristics of the test animals. If the time needed is in doubt, some pretests will be required. Three weeks to one month is a suitable preconditioning time for most experiments. Only sound, healthy animals should be used for extensive environmental studies or results may be

incomplete, unreliable, or unsatisfactory for interpretation. The animals should be examined by a veterinarian and herdsman prior to the experiment, and health records and medications should be monitored closely on animals during the course of the study.

7. Precise schedules are essential after the above factors have been determined. Daily and weekly periods should be established for the entire experiment. Daily schedules in management procedures—measurements of each environmental condition or treatment—are extremely important for a successful experiment and must be adhered to closely so that the measurement of one animal function may not be confounded with others. The cleaning and maintenance of the laboratory should be scheduled for minimum disturbance—not at times when measurements are being made that may reflect this activity. Intervals between animal measurements may be hourly, daily, weekly, and so on, depending on the nature of the treatment and the time period of the environmental condition. Table 6.3 illustrates a daily schedule for an environmental study on cattle.

## ROLE OF VARIOUS DISCIPLINES

A team approach to experimentation in environmental physiology is desirable in most instances. It is unlikely that one scientist would be well qualified in all necessary disciplines to be considered. Desirable talents during a bioclimatic experiment include: (1) a *physiologist and biochemist* to provide knowledge of cellular and organ functions as they relate to acclimation to environment; (2) an *agricultural engineer* or *bioengineer* to provide knowledge of the physical systems involving heat transfer, instrumentation, and application of results to engineering technology; (3) a *behaviorist* to provide understanding of how environmental and other treatment factors may affect behavioral responses; (4) a *pathologist and histologist* to determine the extent of environmental pathological influence on animal tissues; (5) a *nutritionist* to determine ration formulation and requirements for production and maintenance in various environments, role of trace elements, and vitamins that may aid in acclimation to the environment; (6) a *biometeorologist* to interpret environmental conditions that may be encountered and to investigate the energy budget of the animal–environment interface; (7) a *statistician* to provide special insight into rigorous experimental designs and analysis programs for the most ad-

**TABLE 6.3 Illustration of A Daily and Weekly Schedule for an Environmental Temperature Study**

| | Control Period (Thermoneutral) | | | | | | | Experimental Treatment Period (Hot or Cold, etc.) | | | | | | | Control Period | |
|---|---|---|---|---|---|---|---|---|---|---|---|---|---|---|---|---|
| Calendar date | (1) | (2) | (3) | (4) | (5) | (6) | (7) | (8) | (9) | (10) | (11) | (12) | (13) | (14) | (15) | (16) |
| Day[a] | 1 | 2 | 3 | 4 | 5 | 6 | 7 | 1 | 2 | 3 | 4 | 5 | 6 | 7 | 1 | 2 |
| 4:30 a.m. | ← Lights on; clean chambers → | | | | | | | | | | (Repeat) | | | | | (Repeat) |
| 5:00 | ← Brush animals → | | | | | | | | | | | | | | | |
| 5:30 | ← Feed animals → | | | | | | | | | | | | | | | |
| 6:00 | ← Milk animals → | | | | | | | | | | | | | | | |
| 7:00 | ← Animal measurement (thyroid, etc.) → | | | | | | | | | | | | | | | |
| 8:00 | Bl[b] | | | Bl | | Bl | | | | | | | | | | |
| 8:30 | Wt | | | Wt | | Wt | | | | | | | | | | |
| 1:00 p.m. | ← Routine physiological measures → | | | | | | | | | | | | | | | |
| 2:00 | M | | | M | | M | | | | | | | | | | |
| 3:00 | V | | V | V | | V | | | | | | | | | | |
| 4:00 | ← Animal measurement (thyroid, etc.) → | | | | | | | | | | | | | | | |
| 4:30 | ← Clean, feed → | | | | | | | | | | | | | | | |
| 5:00 | ← Milk → | | | | | | | | | | | | | | | |
| 5:45 | Bl | | | Bl | | Bl | | | | | | | | | | |
| 6:00 | | | | Lights off | | | | | | | | | | | | |

[a] Judgment as to the best day of the week for Day 1 is important so that the most intensive measurements are made during the 5-day work week.

[b] Bl = blood sample; Wt = weigh animals; M = energy metabolism; V = total body vaporization. Routine measures are pulse rate, respiratory rates, rectal temperature, etc.

If possible blood samples should be taken on a different day than that for metabolism or other sensitive measures so that metabolism is not adversely affected by the sampling procedure. However, if blood samples are taken properly several hours prior to metabolism, reliable results are usually obtained on trained animals.

vantageous use of animals, facilities, and manpower; and (8) an *economist* to relate the knowledge of techniques and results for evaluation of the feasibility of application to climatological problems.

## PITFALLS IN DESIGN, ANALYSIS, AND EXECUTION OF EXPERIMENTS

Inadequate planning is often the cause of unsuccessful experiments. When designing laboratory studies with animals, the investigator should be aware that such facilities are expensive and require close scheduling in advance of the experimental period to obtain maximum efficiency in use of the facility. With a limitation on number of animals that can be accommodated in a laboratory, a greater frequency and number of measurements should be made, which necessitates precise prescheduling. Replication and adequate control are a must; the former may be done either in space or in time.

Availability of animals meeting the desired qualifications is often a problem of considerable importance. For example, one may need a homogenous group of unbred cows in a certain stage of lactation for a specific experiment. This requires months or even a year of preplanning unless there is a large group from which to choose. Insufficient numbers of rare species, such as eland and musk-oxen, also present a problem.

Investigators should be fully aware of a finite time lag between imposition of the environmental change and some measurable response. This time lag is a function of several factors, including the mass of the animal (as in the case of the mouse vs. the cow, with a considerable difference in storage effect of added heat from exogenous sources), and the rates of change of the numerous physiological functions that occur following alteration of condition (Johnson *et al.*, 1967; Kibler *et al.*, 1965). Time constants involved with each of these physiological factors are not known for most species, although Bond *et al.* (1964) reported on the physiological response times for selected parameters in thermally stressed swine when various cooling media were supplied.

One should be aware of circadian changes that appear in most species because of varying metabolic rates, feeding activity, and so on. The time of day may be a variable in measurement of any rapid-response parameter of an animal. Thus the inherent diurnal rhythm is of importance in terms of interpretation as well as in estimating the repeatability of the observations while evaluating animal differences.

## LABORATORY STUDIES VERSUS FIELD STUDIES

The range in the complexity and precision of control of factors in bioclimatic chambers is similar to the variety of physical factors present in field studies. Laboratory facilities may be as comprehensive as those described for physiological acclimation research (Johnson and Hahn, 1968), where precision control of temperature, humidity, air movement, barometric pressure, and thermal radiation is maintained or as minimal as a single cow chamber constructed to provide only limited temperature ranges with no control over other environmental factors except, perhaps, light and air flow.

Certain advantages over field studies and certain limitations can be listed for the controlled environment facility. Some advantages are high precision of interpretation of data since control is exercised over many environmental factors with more accurate resultant observations; ease of administration of treatments; repeatability of treatments, since environmental factors are repeatable and are not restricted to local and seasonal conditions and other disturbing influences; and ease of studying the more complex aspects of mechanisms of heat-production or heat-loss functions for many phases of basic and applied research. It is possible to separate the environmental factor influences such as temperature, humidity, radiation, and air pollution, which cannot be done in the field. Without control, these influences combined with other indirect environmental factors may result in confounding interactions between treatment and environment. Another advantage of a research laboratory is in the capacity to create an arctic cold or tropic heat, or altitude environment at the convenience of the personnel and students.

Some limitations of laboratory facilities are the shortage of trained personnel to operate the facilities; possible errors in extrapolation of laboratory results to field or biometeorological environs; acclimation of the animals to the laboratory under constant conditions for long periods; and the necessity for working with limited numbers of animals. Seasonal and other adaptive changes that may have occurred among test animals prior to test must also be considered as a confounding factor when animals are introduced to a specified laboratory environment.

## RESEARCH PROBLEMS AND TRENDS IN ENVIRONMENTAL PHYSIOLOGY

Some current problems are summarized below. Some of the views of Findlay (1965) are included in this listing of laboratory environmental physiology and shelter engineering studies that are worthy of intensive study. More specific problems are merely listed and then general environmental research needs and trends are stated somewhat more fully.

1. The relationship of the neural and endocrine system to cold, heat, and related physical and chemical environmental factors.
2. Mode of action and relationship of calorigenic hormones to heat-production and heat-loss.
3. Biochemistry, physiology, and neurophysiology of acclimatization.
4. Diurnal temperature, radiant effects, and interactions with physiological rhythms in domestic animals.
5. Time–temperature effects on partitional heat-loss of domestic animals.
6. A study of thermoenergetics, the change in efficiency of utilization of energy for "productive" processes when the animal is subjected to stress.
7. Simultaneous determination of thermal, energy, and water balance in varied environments.
8. Relationship of body water distribution to heat adaptability of domestic animals.
9. Comparative environmental physiological studies on exotic wild species and domesticated animals are extremely important in the clarification of physiological mechanisms associated with production and climatic adaptability.
10. Is sweat gland activity, due to hormonal or neural control, a principal factor in heat tolerance?
11. What is the role of the peripheral neural stimuli and involved hypothalamic centers in the regulation of appetite, hormones, energy, water metabolism, temperature, etc.?
12. From a confinement housing viewpoint, information on the relative effects of constant vs. varied cyclic environmental conditions on the physiology and production of domestic animals is extremely valuable.
13. Mechanisms of the effects of natural environments (tropical, temperate, arctic), as well as specific environmental factors, on reproduction in the female (LH release) are of vital importance to the livestock industry. The basic studies for these problems should be conducted under carefully controlled conditions.
14. Environmental studies are needed to determine an animal's preference for a temperature, humidity, air flow, or radiation level.

Cyclic variations of these environmental factors would give valuable information for preferable environmental complexes.

15. Efficient measures of acclimatization of animals to the various environmental factors are also needed. Research at the University of Missouri on cattle points to the important effect of acclimatization on water and feed intake and lactation (Johnson *et al.*, 1962). University of California work on oxygen consumption and shivering in chickens (El Hanawani *et al.*, 1970) is in a similar category. We should know the extent of cross-acclimatization, e.g., does acclimatization to cold or heat aid in acclimatization to altitude? Are specific or general acclimatization factors more important? (See also Chapter 8.)

16. The nature of biological rhythms requires investigation (see also Chapter 8). The previous history of an animal can greatly influence its response in bioclimatic research. It has been suggested by Wilson and Siopes (1967) that a conditioned reflex associated with photoperiod may be involved in the release of gonadotropin. If reinforced and new photoperiodic regimens are given, greater gonadal growth may result even though less total light was given.

17. Primarily, from a shelter engineering point of view, environmental equivalence of various climatic factors must be determined on all domestic species. In studies of environmental effects on animals to date, the effects of individual climatic factors, such as temperature, humidity, or light, have been investigated. Although this is an important first step, there is great need to determine the interacting effects of the various climatic parameters. The concept of equivalent environment provides a helpful index for expressing interacting effects. Laboratory tests are made in which animals are stressed by arbitrarily selected combinations of environmental factors, e.g., air movement, humidity, temperature. Any two combinations of variables that result in equal stress as measured by body temperature are said to be equivalent environments. When equivalent environment information is available, it can be used to answer such questions as: "If thermal radiation levels surrounding the animals are increased, what decrease in air temperature is necessary to counterbalance each step increase in thermal radiation load so that the animal maintains the same body heat balance and, therefore, the same elevated body temperature?" The diurnal effects of the factors mentioned above should eventually be evaluated. Such index values are not established for any domestic species with the exception of man.

18. The precise relationship of physiology to performance and its mechanisms and knowledge of the method of control of an animal's

physiological functions for special situations, such as high lactation or growth in a hot or cold environment, have not been adequately determined for most domestic animals. This information is essential if animal scientists are to capitalize fully on environment–animal interactions.

19. The effects of air pollutants $SO_2$, $H_2S$, $NH_4$ oxidants and other products of animal and industry and their interaction with other climatic factors require evaluation on domestic animals, from both the animal and human health viewpoint. The effects of specific environmental factors and climate on animal diseases and parasites are important to the livestock industry.

20. The duration of exposure, especially to temperature extremes, is critical in interpreting physiological and production responses. For example, short-term responses (minutes) to critical temperatures usually stimulate compensating responses in heat production, respiration, and related functions that differ quantitatively as acclimation or adjustment to the extreme is attained.

21. An understanding of the mechanisms of acclimation and acclimatization involves a fundamental understanding of organism, endocrine, and cellular components as they respond to environmental stress. In other words, how are animal stressors expressed at the molecular level, and what are the roles of neural and hormonal compensations on phenotypic actions of the cells?

22. From the viewpoint of ecology and bioclimatology, plant and animal interrelationships should be studied.

23. Animal behavior as it relates to season and shelter design regarding beef, dairy, and poultry management is becoming increasingly important with current trends toward more and more confinement housing.

24. The establishment of valid selection indices for predicting the adaptability of animals to varied geographic environments, such as tropic, arctic, mountainous, or desert, or stressful management systems in temperate environments continues to be extremely important.

## REFERENCES

Bond, T. E., H. Heitman, and C. F. Kelly. 1964. Physiological response time of thermally stressed swine to several cooling media. Proc. VIth Intern. Cong. Agr. Eng. Lausanne, Switzerland.

Brandt, A. E. 1938. Test of significance in reversal or switchback trials. Agr. Exp. Sta. Iowa State College Agr. Res. Bull. 234.

Cochran, W. G., and G. M. Cox. 1957. Experimental designs. 2nd ed. John Wiley & Sons, New York.

El Halawani M. El S., W. O. Wilson, and R. E. Burger. 1970. Cold acclimation in the role of catecholamines in body temperature regulation in male Leghorns. Poultry Sci. 49(3):621.

Findlay, J. O. 1965. Report of first meeting of the FAO expert panel on animal breeding and climatology. pp. 38–44. FAO Mtg. Rpt. An. 1965/2, Rome.

Folk, G. E., Jr. 1966. Introduction to environmental physiology. Lea & Febiger, Philadelphia.

Gill, J. L. 1969. Sample size for experiments on milk yield. J. Dairy Sci. 52:984.

Johnson, H. D., and L. Hahn. 1968. Laboratory facilities for adaptation research. *In* E. S. E. Hafez (ed.) Adaptation of domestic animals. Lea & Febiger, Philadelphia.

Johnson, H. D., A. C. Ragsdale, I. L. Berry, and M. D. Shanklin. 1962. Effect of various temperature–humidity combinations on milk production of Holstein cattle. Mo. Agr. Exp. Sta. Res. Bull. 791.

Johnson, H. D., A. C. Ragsdale, I. L. Berry, and M. D. Shanklin. 1963. Temperature–humidity effects including influence of acclimation in feed and water consumption of Holstein cattle. Mo. Agr. Exp. Sta. Res. Bull. 846.

Johnson, H. D., L. Hahn, H. H. Kibler, M. D. Shanklin, and J. E. Edmondson. 1967. Heat and acclimation influences on lactation of Holstein cattle. Mo. Agr. Exp. Sta. Res. Bull. 916.

Johnson, H. D., and L. Hahn. 1968. Laboratory facilities for adaptation research. *In* E. S. E. Hafez (ed.) Adaptation of domestic animals. Lea & Febiger, Philadelphia.

Kempthorne, O. 1952. Design and analysis of experiments. John Wiley & Sons, New York.

Kibler, H. H., H. D. Johnson, M. D. Shanklin, and L. Hahn. 1965. Acclimation of Holstein cattle to 29°C temperature changes: Changes in heat producing and heat dissipating functions. Mo. Agr. Exp. Sta. Res. Bull. 893.

Lucas, H. L. 1956. Switchback trails for more than two treatments. J. Dairy Sci. 39:146.

Lucas, H. L. 1959. Critical features of good dairy feeding experiments. Preprint, Dairy Sci. Mtg. 23 pp.

Patterson, H. D., and H. L. Lucas. 1962. Change-over designs. North Carolina AES Tech. Bull. 147. 52 pp.

Snedecor, G. W., and W. G. Cochran. 1967. Statistical methods. 6th ed. pp. 56–59. Iowa State University Press, Ames, Iowa.

Steel, R. G. D., and J. H. Torrie. 1960. Principles and procedures of statistics. McGraw-Hill, New York.

Wilson, W. O., and T. D. Siopes. 1967. The conditional gonadal response of Japanese quail to the cyclic photoperiods. Intern. J. Biometeorol. 11:289.

# 7

# DEVELOPMENT OF RESEARCH FACILITIES

## INTRODUCTION

Planning and constructing climatic research facilities for animals requires the combined efforts of the researcher, designer, and administrator. Each person involved in the development of the laboratory must recognize the extent of his own responsibilities and understand the problems of others with whom he will be working.

By way of definitions it should be kept in mind that climatic facilities for animals may vary in size from a small box for a single small laboratory animal to a full-scale production unit for several cattle. The degree of environmental control may vary from a simple sunshade to the very precise simultaneous control of several factors, such as temperature, wind, light, humidity, noise, and air quality. Similarly, the number of researchers and project objectives may vary widely. In some cases the facility may serve the need of a single researcher seeking an answer to a very specific question. On the other hand, it may serve the needs of several researchers seeking answers to a wide variety of questions. The questions themselves may be basic or applied: An endocrinologist may attempt to determine a specific endocrine function as related to the environment, or an agricultural engineer may attempt to appraise the impact of an environment control system on the production efficiency of an animal.

Livestock shelters and other simple structures provide a form of environmental control, but they cannot maintain stabilized conditions for extended periods of time. Failure to hold stabilized conditions can greatly reduce the value of some research studies.

Construction and operational costs are major constraints in all environmental research. Planners of research at all levels must certainly question its value in relation to the expenditures involved. Both parts of this relationship can be improved by careful planning. The cost of the facility can be greatly reduced by setting the design requirements as close as possible to the specific needs of the research contemplated. The value of research with climatic facilities can be greatly enhanced by considering multiproject usage of the facility.

In the case of small animals, the researcher will be well advised to seek ready-made climatically controlled boxes or chambers currently available. Laboratory supply houses and manufacturers of air-conditioning equipment are one source of contact for information on such equipment. In the United States, the "Thomas Register" lists manufacturers and fabricators of environmental cabinets. Air-conditioning journals and periodicals may provide further identification of fabricators of these cabinets and boxes. They may be indexed as environmental cabinets, temperature cabinets, animal cabinets, animal boxes, growth chambers, or by some other similar terms.

It is quite likely that standard prefabricated units will require some modification. This is frequently less expensive than building to a completely new design. The possibility of using or remodeling existing facilities should also be considered. Regardless of the choice, it is well to consult with other researchers and designers who have had experience in an area that is closely related to the activity contemplated by the researcher. The section "Selecting Design Services" should be read before designers are contacted. This section emphasizes that the actual facility design is primarily an engineering activity. It also emphasizes the need to obtain engineering services from those whose expertise is in the design of climatic facilities for animals.

## PRELIMINARY PLANNING

When a researcher determines that his work cannot be performed under field conditions he must turn to laboratory simulations. His first consideration must be what conditions are to be maintained in the facility and with what degree of control. Colleagues should be consulted to determine their interests in joining in the use of the contemplated facility. The requirements of all who conceivably would be working in the facility should be concisely listed and diagrammed, and a simple floor plan drawn to show size, shape, and so on.

In essence, this is a written plan of work and is essential to assure effective communication and understanding among all parties involved. This plan may be put into a brochure, which could be used to advantage in communicating with designers and obtaining financial support.

Generally, researchers are not salesmen and have difficulty in presenting their requirements to the administrator or others who authorize and finance construction. The administrator must understand the needs of the researcher and the value of the research to be performed before he can promote a project. He may use the brochure as a tool in securing funds. The brochure should set forth the need for the facility and the justification of construction at the time and location proposed. It should include some tentative construction and operating budget estimates. The estimates can be developed through consultation with others who have had previous experience with similar facilities, the institution's engineering service, or a consulting engineering service. Depending on the size of the project, complexity, location, and other relevant data, an independent consulting engineer may ask for a preliminary fee. (See "Selecting Design Services" for further guidance.)

Conferences with engineers will enable the administrator and researcher to improve cost estimates and more explicitly set forth the design criteria. (See "Establishing Design Criteria" for further information.) These estimates should be as realistic as possible. Unfortunately, much emphasis is often placed on the "high cost" of a facility and there is a tendency to set cost estimates low or to compromise requirements. This may later result in a facility that fails to fully serve the intended purpose.

The researcher must keep oriented toward his research goal and be careful not to infer or be forced into agreement that his needs can be reduced and therefore fulfilled with less money if indeed that is not the case.

In the initial planning the emphasis should not be on how much the facility will cost, but how important it is to research and how much will be lost in time, money, and validity of research findings if the facility is not made available. Research has often been terminated because of the lack of a closely controlled climatic facility to obtain answers to questions on the effects of climate.

Having a facility available will also reduce research time and make the results of research available at an earlier date. This should not be construed as a recommendation for a climatic facility at any cost: The need must always justify the end. The researcher must not ask for more than he can justify, but he should consider the future in his planning.

Many serious losses of time and money have occurred when funds have been insufficient and the researcher has taken on the entire task of designing and building the facilities. He consulted with suppliers of equipment and materials, but since he was not experienced enough to judge whether their advice was good or bad, he failed in obtaining the desired performance. His own time would have been better used to conduct the research for which he was trained and paid. It is not that a sales representative cannot give good advice nor that such practices cannot provide satisfactory facilities; rather, adequate funding can prevent unwise expenditure of talent and money.

## RESEARCH PLANNING COMMITTEE

To assure the best design for multiple use, present and future, it is recommended that a research planning committee be formed. The committee should be composed of a broad spectrum of academic disciplines, both those who might conduct research and those who, although not involved directly in research in the laboratory, might be called upon to service the laboratory. For instance, veterinary care is usually an important consideration. Veterinarians should be on the committee to assure that the designs are compatible with health requirements of the animals and any medical attention they may require. Specialists in physiology, genetics, engineering, animal management, nutrition, meteorology, and biometrics might also have service functions. If various classes of animals are envisioned as research subjects, there should be expert representation for each.

In operation, each researcher on the committee should submit his specific requirements. These in turn could be acted upon by the group as a whole to determine what phases of each of the proposals might involve multidiscipline effort. Once a list of projects and requirements is compiled, a priority schedule of required laboratory usage can be developed. At this time, also, considerable thought should be given to the possibility of combining projects within the same tests. For instance, in a dairy cattle climatic facility, milk production studies could be combined with engineering studies of heat and moisture dissipation rates. Classes of livestock might also be combined provided there was no disease transmission or other confounding problems.

It will be quickly recognized that a broad array of disciplines will bring forth a variety of terminology that will make communication difficult not only among the researchers but also with engineering and construction personnel. An engineer from within the organization would be helpful toward a solution of the latter problem.

It is suggested that bringing in engineering services from outside the organization be delayed until the committee has its goals and requirements well identified.

The committee should be prepared to meet frequently to review progress. It must be prepared to make many adjustments as cost and engineering design limitations become apparent.

If the research committee is large, it should have a small executive committee delegated to work with the administration and design engineers. This executive committee should be directly involved with selecting the engineering design services.

## SELECTING DESIGN SERVICES

The selection of a designer is a critical consideration for the success of a facility. The problem of designing a climatic laboratory is more engineering than architectural. It is therefore recommended that the leadership of the design of climatic facilities should be given to an engineer rather than to an architect. Should this not be possible, such as in a large construction complex, the design of the climatic facility could be set apart from the architectural contract as a separate item.

Engineering experience in the type of facility being considered is limited, and claims must be carefully investigated. Evidence in the form of successful projects of equal sophistication and complexity should be sought. Visits to such projects are recommended. The engineering design service must be capable of translating the needs of the researcher to facility plans.

The time spent in finding the correct or most experienced organization will be well justified. The engineering services should be impartial ones—preferably without commercial interests in a specific line of products, because their decisions must often be biased toward their products. Such organizations often have qualified engineering sales advisory personnel, who can provide valuable advice, but judgment is required in using this advice.

A design engineer has the responsibility for designing the most economical system and equipment that will adequately and economically perform the functions desired. In addition, he or his design service group must also be capable of supervising the erection of the facility and its final testing and adjustment; some organizations are capable of providing the design but lack operational experience with the systems, equipment, and controls involved. A start-to-finish design organization is the most satisfactory.

Consulting engineering services, like those of doctors and lawyers, are secured through negotiation rather than by any system of price competition.* The negotiation may be for complete engineering services from project initiation to the completed and tested facility, or for one or more of the engineering steps necessary. Determination of his fee may require the engineer to make a feasibility study. The feasibility study is a detailed report that includes all technical information, preliminary engineering calculations, proposed layout, description of operation and control, and cost estimates. The establishment of the design criteria (covered later in this chapter) is essential to the feasibility study. In most cases it becomes an integral part of the feasibility report. The feasibility study can be construed as the concise compilation of all thoughts, ideas, and requirements of the planning committee integrated into a practical engineering concept. Such a study will usually enable a consulting engineer to quote a fee for all of his services in connection with the facility design, construction, and testing.

Consideration may be given to retaining the consulting engineer after the facility is completed to instruct personnel in operation and maintenance. An instruction manual should be provided as part of such a training program. The personnel instructed should include the researchers who will be working in the facility. If research personnel are well informed, they can recognize the indications of malfunctions and have them corrected before serious consequences ensue.

*For complete information concerning consulting engineers in the United States, it is suggested that the Consulting Engineers Council of the United States, Madison Building, 1155 15th Street, N.W., Washington, D.C. 20005, be contacted. They will furnish the "Manual of Practice" giving complete information on how consulting engineers work and an indication of charges for their services.

## ESTABLISHING DESIGN CRITERIA*

Design criteria are the fundamental set of requirements and other pertinent data that will influence and determine the facility design. Table 7.1 outlines some factors that must be taken into consideration. These must be thoroughly reviewed and analyzed from all standpoints—particularly the practicality of construction, degree of importance of each item, and operational constraints. The remainder of this section enlarges on the items capsulized in the table.

The responsibility for identifying the needs of the facility rests with the organization that seeks the facility. The organization may utilize outside assistance but it is usually the researcher who must identify the needs and place priorities on various elements of performance.

**TABLE 7.1 Design Criteria and Division of Responsibilities**

| Item | Researcher Identifies | Designer Determines |
|---|---|---|
| A. *The Facility* | | |
| 1. Over-all design | Location of facility<br>Number of people<br>Number of animals<br>Other animal needs<br>Storage needs<br>Movement of materials, animals, and personnel<br>Employee needs (showers, housing, etc.)<br>Special sanitation or disease control needs<br>Remote instrumentation needs<br>Special considerations with respect to other facilities<br>Possibilities for future changes | Layout, design, and specifications:<br>Coordination of space, sound, lighting, decoration, etc., to needs of animals and personnel<br>Provisions for installation and operation of instrumentation<br>Provisions for disease protection (i.e., bacteriological filters)<br>Provisions for future expansion or modification |

*This section envisions a major construction project. However, it is applicable to any facility. No matter how small the project, researchers and designers should check details set forth in this section to avoid oversights.

**TABLE 7.1 (Continued)**

| Item | Researcher Identifies | Designer Determines |
|---|---|---|
| A. *The Facility* (cont.) | | |
| 2. Space allocation | Numbers, kinds, and areas of offices, support laboratories, and other spaces | Layout and dimension of rooms and selection and organization of fixed utilities and equipment |
| 3. Test rooms | Test conditions, subjects, and measurement program:<br>Climatic factors<br>Degree of precision<br>Required variability of conditions<br>Description of animals to be used<br>Special features of animal care and feeding<br>Kinds of measurements to be made<br>Instrumentation<br>Number and talents of people involved in conducting tests and operating laboratory<br>Special considerations such as toxic chemicals, radiation<br>Need for protection from electrical or other interference<br>Possibilities of future program changes | Design and specifications for:<br>Construction<br>Environmental conditioning equipment and control systems<br>Built-in instrumentation<br>Fixed animal care and handling equipment<br>Utility services |
| 4. Special equipment | Needs for safety hoods and alarms for failure in lighting, air flow, temperature, humidity, electricity, etc. | Selection and arrangement for quantity, quality, and performance |

**TABLE 7.1 (Continued)**

| Item | Researcher Identifies | Designer Determines |
|---|---|---|
| B. *Service to facility* | | |
| 1. Electric service | Kinds and location of electric equipment and instruments needed for research including special current and phase needs | Electrical service and distribution: Total connected load; Instantaneous load; Standby generation where needed |
| 2. Water | Location of use; Approximate rate of use; Quality requirements | Systems for: Treatment; Distribution |
| 3. Transportation and access | Animal sources; Methods of delivery for: Animals; Supplies; Special vehicular needs; Personnel transport | Roads, walkways, parking, entrance doors, gates, animal handling equipment |
| 4. Communication | Type and location of: Telephone service; Radio service; Intercom service; Closed circuit TV; Sound recording, etc.; Alarm systems | System, installation, and performance |
| 5. Emergency services | Need, kind, and critical location for: Light; Heat; Water; Fire protection; Equipment failure alarms; Police protection; Vandalism protection | Location, performance, and installation specifications |

**TABLE 7.1 (Continued)**

| Item | Researcher Identifies | Designer Determines |
|---|---|---|
| B. *Service to facility* (cont.) | | |
| 6. Feed | Kinds and sources of feed, rates of use, and quality requirements<br>Processing needs and special care requirements | Processing system, distribution system, and storage design:<br>Rodent and insect control<br>Environmental protection |
| 7. Spare parts and idle equipment | Equipment requiring storage | Repair parts needs and design of storage areas |
| 8. Fuel(s) and other volatiles | Needs for:<br>Laboratory purposes<br>Vehicles and special equipment | Source<br>Storage<br>Distribution<br>Service access |
| C. *Service from the facility* | | |
| 1. Waste handling | Limitations involving the contact of animal with its waste<br>Limitations involving contact of waste products with objects external to the facility | Methods and systems for handling, treatment, and disposal |
| 2. Drainage | None except for possible limits of interference when animals are kept outside of the facility | Topography, soil absorption, and system layout |
| 3. Animal disposal | Condition of animals<br>Frequency and amount of removal<br>Any known restrictions | Alternate methods of disposal<br>Equipment design, selection, and installation |

**TABLE 7.1 (Continued)**

| Item | Researcher Identifies | Designer Determines |
|---|---|---|
| D. *Economics* | | |
| 1. Construction cost vs. functional value | Priority and final choice for major facets of program: Difficult test conditions Degree of acceptable control | Cost and ability to meet performance desired |
| 2. Construction cost vs. budget | Acceptable reductions in performance | Loss of performance when reducing costs |
| 3. Special vs. standard commercial components | Degree to which rigorous requirements must be met | Relative difference in performance and costs |
| 4. Operational costs | Anticipated annual operating budgets and labor skills available for maintenance and operation | Anticipated operation costs for fuel, utilities, labor, operation, and maintenance materials Design for optimum mix for construction costs, operational costs, and performance |
| E. *Site planning* | | |
| 1. Placement | Relationship to other buildings and supporting facilities Special terrain and exposure considerations for animal management, etc. | Layout to suit terrain and site development |
| 2. Legal | Sources of assistance with legal and administrative details | Permits, zoning boundary restrictions, public health restrictions, standards, and codes |

**TABLE 7.1 (Continued)**

| Item | Researcher Identifies | Designer Determines |
|---|---|---|
| E. *Site planning* (cont.) | | |
| 3. Public relations | Nature of surrounding property use and occupancy | Ways to minimize nuisance to surrounding community by landscaping, controlling air and water pollution, flies, etc. |

## The Facility

### OVER-ALL DESIGN

The overall design of the structure and climatic control systems are the designer's responsibility. He makes all thermodynamic calculations, determines the heating and cooling system needs, decides on the control and monitoring systems, determines the applicable structural loading and heat transfer characteristics, and designs the complete facility accordingly. The designer should not rely solely on equipment manufacturers' promises to provide him with equipment to meet the specified test condition. He must make his own thorough analyses. Performance testing of environmental facilities is difficult in itself, and proving the lack of performance in a legal suit is even more difficult. Geographical location is a factor. Designs will need to be altered to fit the climate. An extremely warm and tropical humid climate presents problems quite different from an arctic climate.

The movement of animals, materials, and personnel is of extreme importance. An environmental study may be upset when access doors are opened unless adequate provisions are made to avoid drafts or rapid climatic factor changes. A helpful practice is to build the climatic chamber within a similarly controlled space. The climatically controlled area around the chamber can serve as a buffer area; a service area; an acclimitization area for tempering instruments, water, and feed to the climatic conditions area; and a work area outside of the room for weighing or animal treatment.

Sanitation and protection from contamination are important. A positive ventilation system that is always operative through biological filters is usually essential for disease studies. Isolation of one animal or animal group from another may also be needed. These needs must be identified by the researcher. In some cases, outside air may contain contaminants, pathogenic or otherwise, that might upset the climatic study. Generally, this has not been a problem for most climatic facilities, but it must be included among the items to be checked. Both the durability of materials and ease of cleaning and sanitation must be considered.

### SPACE ALLOCATION

Space allocation is primarily the designer's responsibility, but the researcher will need to identify his requirements—particularly instruments and the fixed equipment, such as cabinets, water still, surgical fixtures, and animal weighing scales, as applicable.

### TEST ROOMS

The test rooms or climatic chambers require environmental control. They must provide the required controlled environmental conditions; facilities for holding, feeding, and otherwise caring for the animals; and access for man and his instrumentation in conducting the tests.

The range of controlled climatic conditions, the degree of precision required for controlling each condition, the rapidity of change from one test to another, and the nature of conditions within a test—single or multiple fixed conditions or variables—are important early decisions. These elements will influence the costs of the facility. High degrees of precision, wide ranges of controlled climatic conditions, and rapid fluctuation of temperature escalate costs. Many research purposes are adequately served with air temperature fluctuations of ±2°C from the scheduled test condition.

The cost of providing equipment and controls for such tests is much less than that required for finer degrees of control. The capabilities of laboratories may be extended and operating costs reduced by scheduling cold tests in the winter. Climatic tests that require defrosting of cooling coils will greatly increase the cost and complexity. Extremely low temperature conditions are even more costly. Continued operation at freezing temperatures within the laboratory will require special provisions to avoid frost buildup and upheaval of the floor.

Establishing the range and combinations of climatic conditions requires knowledge of psychrometry and the ability to read a psychrometric chart (see Chapter 2). Unless one has this fundamental knowledge he may be asking for conditions of temperature, humidity, etc., that are unreasonable, unrealistic, or even impossible. A knowledge of psychrometry will also make communication with the designer much easier.

Permissible air velocities among the animals are required data for designers. Since normal thermal convection currents around large animals may be as great as 15 meters per minute (about 50 feet per minute), there is little point in specifying lower velocities than this for large animals—lower velocity limits will usually increase the complexity of design, and therefore costs. Velocity tolerances are essential, but setting them lower than necessary will be expensive.

Acceptable limits for air quality should be identified. Ventilation systems for animal shelters are usually designed to remove excess moisture through regulating the quantity of outside air circulated through the room. When this is done, the levels of gas contamination are usually well below the toxic limits. However, in climatic facilities under conditions where outside air is greatly restricted, the gases may build up to toxic levels. Excess carbon dioxide is a factor. In cattle facilities, respiration rates were noted to rise when the fresh air intake was dropped to 10 cfm (0.28 cubic meters per minute) per animal (average body weights were about 454 kg) but returned to normal when 20 cfm (0.56 cubic meters per minute) of fresh air was used. Ammonia in excess of 30 ppm may blind chickens.

Light affects maturity and production of fowl. The researcher should check whether sound, atmospheric pressure, and precipitation have significant effects.

Psychological considerations have often been given little attention in room design. In the higher orders of animals, particularly primates, this can become very important. The researcher has the responsibility for specifying applicaple tolerances. The constructive value of pleasant surroundings and convenience to the personnel working in the laboratory has been well established.

The range of testing programs must be identified. This includes: the kinds and numbers of animals; the measurements to be provided for; the most demanding testing schedule; and the people involved in the care and testing of the animals. The needs of the animals usually are easy to establish. Frequently overlooked, however, are explanations of special features such as individual feeding; special requirements for access; maximum permissible noise levels; floor drains (that will not freeze); need for isolation of animals; and precautionary measures associated with the use of toxic agents and avoidance of entry by disease, insect, and parasitic vectors.

Instrumentation needs should certainly be identified by the researchers. An access port should be provided to climatic rooms for lead wires or tubing for remote indicating instruments that might be installed at a later date. Frequently, researchers are frustrated when they wish to install additional instrumentation or update existing instrumentation. Removable wall sections are helpful. This concept could well be incorporated with a need for an emergency exit.

The sensitivity of anticipated instrumentation to humidity, dust, vibration, electrical disturbance, and voltage change must certainly be identified by the researcher, and the designer must be sure that this factor is given adequate attention. In practically all experimentation, the instrument is the eye of the researcher. The finest facility would be useless if the instrumentation did not adequately record the events—both the environmental conditions and animal reactions. There can be no compromise with the reliability of the instrumentation.

#### SPECIAL EQUIPMENT

Most facilities depend on outside air for maintaining satisfactory air composition. A warning system to alert operators to the failure of the ventilation system is essential. A simple device may suffice when someone is on duty at all times. A recorder of the air velocities or air volume moving through the ventilation ducts will indicate failure. An alarm bell and/or light is of further benefit. In general, refrigeration compressors will have gages and an alarm bell. Alarm systems that may be coupled with personal telephones, a central telephone office, or a central security office are available. Chemical hoods and facilities to handle radioactive isotopes may be needed. The designer should include the necessary fire, electrical, and mechanical equipment alarm systems in his design. Fire safety is usually handled quite well through following standard practices for safe construction. However, one may wish to include a detection system for smoke that will provide an early alert and permit removal of animals in case of fire. Generally, a good fire prevention program coupled with safe construction will be adequate.

### Service to the Facility

Services to the facility are closely related to the design of the facility itself and, indeed, may be considered by some as a part of the basic facility design. Remoteness of location and degree of development within the general geographical area will greatly affect design decisions as to needs and costs.

### ELECTRIC SERVICE

The electrical service requirements will be mostly determined by the designer on the basis of the electrical equipment used in the design of the facility. The researcher will have to identify special requirements (generally for instrumentation) such as levels and kinds of voltage, the phase (single and three phase will probably be normally available), and voltage stability. The designer will need to determine the availability of electrical energy to the site and develop the design. The researcher may realize some savings in working with the local electrical utility in obtaining rate concessions and some of the "on site" transformers and other equipment. Standby electric generation is worthy of consideration. Experiments and sometimes animals may be lost without it. The researcher will need to check that this need has been explored. Often standby electricity is provided only for certain facets of the operation, such as ventilation, alarm systems, and lighting.

### WATER

Adequate water supply is often taken for granted but this needs checking. Both quantity and quality need consideration. The rate of use is often more important than total daily consumption. Animal facilities frequently have very high peak demands. The water consumption of animals will be greatly increased at high temperature conditions. Treatment of water may be required for both animal and equipment use. The researcher should assist the designer in appraising demand and treatment requirements.

### TRANSPORTATION AND ACCESS

Transportation and access are also frequently taken for granted and may not be part of the original design contract. However, if they are not, the researchers must be sure that these needs are met. Separate quarantine facilities, animal corrals, personnel parking, and access for feed service are sometimes overlooked. In developing countries, the availability of freight handling services is of great importance. The dependence on air freight may require a landing strip and would certainly require consideration in the design of the facility itself.

### COMMUNICATION

Communications systems are closely related to the special equipment within the facility. For example, alarm systems may be tied into the telephone system. The designer and researcher should work together in ascertaining telephone needs and making provisions for installation. Closed circuit TV has great potential in monitoring animal behavior. A radio station may be important in remote locations. The researcher will need to identify these needs but they might well be "add alternate" items in the construction contract that could be added later.

### EMERGENCY SERVICES

Emergency services include fire and other security protection devices. Security fences are recommended. Remote locations may require a guard and fire house. Often the police function can be handled in conjunction with the care of the animals.

### FEED

Feed is an essential consideration in any animal research facility. The research planning committee will usually be well acquainted with the kinds, quality, and amounts required. They will be able to advise the designer on storage, processing, and distribution needs and also on commercial equipment that might do the job. An agricultural engineer is an excellent source of consultation for the designer. The designer will also find consultation with manufacturers of feed processing equipment helpful. Generally, the problem of dust and deterioration of feed make it desirable to provide a separate facility for storage and processing. The designer and researcher should consider feeding of pellets in order to reduce dust within the environmental facility. The construction of separate processing facilities permits alterations in the processing plant without upsetting the climatic laboratory. Feed damage through caking by high humidity, pocketing in distribution systems, and rodent and insect infestation must be considered.

### SPARE PARTS AND IDLE EQUIPMENT

Storage space for repair parts and idle research equipment is often overlooked. Space for this need may be provided inside the facility or in an adjacent building. There is no level of supply that is generally applicable to all facilities. The remoteness of the facility from sources of supply, the adverse impact of shutdown on an experiment, the availability of backup or alternate climatic control systems, the reliability of the equipment, and stocking characteristics of sources of supply will all have a bearing on repair part needs.

#### FUEL(S) AND OTHER VOLATILES

The choice of fuel is generally contingent upon its availability and the economies of its use. However, in animal facilities, consideration must be given to possible toxic effects of products of combustion—particularly those that might enter the space where the animals are held. Even electrical heating equipment may create toxic gases, particularly when energized after being out of use and having collected considerable quantities of dust and other contaminants. To date, this problem has only been observed as detrimental in some poultry environmental facilities.

Trucks and other service vehicles are needed in conjunction with animal care. Fuel, paints, and other volatile materials require special considerations for storage.

### Service from the Facility

Services from the facility should be considered in basic facility design, but may not be a part of the design or construction contract. Local practices and regulation codes will determine the appropriate technology and designs.

#### WASTE HANDLING

The technology for the disposal of wastes from man are well established. Community sewerage systems should be used whenever available. A properly designed septic system is a satisfactory alternative, and in some cases a sanitary privy may be all that is justified. The problem of handling wastes from animals is much more complex. Animal wastes differ markedly from human sewage. With proper precautions, animal wastes can be put into sewerage systems, but generally some other means of disposal is more economical. The application of animal wastes to agricultural land continues to be an acceptable practice in most areas of the world. However, provisions must be made to collect, transport, and spread it. Often storage is required; volume may not be great enough to justify daily hauling or there may be times at which the waste cannot be applied to the land.

#### DRAINAGE

Drainage is a primary consideration for septic systems, important for roads and parking areas, and a factor in the design of animal holding areas. Flood problems should be avoided by locating facilities above the flood plain; most climatic laboratory control equipment would be ruined by flooding.

#### ANIMAL DISPOSAL

Often disposal services are available elsewhere on the research station. Sometimes commercial animal rendering plants may be used. Generally, burial is expensive and land unavailable for that use. Some poultry operations have utilized heated septic tanks. Incineration is used by many, particularly where diseased carcasses are involved. The design criteria should contain advisory information and any research constraints.

### Economics

The design of a facility is a balance between that which might be considered the researcher's dream and the money available for the project. It is important that the researcher and the designer come to an early understanding as to financial limitations and other economic considerations before the designer goes to great expense in developing a design.

#### CONSTRUCTION COSTS vs. FUNCTIONAL VALUE

The cost of providing various combinations of climatic conditions within the laboratory must be studied in relation to their functional value. The research planning committee must identify priorities and be prepared to consider possible alternatives. A cost–benefit analysis is helpful. This analysis should be subdivided according to the questions that the research is expected to answer. Functional concessions may be required.

#### CONSTRUCTION COSTS vs. BUDGET

If functional concessions cannot be made, cost-cutting techniques in construction and equipment must be sought or the budget increased. Functional value may suffer as the precision is usually less than originally sought. The time delay and added costs of making something meet acceptable alternative performance levels are also problems. Generally, these adjustments should be avoided.

#### SPECIAL vs. STANDARD COMMERCIAL COMPONENTS

Attempts to make standard commercial air-conditioning or refrigeration equipment meet specifications for which they were not originally designed usually result in failure. This should not be interpreted to mean that standard commercial equipment should not be used but rather that care should be exercised to use it only for the conditions for which it is designed. Many climatic facilities will require refrigeration equipment that is much more complex than that for domestic and industrial air-conditioning practice. The researcher will need to be prepared to justify to his administrators that his requirements (and therefore the costs) are more stringent (and therefore higher) than commercial equipment will provide. NOTE: COST ESTIMATING VALUES FOR AIR-CONDITIONED SPACE OR EVEN COLD STORAGE ROOMS ARE NOT APPLICABLE TO MANY CLIMATIC LABORATORY SITUATIONS.

#### OPERATIONAL COSTS

Operational costs are frequently overlooked in planning research facilities. Often cost savings in construction result in greatly increased operational costs. The researcher and his administrators must fully understand and communicate to the designer the annual budget within which they expect to operate. The designer, in turn, must be prepared to advise whether or not this is realistic. Unfortunately, the ideal condition of establishing the operating costs and then obtaining the funds specified by the designer are seldom realized.

Often, skilled labor is not available for operation and maintenance of equipment. Maximum reliability and automatic control systems must then be emphasized. This will increase the construction cost. The economic considerations of Table 7.1 refer primarily to construction and operating costs of the facility itself, but the researcher must allow for animal feed and animal care as well as other research support costs. Because of the specialized nature of research, these may be considerably higher than in a regular animal production unit.

Equipment maintenance, whether from labor within the organization or through outside repair services, is another consideration. In many localities, even within the more fully developed nations, the usual refrigeration repair and maintenance services will be inadequate.

## Site Planning

Site planning will be applicable where an existing building is being remodeled as well as in new construction. The flow of people, materials, and animals must be given major consideration. Esthetic acceptance by the laboratory workers on the job and the community are also important.

### PLACEMENT

Placement of buildings on the site requires a topographic map, soil load-bearing capacity tests, and knowledge of available utilities. Generally these are obtained under the guidance of the design engineer. Since there is often a relationship to other facilities, such relationships must be identified by the researcher. In some cases, special attention must be given to protecting animals from exposure, particularly those animals being held outdoors. Exposure of the facility to the environment will also be a matter of concern as high winds or extreme thermal or solar radiation may upset the heat balance calculations that were made in accordance with "standard" conditions assumed for much of the tabular design handbook data.

### LEGAL

Consideration of local building codes and restrictions is standard practice for designers. Legal clearance of property ownership and special zoning authorizations are usually not the concern of the designer but are the responsibility of the institution that seeks a facility.

### PUBLIC RELATIONS

The organization seeking the facility and the researchers operating it will be faced with maintaining good public relations throughout the life of the facility. It is important that the design criteria identify the nature and occupancy of surrounding property. The design criteria are also useful to an information specialist or reporter who may wish to initiate action for a friendly reception and acceptance by the community.

## GENERAL DESIGN PROCEDURES

This section describes steps in the design, layout, and writing of specifications for a specific climatic chamber. It is written to acquaint the researcher with design procedures and serve as a checklist for the engineer.

Regardless of size or complexity of a climatic chamber, a complete orderly design and specification, including performance requirements and guarantee, is important. The construction contractor must be carefully selected for competence in carrying out the engineer's design and layout. Piecemeal construction and assembly of equipment and materials should be avoided. The design criteria must be followed.

### Calculations

#### HEAT LOAD CALCULATIONS

The heat load calculations are made to determine the heat required to maintain high temperatures, as well as the heat to be removed to maintain the low temperatures. Similar calculations must be made to determine moisture removal or addition to maintain the desired humidity.

The engineer takes into consideration in his calculations the following:

1. All conditions of temperature, humidity, air motion, ventilation, pressure, and natural weather phenomena that are to be reproduced in the test chambers.
2. Heat transfer through walls, floor, and ceiling at the temperature differentials that will exist. Materials of construction, insulation, and air motion over the inside and outside wall surfaces are important factors.
3. Metabolic sensible and latent heat dissipation by people and animals who will be in the chamber. Duration of occupancy, activity, and climatic conditions are significant.

4. Moisture losses from animal waste, cleaning, feeding, and watering (including water wastage). The test conditions will alter these losses.

5. Heat from lights, electric appliances, motors, and other heat-producing items in the chamber.

6. Sensible and latent heat from open gas flame or similar combustive process if used in the chamber.

7. Heating, cooling, humidifying, and dehumidifying of outside-ventilation air.

8. Heat from periodic defrosting of equipment.

9. Heat given off and absorbed when chamber floor is heated and/or cooled.

10. Effects of rain, sleet, snow, wind, and altitude.

#### STRUCTURAL CALCULATIONS

The engineer takes into consideration the following factors in making structural calculations:

1. Structure size, shape, and arrangement
2. Type of construction and materials
3. Insulation and vapor-proofing
4. Wind and snow loads
5. Soil load bearing capacity and water table
6. Earth shocks, sonic booms, and vibration from equipment
7. Expansion and contraction of building structure under temperature changes and frost action
8. Space allocation for climatic conditioning and other equipment
9. Attachment to adjacent structure
10. Abuse from animals or other sources both inside and out
11. Effect of operations inside and outside of the structure
12. Need for negative or positive pressure control as required for prevention of drafts on animals, disease control, sanitation, or pollution control

### Development of the Design

The development of the final design is the result of successive preliminary designs in which tentative decisions are made and reviewed for optimum compatability of equipment, control systems, and structural design. Costs must be continually considered.

Key design considerations are as follows:

1. *Utilities*: These include steam, hot water, gas, water, electricity, sewage.

2. *Type of Heating*: This will include a source, such as steam from a central plant, fuel fired steam or hot water boiler, direct fired furnace or unit heaters. Some systems will require nonfreeze secondary heating mediums of various types to suit the conditions.

3. *Type of Fuel*: The type used will depend on availability, dependability of source, storage, capacity required, control, maintenance, and cost.

4. *Air Handling System Design*: The engineer takes into consideration the method of distributing the air in the chamber, including air motion in the chamber, and conducting it by a system of ductwork through the air handling equipment. The equipment will consist generally of filters, ducts, air diffusers, heating coils, cooling and dehumidifying coils, solid desiccant or other types of dehumidifiers, humidifiers, and circulating fans. Air sound traps are frequently used to reduce air noise. All equipment must be selected for its adaptability in the conditioning system and for operation with the heating and cooling mediums that are selected. The air handling systems will bring in outside ventilation air, which must be exhausted by a fan or fans. Where chambers are to be operated to simulate various altitudes, the supply and exhaust fans are selected for type and capacity to produce high or low pressures as the case may be.

5. *Refrigeration System Design*: These can range from the simple direct expansion type, wherein refrigerant is expanded directly in the tubes of the cooling coil, to the more extensive secondary cooling medium or brine system. Actually, calcium chloride or sodium chloride brines are not often used, but the term "brine" is commonly applied to a nonfreeze solution such as glycol and water. Where very low temperatures are involved, one of the low boiling point halocarbon refrigerants, such as refrigerant 11, is used as a brine for circulating through the system. Direct expansion systems are usually less expensive than brine or secondary cooling medium systems, but their adaption to meet all conditions of a climatic system can be complicated by numerous refrigerant control devices. The brine system, in fundamental operational respects, is simple and provides a basis for far more precise control. Each refrigeration system has to be provided with suitable condensers of the shell and tube, air cooled, or evaporative type. The selection is based on climatic conditions, availability of water, and local ordinances or codes. One of the engineer's major considerations in his refrigeration system design is the variation in load on the system.

His selection of machines, capacity control, artificial loading, together with all automatic controls for same must be carefully studied and the system arranged to meet all conditions at all seasons of the year without injury to the equipment.

6. *Equipment Selection*: As can well be visualized from the preceding paragraphs the engineer selects equipment and machines as he progresses with his design. He usually incorporates commercially available proven equipment, as this usually provides the greatest dependability, ease of maintenance, and lowest cost.

7. *Automatic Control System*: In designing his controls, the engineer considers recorders and indicators for monitoring all the functions of a system. Frequently these are incorporated together with the controller in a single instrument. Preferably the controls and all monitoring devices are furnished by a single supplier to eliminate divided responsibility and provide the other advantages of a single source of supply.

8. *Structure*: In addition to the test rooms, the structure must provide for space and protection, for the air-conditioning equipment, control devices, and any special needs of the researcher for storage, instrumentation, animal holding, tools, and other items. The design will include all details of construction, including those for auxiliary buildings.

9. *Electric and Plumbing Systems*: Electric and plumbing systems for the operation of equipment, chamber lighting, animals, attending personnel, cleaning, sewage, washing, etc., are designed and laid out completely by the engineer. He also takes into consideration auxiliary utilities, such as a standby electric generator for use in case of failure.

## Final Plans, Specifications, and Cost Estimates

After much review, the design engineer completes all plans in detail. Of utmost importance: The engineer must supply the researcher with complete diagrams of all systems even though the piping, lines, controls, wiring, etc., may be indicated on the construction plans. Such diagrams are not usually furnished for ordinary construction, but it must be remembered that a climatic facility is a highly specialized and exacting project. All the requirements must be clearly conveyed to the contractor and should be retained at the facility after completion of construction.

In addition to the plans, a complete set of specifications is drawn up by the engineer. The specifications must include a description of how the entire facility is to operate, the conditions that must be maintained, and a description of how each subsystem (heating, refrigeration, etc.) must operate and perform. The specifications are legal documents. They are used by reputable contractors bidding on the work and form the basis of the contract and guarantee of performance and equipment by the contractor. Table 7.2 indicates items that are frequently included in climatic chamber specifications.

**TABLE 7.2 Specification Index—Typical for Major Climatic Facilities**

*General Provisions*
Storage areas, temporary heat and utilities, progress charts, subcontractors, quality of materials, protection of structures and work, bonds, inspection, responsibility, accidents, and many other legal items
*Special Conditions*
Scope of the work, list of contract drawings and specifications, shop drawing requirements, layout of work, contractors liability insurance, time extensions, record drawings
*Special Provisions*
Special requirements for selection and approval of qualified contractors and special equipment peculiar to the job
*Applicable Minimum Hourly Rates of Wages*
Usually required on federal government work or work subsidized by federal government money
*Construction Engineer's Office*
Required for resident inspector
*Samples*
Describes how construction material samples shall be submitted
*Demolition*
Required when facility is to become part of or replace an existing facility
*Earthwork*
Grading, excavation
*Topsoiling, Seeding, and Sodding*
*Landscaping*
*Bituminous Pavement*
*Damproofing and Membrane or Metallic Waterproofing*
For walls below grade, slabs
*Concrete*
For footings, walls, columns, slabs, beams including control and tests, curing requirements
*Structural Steel, Stainless Steel Plate, and Sheet Lining of Chambers*
Columns, beams, and miscellaneous steel framing for building and chambers; stainless steel lining of chambers with detailed requirements for fabrication
*Sheet Metal*
For building construction such as gravel stops, rainspouts
*Mortars*
For masonry work
*Calking and Sealing*
For general construction
*Masonry*
Brick work, concrete block, glazed structural tile

**TABLE 7.2 (Continued)**

*Roof Insulation*

*Roofing Systems*
Requirements for built-up roof construction, i.e., felt, asphalt, application, gravel, flashing

*Windows*
For buildings

*Miscellaneous Metal*
Building expansion joints, handrailing, steps, ladders, louvers

*Hollow Metal*
Metal doors and frames for building construction

*Plate Doors and Frames*
Chamber doors where evacuated or pressurized spaces will be required

*Furring, Lathing, and Plastering*

*Acoustical Tile*

*Woodwork*

*Resilient Flooring*

*Ceramic Tile*

*Finish Hardware*

*Painting and Finishing*

*Glazing*
Laboratory equipment including exhaust hoods, cabinets

*Mechanical and Electrical Equipment*
General requirements, general equipment requirements for motor characteristics, belt drives, equipment bases, hangers, emergency services

*Chamber Insulation*
Construction required for chamber floor, wall, and ceiling insulation including vapor barriers, expansion joints, installation and protection; materials of construction

*Chamber and Anteroom Doors*
Detailed requirements for specially constructed doors to withstand extreme differences in temperature and pressure

*Chamber Conditioning and Exhaust System*
Special make-up air and exhaust systems for chambers including absolute filters, positive displacement blowers, cooling coils, heating coils, heat exchangers, ductwork, access doors; chamber recirculating air systems including special insulated dampers, casings, cooling coils, heating coils, humidifiers, filters, access panels, fans, and drives

*Brine Circulating System*
Special brine pumps, heat exchangers, headers, valves, piping, expansion tanks, pressurizing system, charging, and testing

*Low-Temperature Refrigeration System*
Central refrigeration plant and auxiliaries. May be reciprocating compressor with booster; centrifugal cascade system or reciprocating cascade system; cooling tower, pumps, drainback tank, condenser water piping, and valves

*Chamber Control, Recording, and Indicating System*
Complete description of individual system, how it operates, tolerances expected, quality, features of components and description of all components, method of installation, testing, and adjusting

*Control Panels*
Custom fabricated panelboards containing variable speed motor controls, other electrical power control devices, automatic temperature and humidity control equipment, and all interlock wiring

TABLE 7.2 (Continued)

*Sheet Metal and Air Handling Apparatus*
All comfort cooling and heating equipment serving auxiliary personnel spaces, laboratories, instrument rooms, corridors, offices, animal holding areas including air handling units, ductwork, grilles, and automatic controls
*Air Conditioning for Adjacent Spaces (Offices, Animal Holding Areas, Laboratories, etc.)*
*Chilled Glycol Solution System*
Nonfreeze solution chilling system for outside air make-up coils. Chiller, glycol tanks, pumps, heat exchangers, expansion tanks
*Heating Apparatus*
(Steam and hot glycol solution systems) heat generation equipment and distribution system. Steam boiler, condensate return units, piping and heat exchangers. Hot glycol piping to outside air heating coils, chamber radiant floor panels, trench and floor heating coils, pumps, heat exchangers, piping, valves
*Underground Heat Distribution System*
Used if steam or high temperatures of hot water is available from existing central plant. Piping, manholes, underground conduit, valves, anchors, insulation
*Plumbing*
Plumbing fixtures, drains, soil waste, vent, hot and cold water piping
*Insulated Pump and Header Rooms*
Used where pumps, valves, and headers can be grouped and placed in a small insulated room to facilitate service and maintenance
*Thermal Insulation (Mechanical)*
Pipe, equipment, and duct insulation for all items other than chamber construction
*Balancing, Testing, and Adjusting*
Requirements for initial testing and balancing of air flow, brine flow, glycol flow for points in system as specified. Full operational tests of all equipment for 72-hr periods on predetermined control sequences or test programs to determine accuracy of control and equipment performance
*Electrical Systems*
Panelboards, feeders, instrument connectors, special wiring cabinets, chamber panic switch and buzzer system, seal-off wiring boxes
*Lighting Fixtures*
High intensity specialized lighting cycle and dimmer controls, control panel lighting
*Intercommunication System*
Two-way communication between chambers and control area in gallery
*Closed Circuit Television System*
Electrically operated swivel cameras in chamber with monitor screens on gallery floor control area
*Lightning Protection System*
Air terminals, ground rods, and down cables on exterior chamber roof and walls
*Standby Electric Generation*

The engineer often prepares a complete cost estimate based on the final plans and specifications. This serves as a final check on the sufficiency of available funds, as a basis for deciding whether the inclusion of deductible alternatives is necessary, and as a check on the bids by contractors.

## SPECIAL DESIGN CONSIDERATIONS

Although heat load calculations for climatic control chambers generally follow the procedures for heating, air-conditioning, and refrigeration systems, they are necessarily a great deal more extensive and exacting. Chamber construction and conditioning systems require special consideration and design, especially when the temperature, humidity, and other conditions to be produced in the chambers are extreme. Variations in these conditions, such as diurnal simulation, add to the complexity. When climatic conditions are changed in a chamber, undesirable conditions, especially in regard to humidity, can occur during the period of change. They cause unusual control problems that must be solved.

Sources of technical data and information are listed at the end of this chapter.

### Heat Load Calculations and Climatic Control Techniques

Calculations must be made for not only maximum heat loads but also for minimum loads and several points in between. When conditions are to be cycled in accordance with programs, special analysis of load-related factors must be made.

Humidity and dew point changes must be plotted, especially when conditions of temperature and humidity are to be cycled. This will indicate points of oversaturation where condensation will appear on coils, ducts, test room surfaces, or even as precipitation in the test rooms.

All psychrometric conditions must be thoroughly analyzed so that the optimum method of producing the required conditions of temperature, humidity, air circulation, vapor pressure, and so on, can be decided upon. Where conditions of high altitude or high pressure are involved, the appropriate psychrometric charts or data must be used, since psychrometric values may be substantially different than at the sea level atmospheric pressure at which many charts are standardized.

Where cooling is required to prevent overheating in chambers and high humidity is desired, humidity must be added directly into the chamber. Humidity control is usually not provided below 2°C dew point; however, if humidity control below 2°C dew point is required, it can be provided. Calculations become very involved, and many variables are introduced, all of which must be resolved. Automatic control necessarily becomes complicated, and special techniques are required. Where temperatures below freezing are to be run in the chamber, a coil defrosting system is needed. Dehumidification by desiccants can be employed to reduce frost build-up on cooling coils and provide better control of conditions.

Changes in vapor pressure created by temperature and humidity conditions must be considered when extremes are to be produced.

Frequently special techniques must be applied in the conditioning and handling of outside air (fresh air) for ventilation, respiration, and odor control, especially where extreme conditions are to be produced. Sometimes outside air must be reduced to a minimum because of the high heat load required for conditioning, and consideration must be given to reducing odors through means other than ventilation.

Many chambers require heating and/or cooling of walls, floors, and ceilings to control heat radiated from large animals, to simulate heat radiation, or to maintain surface temperatures in equilibrium with air temperatures when air temperatures are being cycled. Low chamber mass will aid in attaining rapid changes in chamber conditions.

Frequently separate consideration must be given to varying latent heat loads to ascertain the heating system and equipment that can handle them.

## Systems and Equipment Selection

Climatic control systems frequently require excessive air volumes for heat transfers, humidity control, and other functions. These must be handled in a manner that will not produce excessive air motion in the chamber.

Where extremes of temperature, humidity, or pressure are required, conditioning equipment located outside the chamber can impose extensive problems. Such problems include the necessity for thick insulation, vapor-proofing, awkward duct connections, difficulty of access for servicing, and possibility of leakage around duct penetrations through chamber walls.

Usually climatic chambers must be capable of operating under a wide range of conditions. This means that the equipment may be oversized for the lighter loads. Contrary to common belief, control of light loads may present as many problems, is not more, than control of heavy loads. The performance capability of equipment and its control system must be considered, not only at extreme conditions but at many intermediate conditions.

The direct expansion refrigerant system is not generally adaptable to a wide variation in control and loading without special refrigerant control devices, which are potential sources of maladjustment and malfunction. This is especially true with high-speed small-displacement refrigeration compressors. Extreme care must be used in the application of such devices. They can upset the normal lubrication circulation through a system and create operation that will permit liquid refrigerant to enter the return or suction gas side of a compressor and cause complete breakup of the machine. They can also produce operation under very light loading wherein compressor compression ratios are increased to the point where overheating occurs or injurious short cycling occurs.

The brine-type system is generally preferred. The brine can be glycol and water or one of a number of liquid chemical compounds having suitable low freezing points. For chemical compounds, special pumps must be used and, particularly in the case of low-boiling-point chemicals, the system must be artificially pressurized with an inert, insoluble, and noncondensing gas at the system temperatures. The brine-type system lends itself to close temperature control and provides satisfactory operation over a wide range of loading including frequent load variation.

All equipment must be selected for the atmospheric conditions under which it must operate. Materials used must have suitable expansion and contraction properties and resistance to corrosion, electrolytic action, and injury from freezing.

Air filtering can become a great problem because of hair and dust from animals. Frequently filter staging must be used. Where diseases must be considered, it is frequently necessary to use absolute or bacteria filters before exhausting air from chambers; and in some instances recirculated air must be passed through such filters. Coarse filters are placed first in the direction of air flow and are followed by one or more stages of finer filtering media. The coarse filters obviously must be replaced more frequently, and they are generally of a type that can be easily and inexpensively replaced. It is advisable to monitor the air pressure drop through each filter stage. Many good commercial devices are available for this.

The selection and arrangement of refrigeration machines, especially for the more extensive facilities, are extremely critical because of the serious consequences of failure. For the smaller installations up to machines requiring a total horsepower of 125, the reciprocating machine is generally used, but there is no hard and fast rule that can be followed. Each design must be completely scrutinized for the best selection. More than one machine is generally advisable, each with a completely separate refrigerant circuit. Interconnected crankcases for oil equalization, suction discharge, and crankcase gas equalization can readily nullify the advantage of multiple compressors. If a machine breaks up internally, metal and other particles will readily contaminate the entire system, making it inoperative.

Centrifugal machines are generally used for the large installations, although in some installations the absorption type of refrigeration unit has been preferred. Absorption machines are not capable of economically producing the extremes of temperature provided by reciprocating or centrifugal machines. The centrifugal machines have the advantages of size, rotating rather than reciprocating internal parts, efficiency, simplicity, a wide range of capacity control, and long life. They are available in multiple stages where extremely low temperatures are desired.

Extremely low temperature and humidity conditions usually require that refrigeration compressors be staged. Two alternatives are available, a compound system and a cascade system. The compound system is a multistage application where compressors are interconnected in series and use the same refrigerant system. The cascade system involves the interconnection of separate refrigerant systems in a matter that one provides a means of heat rejection (condenser) for the other. Cascade systems permit the use of a different refrigerant for each cycle of the cascade. It is most often used where the range of pressures from evaporating to final condensing conditions is greater than can be handled in a single commercial compressor. It might also be more economical than compound systems as the size of the lower temperature stages components can be reduced and the problem of balancing oil pressures between compressors is eliminated.

The heating medium for chamber conditioning must be considered from the standpoint of control and the effects of chamber conditions on the heating medium. Circulated hot water, nonfreeze solution, or liquid chemical, as the requirement may dictate, usually provide the greatest opportunity for close accurate control and freedom from injury to equipment through changes in conditions. Steam can be used but, as with water, there is the danger of freezing. Further, it is not possible to achieve as great control accuracy with steam as with a liquid heating medium.

Electric heat can be used when available in sufficient quantity; it provides many advantages including freedom from freezing. Modulation of temperature can require a difficult and expensive control system, however.

Direct-fired, hot air heating systems are seldom worth considering except for minimal facilities of the crudest type.

Humidification systems that can be used include steam injection, compressed air injection, heated pan evaporation, and water spray.

The two basic systems of control are pneumatic and electric. Electronic parts and devices are used in both basic systems, and a given system of controls may include both pneumatic and electric. The selection of the control system is of utmost importance and of necessity very complicated for an installation of any appreciable size. It must be remembered that the automatic control system cannot do the impossible and given controls will not operate with the same degree of accuracy over their entire range unless the system and equipment are readjusted for each change in controlled conditions.

### Chamber Design and Construction

Prefabricated chambers or refrigeration boxes are commercially available. When such a selection seems practical, the supplier must provide the purchaser with all engineering data covering the conditions of operation and performance.

For larger installations, all the factors covered earlier in this chapter must be given adequate consideration.

All forces acting on the chamber should be considered including outside weather air exchange, pressure differentials, changes in pressure created by changes in air conditions within the chamber, low and high pressures to simulate high or low altitudes, expansion and contraction created by temperature changes, mechanical shock, and temperature shock.

Consideration of the effects of vapor pressure on chamber insulation is of great importance. Frequently only the outside wall of a refrigerated enclosure is considered to require vapor-proofing. This contemplates that the enclosure is always in warm or heated surroundings. Climatic chambers, on the contrary, frequently are required to operate under very different conditions and require the best possible vapor-proofing, both inside and out. An example is the chamber that has been controlled at a low temperature for an extended period of time and the insulation itself has become cold. Subsequently, if the temperature and humidity are raised and an inside vapor barrier is not provided, moisture will be drawn into the insulation saturating it. Once saturated, it is extremely difficult to dry out the insulation. The matter of vapor-proofing must not only be considered for the chamber structure but also for air handling equipment, such as ductwork and cooling coil enclosures.

Chambers suffering saturation of insulation are subjected to extreme damage and structural failure when the temperature is reduced below freezing and the moisture in the insulation freezes and expands. If a chamber is built with its floor insulation on the ground and subfreezing temperatures are maintained inside over long periods, the floor insulation temperature will be reduced below freezing. This will attract ground moisture that will collect, freeze, and expand to heave the chamber floor upward. Consideration must always be given to providing heat under such floors. Heater stripes or their equivalent would also be needed around doors and other openings to prevent frost from sealing them closed.

## ARRANGING FOR CONSTRUCTION

After all plans and specifications are completed for a climatic facility the next obvious step is to arrange for its construction. Large projects will usually be constructed under contract with a well-established contractor experienced in climatic facility construction. The average building organization, although capable of constructing a number of different types of buildings including their heating, plumbing, air conditioning, refrigeration, and electric wiring, may not be experienced in, and thus cannot recognize the many pitfalls in, the construction of climatic facilities.

The design engineer can be of great help in selecting a construction organization and/or in getting together a list of competent bidders. He will be able to assess the quality of an organization's past work, experience, ability, financial stability, reputation, and follow-up reliability. Important considerations are the skills of supervising personnel who oversee specific phases of construction. However, the design engineer should be retained for continual supervisory cognizance over the construction organization and the total project.

When the research establishment itself maintains a construction department, an "in-house" construction arrangement may be possible. However, the same strictures must apply as for any commercial organization. An in-house project is usually simpler and less complicated than one for which "outside" services are desired. The engineer working with an in-house department has the responsibility for the acquisition of equipment and materials to meet his design and performance requirements. Also, the research institution must assume full responsibility for facility performance and for administration of the guarantee on equipment purchased from commercial suppliers. Such guarantees must be well defined and established at the time of purchase; too frequently performance and mechanical stability as are not clearly defined in writing. When the researcher finds himself faced with a malfunction due to equipment inadequacy, he must know to whom to turn. His procurement representative should be warned of this need.

On occasion, but certainly infrequently, construction contracts are awarded on a time-and-material (T&M) basis. With extremely close supervision of the contractor and control over his operations, this can be satisfactory, but it invariably results in questions of responsibility for equipment and system performance when malfunctions occur. Such questions can cause long delays in correcting difficulties. It is strongly recommended that, if a T&M contract is entered into, that the contractor's responsibility and guarantee be clearly understood and agreed upon in writing. A T&M contract can result in the very best in construction because the contractor does not find himself working under a monetary limitation as is the case when the contract is based on competitive bids. Time-and-material contracts are also entered into for the sole purpose of assuring that a particular highly specialized contractor does the work. This can be an important consideration in foreign and remote locations or under circumstances that limit the method of procurement. T&M contracts can be entered into on an upset price (a set maximum) basis, which approaches the fixed price type of contract. The fixed price contract guarantee is usually more closely defined. It is not often, however, that a research organization is permitted anything except competitive bidding.

## CONSTRUCTION SUPERVISION AND FINAL TEST

Once the plans and specifications are completed and a contract entered into, the engineer's role in the project should not cease. Instead, he should be retained to supervise the construction, conduct all final tests and adjustments, and perform the many other functions associated therewith. The engineer knows more about the construction of a project than anyone else and is responsible for its performance to suit the needs of the researcher. Frequently also, during construction, the engineer should inspect the construction with the researcher and resolve any problems of visualization difference. If changes are necessary, it is far better to make them at the earliest possible stage of construction.

Catalog cuts, descriptive literature, drawings, rating sheets, etc., covering all materials, equipment, devices, and work that the contractor proposes to furnish should be submitted to the engineer for review and approval prior to procurement or fabrication by the contractor. If they do not meet the requirements of the plans and specifications they will be rejected by the engineer and the contractor must make new submittals. Copies of all such submittals and detailed construction drawings are retained in the engineer's file for future reference. They frequently prove invaluable during the immediate post-construction period and in maintaining and operating the facility. Payment invoices are usually submitted by the contractor at the end of each calendar month, and they are reviewed and approved by the engineer, usually less a 10 percent retainer, to be paid upon satisfactory completion and acceptance of the work.

The engineer should carefully inspect all work, materials, and equipment at regular intervals. Interpretations of plans and specifications by all concerned are not always the same, and the engineer must act as required to straighten out any misinterpretations.

After any phase of the work has reached the stage where it is ready for test, the engineer should make sure that all tests are complete and conclusive. This applies to pressure tests on piping, electric circuits, performance of individual units, controls, construction seals, and so on. When the entire facility is completed, all acceptance tests performed by the contractor must be witnessed by the engineer, who should keep his own log, in addition to that kept by the contractor. During the test period adjustments are made and tests rerun with additional adjustments until the performance meets the plans and specifications. *Whenever possible, it is highly desirable to have operating and maintenance personnel inspect the facility during the construction stage and be on hand during all tests.*

When the facility is completed, the contractor and the engineer should thoroughly instruct all operating and maintenance personnel over a sufficiently long period of time to make sure that they are proficient in the operation and servicing of the complete facility. Written data, diagrams, rating information, instructions, parts information, and final "as built" drawings should be provided.

## OPERATION AND MAINTENANCE

Operation and maintenance effort and study should not wait until the completion of a project. Rather, the personnel who are to operate the chamber systems and the personnel who will maintain the equipment should be on hand during the construction stage, at least occasionally. During final tests and adjustments they should be on hand at all times. These observations plus copies of construction plans and specifications, copies of contractor's submittals to the engineer for approval during construction, and the complete set of equipment operating and maintenance instruction (usually furnished with parts lists, etc., after completion) are valuable. Arrangements should also be made for instruction of operating personnel. The design engineer and contractor usually are anxious to assist in such instruction as the competency and experience of the personnel who will be working in, operating, and maintaining the facility are important factors in attaining satisfactory performance.

In the simpler type of climatic facility, what little operation will be required can be done by the researcher and those who will be involved in the maintenance. It is assumed that, in the engineering and design stages, due consideration will have been made for the care of the facility after construction.

For the smaller research institutions and facilities in remote locations, it is probable that experienced operating and maintenance personnel will not be on hand and experienced personnel from other organizations will be used. An extensive facility can require round-the-clock attending personnel, but not necessarily require their continual presence in the laboratory. Frequently large institutions employ round-the-clock operators for heating, air conditioning, and other specialties who may be able to attend the climatic facility along with others of the institution.

Where facilities are very complex and involve very large refrigeration machines (500 horsepower up), an operator is usually required to check all equipment periodically and keep a continuous log. His log should include temperature, pressure, flow, humidity, and electric readings at various points that require monitoring. In many instances round- or strip-chart recorders are provided, either separate from the controllers or as integral parts thereof. Recorder charts become a part of the operator's log and are filed for reference. The operator sets up the conditioning system and all parts thereof to produce the particular conditions that are required by the researcher. In very highly sophisticated systems a large degree of automation is provided so that changes in conditions and programs can be attained by simple adjustment of the master controllers and programmers. Such adjustments are usually done by the researcher and checked by the operator. On occasion the operator must make system adjustments to attain the closeness of control that may be desired. Also, when conditions must vary over an extremely wide range, for instance from −50° to +50°F, some adjustment is usually necessary to keep controls in calibration and maintain close differentials.

The necessity for highly trained and competent operating and maintenance personnel is emphasized. Incompetent personnel can cause as much trouble, and even more, than ordinary wear and tear. It is best to have one or more of the laboratory technicians trained in and responsible for routine maintenance and adjustments of controls.

The researcher will probably have an excellent knowledge of the procedures for animal care. Animal care has a great effect on the mechanical systems. For instance, if clean conditions in the chamber are not maintained, the concentration of ammonia from animal waste can cause detrimental corrosion. Animal care is usually provided by animal caretakers. They rarely are the same as the mechanical operating and maintenance personnel as their training is very different. But the two types of personnel should assist each other.

Many institutions have large maintenance departments that can generally take care of repairs to almost any type of mechanical equipment and system. Where their knowledge of highly specialized items, such as centrifugal refrigeration machines, may be somewhat lacking, outside service organizations are called in. Frequently such organizations are operated or at least authorized by the manufacturer of the equipment. Many climatic facilities are maintained entirely by outside service organizations, either on a yearly contract basis or on "on call" basis.

Of utmost importance in maintenance is what is usually known as "preventive maintenance." This includes periodic inspections, lubrication, cleaning of all kinds, belt tightening, and adjustments. As the term implies, "preventive maintenance" will assure the best in operation by preventing breakdown or malfunction.

A supply of spare parts and items such as lubricating oil, grease, replaceable filters of all types, refrigerant gas, solution additives, and cleaning fluids should be kept on hand. A complete list of such items should be prepared by the operating and maintenance personnel. The researcher and engineer will problably need to assist with the preparation of this list. An important parts stocking consideration is the source of supply. It is impossible to maintain a stock of spare parts sufficient to take care of every eventuality. Accordingly, a concise list should be prepared that identifies where and how the parts may be procured.

## BIBLIOGRAPHY

American Society of Agricultural Engineers. 1966. Management of farm animal wastes. ASAE Publication #SP-0366. St. Joseph, Michigan.

American Society of Agricultural Engineers. 1969. Design of ventilation systems for poultry and livestock shelters. *In* Agricultural engineers yearbook. ASAE, St. Joseph, Michigan.

American Society of Agricultural Engineers. 1969. Effect of thermal environment on production, heat and moisture loss and feed and water requirements of farm livestock. *In* Agricultural engineers yearbook. ASAE, St. Joseph, Michigan.

American Society of Heating, Refrigerating, and Air-Conditioning Engineers. 1965. Environmental control for animals and plants–Physiological considerations. pp. 117–126. *In* ASHRAE guide and data book: Fundamentals and equipment for 1965 and 1966. ASHRAE, New York.

American Society of Heating, Refrigerating, and Aid-Conditioning Engineers. 1967. Environmental control for animals and plants. pp. 405–422. *In* ASHRAE guide and data book: Applications for 1966 and 1967. ASHRAE, New York.

Barker, E. V. 1960. Design and construction of animal quarters for medical education and research. J.M.E. 35(1):15–23.

Decker, H. M., L. M. Buchanan, L. B. Hall, and K. R. Goddard. 1962. Air filtration of microbial particles. Joint Publication of Dept. Health, Education and Welfare, and U.S. Dept. of the Army. P.H.S. Pub. No. 953. Government Printing Office, Washington, D.C. 43 pp.

Esmay, M. L. 1969. Principles of animal environment. A.V.I. Publishing Co., Westport, Connecticut. 325 pp.

Gardner, J. A. 1963. Considerations in waste disposal. Lab. Anim. Care 13(3): 357–368.

Holbrook, J. A., and N. R. Brewer. 1963. Planning and equipping animal operating facilities. Lab. Anim. Care 13(3):424–430.

National Academy of Sciences. 1962. Animal facilities in medical research (A preliminary study). Institute of Laboratory Animal Resources, National Research Council, Washington, D.C.

U.S. Atomic Energy Commission. 1948. Qualitative specifications for animal laboratories for experimental work with radioactive materials to Brookhaven National Laboratory. Report by Arthur D. Little, Inc. for U.S. Atomic Energy Commission, Technical Information Service, Oak Ridge, Tennessee.

U.S. Department of Health, Education and Welfare. 1963. Guide for laboratory animal facilities and care. Public Health Service. Publ. 1024. Government Printing Office, Washington, D.C. 33 pp.

# 8

# RESEARCH UNDER FIELD CONDITIONS

## INTRODUCTION

The response of an animal to a given physical environment—be it measured in units of performance or change(s) in certain physiological traits—is a reflection of the animal's reaction to both the current environment and that to which it was exposed previously (McDowell, 1968; McDowell and Cundiff, 1967). Recognition of this fact becomes very important in the design and execution of experiments under field conditions. Under the control of a laboratory, a standardized preexperimental period, which allows some degree of adaptation to a set of environmental conditions, is desirable and often practiced; but this is not possible in the field. A partial degree of stabilization of the physical environment, i.e., uniform feeding and housing, is desirable but not always feasible; therefore, greater precision in planning is often required for experiments in the field than in the laboratory.

The physical environment is typically referred to as the prevailing climatic conditions, but the climatic elements constitute only a portion of the total environment. Nevertheless, those grouped under "climate" are frequently regarded as the most important of the environmental variables, because they influence the major portion of the others in some fashion. The elements of climate not only have direct effects on the animal but also influence the growth and quality of feedstuffs, the incidence of diseases and parasites, the efficiency of labor, and many other factors. Therefore, to obtain maximum economic returns from animals, man seeks means of ameliorating the adverse effects of climate. Many examples of successful efforts in this direction may be cited in the temperate zones of the world. The development of shelters that protect both man and his animals from severe winter cold is one. Much less progress has been made in the tropical and subtropical zones, but evidence is accumulating to show that the problems of these zones are no more insoluble than those of other regions.

The preceding chapters cover many of the factors that must be considered in determining how animals respond to various environmental conditions and some of the methods of measurement. Much of the information is oriented to laboratory studies conducted under well-controlled conditions in psychrometric chambers or similar facilities. But in the end, almost all physiological studies must be related to animal productivity that can be measured under field situations. While basic research dealing with the response of animals to various climatic and other environmental variables is important in advancing man's knowledge, the ultimate objective of most research with domestic animals is the improvement of performance under practical commercial conditions. Results from laboratory studies have contributed to an understanding of how animals respond under field conditions, but for the most part extreme caution must be exercised in interpreting the laboratory findings in terms of expected responses of animals when they are exposed to less controlled environments. Even so, knowledge of the responses of animals to various climatic conditions should be utilized in planning field studies; but the uncontrollable factors that are inevitable in field work must also be given due weight. The results obtained from field studies are seldom as repeatable as those in laboratory experiments.

"Field," in this context refers to conditions in which there is little effort to control the individual environmental variables. Field experiments can, however, range from simple to complex. In the simplest experiments man usually makes no effort to ameliorate the situation that the animal experiences, e.g., open range grazing. On the other hand, when animals are kept in relatively close confinement, with some effort made to modulate or decrease the direct impact of many elements of the environment, the design may be more sophisticated. The restriction of cattle to small lots or confined housing as often practiced in rearing of swine and poultry would be examples. Although findings from experiments conducted under well-controlled conditions in laboratories and observations made relative to animal performance in different climates are beneficial aspects of technology, extensive research in "practical situations" is also required. This refers to the need to identify more fully the elements involved in the functioning of animals, as well as those that influence the optimum efficiency of animals in various environments, in order that man may continue his efforts to make adjustments to ensure maximum economic return.

The objective of field research as discussed here is to provide guidelines for possible experimentation toward determining means of increasing the economic efficiency of animals in various situations. Within this context are two major areas: (1) determining the relative importance of the various bioclimatic factors as restraints to animal performance; and (2) development of methodology for environmental modifications that will result in more nearly optimum productive efficiency for specific environments. Field research is, therefore, required as a means of determining the constraints of environment relative to productive performance. Performance is defined as the products that man desires from his animals and the economic returns they will yield. Since modifications of the environment usually require high inputs of capital, they must be justified in terms of measurable changes in productivity resulting in increased net economic returns (McDowell, 1967).

It is important to recognize that environmental–animal functional relationships are more difficult to assess precisely in the field than in a laboratory with controlled conditions. Therefore, experiments designed to elucidate the interrelationships among physiological functions and prevailing climatic conditions may not be meaningful or even warranted in the field. An example might be the association of ambient temperature to adrenal cortical secretion rates. This type of work would be more appropriate for a laboratory since field results would not be specific enough to give a thorough understanding of cause-and-effect relationships in this type of study.

This does not imply that the field researcher should not approach his tasks without a sense of pride. The mere fact that the studies are carried out in natural environments or where only limited modifications have been made does not mean that it is not important nor a challenging undertaking. In the case of laboratory work, using climatic chambers, the emphasis is often on equipment and experimental design, whereas in the field sophistication is required in the design of the experiment, the experimental procedures, and the methods of carrying them out. Under most field situations, a wide range of environmental variables must be either controlled or measured or the animals must be exposed in such a fashion that the environmental variables not under study do not influence them or affect all in the same fashion. The necessity for close supervision of experimental procedures and a consideration of a wide range of environmental variables means that the field research worker must be extremely cautious in his design and equally careful in collecting data, supervising labor, and other factors that determine whether or not the results can be effectively interpreted.

Field research may be either basic or applied. The normal definition of basic research in this case would be the testing of laboratory technology in the field without indications of immediate applicability. It is often necessary to take laboratory findings and test them under field conditions for determination of their reproducibility under a wider range of environmental variables. Applied research refers to that designed to improve the efficiency of animal functioning. Normally, this type of work can find immediate application to procedures for animal management, construction of housing, or techniques used for modification of environment and expected to result in improved economic return from the animal.

A frequent fallacy in the conduct of field research is the inclusion of too many variables. Emphasis should be placed on a design as simple as possible in order to provide the information required. A minimum would be the study of a single climatic variable and a single physiological variable. The response to any single environmental factor is likely to change if other environmental factors change. Therefore, field studies are usually carried out in the region where the results are to be applied.

## CLIMATIC CONDITIONS IN THE FIELD

For our purposes, climate is defined as the sum total of the weather conditions that occur in a specific location or within a region over a designated period of time. Even in a local area, weather conditions vary by month, season, and year. However, if sufficient data are available, it is possible to define the most probable ranges in the climatic variables that will be experienced in a given area.

The overriding climatic factor that affects the physiological function of animals is temperature. Local climates are characterized by variation in temperatures that may range from a minimum of approximately 6°C to as much as 16° or 20°C under some very arid conditions. To simplify the description of climates, frequently a mean temperature is derived. The values can be used to select or designate climates that are likely to have adverse effects upon the function and productivity of animals. For most domestic animals, a mean daily temperature, ranging between 10° and 20°C, would be unlikely to place them under thermal stress. This is based on the premise that it is usually difficult or impossible to differentiate between the responses of different animals to climatic variables within this temperature range. The range is sometimes referred to as the "zone of thermal neutrality," since it is assumed that within this range of temperatures animals are not required to expend appreciable amounts of energy to keep their gains and losses of heat in balance. Conditions within the 10°–20°C range are not likely to impose sufficient stress for animal differences to be detectable with any degree of reliability (except in the newborn, or in a very young animal with its much narrower zone). If the temperature falls below a daily mean of 10°C, it is likely that minimum temperatures, at least, will drop below the comfort zone for most adult animals and far below for young animals; thus efforts of some type will be required by the animal to maintain its body temperature. Contrarily, temperatures above a mean daily of 20°C are likely to result in conditions where the animal must exert efforts to rid itself of excess heat. The range in temperatures from 20° to 30°C may be described as a zone of warm temperatures where conditions will be difficult for the animal but not necessarily intolerable. When mean daily temperatures exceed 30°C, conditions may become such that animals are unable to maintain satisfactory physiological functions, therefore, deterioration in the status of the animal and its performance are likely to occur.

As mean daily temperatures fall outside the comfort zones described, other climatic variables assume greater significance in the homeostasis of the animal. The water vapor content of the air is frequently associated with temperature. In relation to animal responses under field conditions, the amount of water vapor in the air is best described by dew point, wet bulb temperature, or vapor pressure. These are deemed more satisfactory for characterizing the field environment than relative humidity since relative humidity expresses a ratio to a given dry bulb temperature. Hence, when the air temperature is in a continuous state of flux, relative humidity does not provide a suitable measure of average humidity conditions. The other measures are better for estimating the temperature–humidity conditions because they may influence animal performance.

As ambient temperature either rises or falls outside the zone of comfort, the effect of humidity is increasingly important as an interacting factor (Barrada, 1957). The reason is that, at high temperatures, high humidities tend to restrict heat loss and, therefore, make it more difficult for the animal to maintain thermal balance. At low temperatures, a high humidity tends to increase the ability of the air to carry heat away from the animal body; hence, increasing the rate of heat loss also makes more difficult the animals maintenance of thermal equilibrium.

Other climatic factors that interact with ambient temperature to impose stress on the animal are wind velocity and solar radiation (Bond, 1967). Solar radiation at low temperatures will tend to raise the temperature of the animal's surroundings, thereby decreasing the probability that the animal will be in thermal difficulty. The heat gained from the solar radiation will partially compensate for any tendency toward excessive heat loss to cold air. At high temperatures, solar radiation serves to increase the severity of reaction to a given ambient temperature. Wind velocity may have a reverse effect to solar radiation since at high temperature, increased wind velocity tends to increase the rate of movement of air away from the animal body, accelerating the rate of evaporation from the body surface. This holds true until ambient temperature exceeds skin temperature, after which the gain from the heated air may more than compensate for the additional evaporative loss. At low temperatures, high wind velocities tend to accelerate heat loss, thereby making any given cold temperature more severe (Williams, 1967).

Another major climatic factor influencing the natural environment is precipitation. If animals are exposed to rain at high air temperatures, the rate of heat loss may increase as a result of evaporation following the rain. When the rain is cold, the chilling effect may be detrimental to the animal, particularly if the animal has elevated skin and body temperatures. At low temperatures, rain tends to chill the animal and increases the severity of response to cold conditions. Snow does not usually create difficulties as serious as rain since it frequently does not melt on striking the pelage of the animal.

## CHARACTERIZATION OF THE ENVIRONMENT

For general purposes, temperature, humidity, solar radiation, barometric pressure, air movement, and precipitation define the climate adequately to assess the stress imposed on animals. Measurement of these parameters is simple and needs relatively simple instruments in the field.

The frequency of observations required depends on the duration of the anticipated experiment and the precision of measure, both in relation to animal response and the prevailing climatic conditions. If the association of a variable, such as milk yield, to climatic conditions is determined on a periodic basis, i.e., weekly or monthly, some general description of the climatic conditions as they exist for an average day would suffice. In contrast, if the animal response is measured as body temperature, respiration rate, and so on, then recordings of the climatic variable should correspond to at least the frequency of the observations.

If there is interest in studying the relationship of extreme high or low temperature conditions to animal response, the exploration should be initiated so as to correspond to these periods. The extent of the contemplated extremes and the precision with which the animal variables are to be measured determines the number of climatic variables that should be considered. In general, the number will vary but is correlated with the extremes. If the dew point of a given area is consistent throughout the day or season, this might be characterized in a very general way, such as a single observation per day or a mean monthly value. The same holds for solar radiation and often also for air movement. Of course, if adequate data are not already available for such determinations, then information on the climatic variables must be recorded along with the animal observations.

The necessity of advanced planning for characterization of the environment cannot be overemphasized, particularly with respect to the climatic elements. In the past, most researchers conducting field experiments have tended to use the approach of one of two extremes—either too few data or many more than necessary. On the extreme low side the tendency has been to describe climatic conditions by some clasic systems, such as that proposed by Köppen (Köppen and Geiger, 1936) or Thornthwaite (1948). These classifications have been found to apply fairly well to plants if general cultural practices are being tested, but anything finer than the major subdividions does not accord well with the direct climatic significance for animal management. Also, experiments with animals are usually of short duration—hours, days, or perhaps weeks—which further restricts the usefulness of these classifications for studying the relationships to animal response.

At the other extreme are the researchers who make continuous recordings of certain climatic elements. Large volumes of data are amassed, but if the animal trait(s) is (are) measured once or twice per day, less frequent recording or data from the local weather station may prove satisfactory (Hahn and McQuigg, 1969; Johnson and Givens, 1962). It is desirable, however, that the meteorological data be collected as near as possible to the area where the animals are to be observed (Lee, 1953).

In most cases, it is best to measure both the dry and wet bulb temperatures since from these may be determined the relative and absolute humidity. The measurement of solar radiation requires more elaborate equipment (see Chapter 2) but, for reasons that will be discussed below, it will usually be satisfactory to use data from a meteorological station situated in the area. If such is not available, then hours of sunshine are sufficient and these may be recorded with a sunshine recorder. Air movement obtained with a vane or cup anemometer located at the height of the animal under investigation will suffice. Rainfall may be measured with a simple rain gauge and the precipitation for 24 hr recorded.

Dry and wet bulb temperatures, solar radiation or hours of sunshine, air movement, and precipitation patterns for an area near the animals will suffice for assessment of climatic stress. If equipment for collecting meteorological data is restricted, priority should be given to that needed for recording dry and wet bulb temperatures since air temperature and air humidity are closely related in terms of causing physiological stress. Table 8.1 illustrates the relative importance of wet bulb and dry bulb temperatures in creating thermal stress conditions for different species. The weighting required varies with the species; nevertheless, both temperature and humidity are important to all species. Even though both air temperature and humidity data are collected, the problem of expressing their combined effect in a single figure remains. One

**TABLE 8.1 Relative Importance of Wet Bulb and Dry Bulb Temperature in Causing Thermal Stress in Different Species[a]**

| | Weighting Factor (%) | | |
|---|---|---|---|
| Species | Wet Bulb | Dry Bulb | Ratio Wet: Dry |
| Man | 85 | 15 | 5.7 |
| Cattle | 65 | 35 | 1.9 |
| Young pig | 35 | 65 | 0.5 |

[a]Adopted from Findlay (1968).

possibility is the construction of lines of equal effect on psychrometric charts, for rectal temperature and respiration rate of cattle (Barrada, 1957). Bianca (1962) recommended the equation ($db \times 0.35$) + ($wb \times 0.65$) to derive a weighted value for correlation studies, where $db$ = dry bulb and $wb$ = wet bulb, both in degrees centrigrade. Others (Hahn and McQuigg, 1967; Kibler, 1964) have related the U.S. Weather Bureau temperature–humidity index (THI) to some physiological responses. Preliminary information suggests that THI is reasonably suitable for estimating changes in performance for man and cattle. The THI values may be calculated from dry bulb temperature plus a measure of humidity, such as THI = $0.72\ (C_{db} + C_{wb}) + 40.6$.

Research with man showed little discomfort from temperature and humidity when the THI was 70 or below, but half the humans sampled were uncomfortable at 75 depending upon the level of physical effort. All felt some discomfort at a THI of 79. The efficiency of workers declined when the index in the working area reached 80 or above.

The use of THI in relation to animal responses has not been extensively investigated, but studies conducted at the University of Missouri (Hahn and McQuigg, 1967; Johnson *et al.*, 1963; Kibler, 1964) showed losses in milk production for lactating cows were clearly related to changes in THI. The cows seemed to have little discomfort at 70, but milk yield and feed intake were depressed at 75. Cattle of all ages showed measurable degrees of discomfort with THI 78 or above. The discomfort became acute as the index increased. The variation in the importance of humidity for different species (Table 8.1) suggests the best constants for weightings of temperature may fluctuate, depending on the species.

Figures 8.1 and 8.2 demonstrate utilization of THI for scaling the temperature–humidity conditions in both the arid tropics—Jacobabad (Pakistan) and Khartoum (Sudan)—and the humid tropics—Calabozo, Caracas, Maracaibo, and Maracay (Venezuela). When viewed initially, the wide contrast in the temperature and relative humidity values for Khartoum and Jacobabad (Figure 8.1) make the expected influence of these elements on animals difficult to perceive. The mean monthly temperatures of Jacobabad, Pakistan, indicate air temperature would not be likely to have a direct relationship with animal function that could be detected with confidence. On the other hand, when the high humidity at this location is combined with a fairly warm temperature, there are indications that the climatic conditions would have a measurable influence on most animals for about 8 months of the year. The relative humidity, coupled with the warm temperature at Jacobabad,

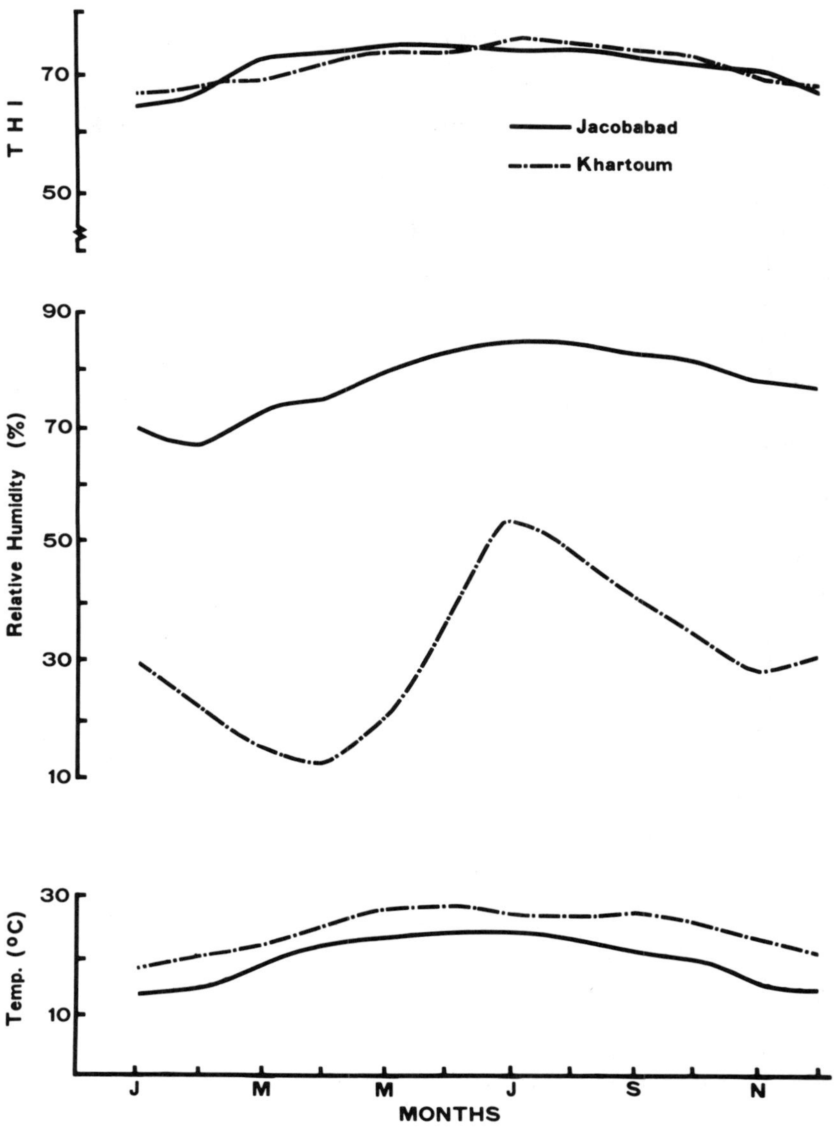


FIGURE 8.1. Comparison of monthly temperature, percent relative humidity, and temperature-humidity index values (THI) for Jacobabad (Pakistan) and Khartoum (Sudan).

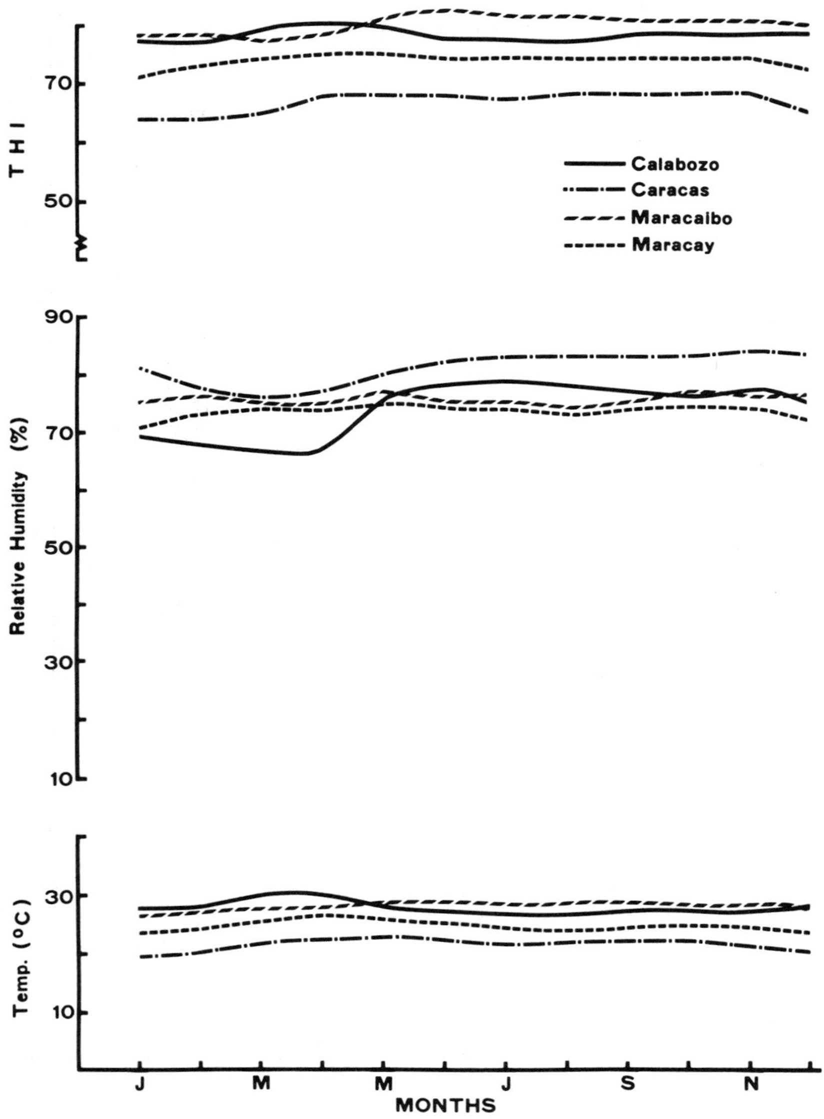


FIGURE 8.2. Comparison of temperature, percent relative humidity, and temperature-humidity index values for four locations in Venezuela.

also suggests the problems of ecto- and endoparasites could be high. Although the relative humidity for Khartoum is lower and much more variable than for Jacobabad, the monthly THI's show that conditions at this location are likely to have a detectable influence on animals for 6 months or more of the year. The extent of the effect would, of course, be dependent to a degree upon age, level of feeding, and general level of performance of the animals.

As might be expected, THI's do not modify the picture of the prevailing temperature and humidity conditions in the humid tropics (Figure 8.2). Even so, the use of THI provides an additional refinement to the expected temperature–humidity conditions, particularly in locations with definite wet and dry seasons, such as Calabozo, Venezuela (Figure 8.2). The calculation of THI seems worth the effort in (1) determining the types of research that may be useful toward changing animal performance, and (2) determining the likely application of results of research among locations.

It has been suggested that still a more satisfactory method for classifying climates for purposes of field research would be to derive an "equivalent temperature" (ET) scale, which would be for animals somewhat like the evapo-transpiration index for plants (Thornthwaite, 1948). The proposed equation for ET is:

$$\mathrm{ET} = K_a\,(db) + K_b\,(rh) + K_c\,(sr)$$

where
ET = equivalent temperature
$db$ = dry bulb temperature
$rh$ = relative humidity
$sr$ = solar radiation
$K_a$, $K_b$, and $K_c$ = constants depending on the units employed

The expected advantage of the ET scale over THI would be an additional weighting for incorporation of solar radiation. The disadvantage is in obtaining suitable evaluations of the amount of thermal radiation received by the animal. The standard instrument for collection of solar radiation data is a pyranometer (Chapter 2), which measures the radiant energy falling on a horizontal surface in langleys. The magnitude of the langleys may be adequate for general classifications of intensity, but more often than not the radiation load indicated by the pyranometer does not give a satisfactory estimate of the total radiant energy received by the animal. This is because the entire body surface does not receive the full intensity of the radiation; therefore, it is difficult to determine the net load of direct radiation to which the animal is subjected. In addition, the reflected radiation load (radiation received from the ground or surrounding objects) is often an important source of thermal radiation, yet this segment is not measured directly by the pyronometer (Bond *et al.*, 1967). There are instruments that give a better evaluation of the total radiant energy (Chapter 2), but these are not in common use. Thus, the current state of technology is such that attempting to obtain precise ET values for a location does not seem warranted.

It would be desirable to have a scale for proportionate weightings of the climatic variables enumerated as well as wind velocity, but unfortunately the constants or values for proportionate weightings that would be needed are not available. Hence, many times researchers have used rather elaborate equipment to collect meteorological data only to find that, in using them they must resort largely to generalizations (Johnson and Givens, 1962; Johnson *et al.*, 1962; Johnston *et al.*, 1959), e.g., mean daily dry bulb temperature, mean daily dew point, total langleys of radiation per day, and average wind velocity for a 24-hr period. Caution should, therefore, be exercised to ensure investments in equipment and time for collection of extensive meteorological data are fully warranted.

## Environmental Profiles

Since the total physical environment influences the responses of animals, the assessment of the likely impact should be more extensive than just the climatic elements. Physical environment as used here refers to the elements that man usually has little control over, i.e., rainfall distribution, soil fertility, soil pH, and to some extent the prevailing diseases and parasites, both endo and ecto.

Indeed, it would be highly desirable to have as complete a picture as possible on all elements of the physical environment but, more often than not, less must be accepted. Even rather crude assessments would be beneficial in the planning and conduct of research. The profile that can be developed for any location will depend on the extent of the information sources. If climatic data are unavailable for an area, examination of world climatic maps in one of several good atlases would be a start. Soil classification maps and charts on patterns of rainfall distribution would also be helpful. Classification of the area by Köppen's system (Köppen and Geiger, 1936) of classification would be useful. On-the-spot surveys to include observations on the types of animals, the uses being made of them, the degree of fatness, and the quality of the crops could be quite useful in identifying some of the fluctuations in various environmental factors that would impose some degree of stress on the animal. Interviews with local residents regarding animal diseases and parasites, age of marketing of the livestock and cropping patterns, identifying grasses native to the area and the approximate crop yields add further to the profile (McDowell, 1966a, 1971). The elevation, latitudes, the proximity to mountains and their orientation, and the distance from a large body of water are additional factors that would be helpful in extrapolating information from other areas that have been more extensively characterized.

But, to pinpoint likely problems influencing the performance of animals, much more information is highly desirable. The approximate sequence of priorities that should be used in attempting to gather further refinements of the profile follow (McDowell, 1971):

1. Mean monthly temperatures will aid in determining the variations as well as the extremes that are likely to occur during the year.

2. Mean monthly maximum and minimum temperatures will serve to further characterize the extremes as well as to derive estimates of the magnitude of diurnal heating and cooling.

3. Monthly precipitation, coupled with temperature data, will aid in determining the likely growing season for plants, estimating the rate of plant growth and maturity, determining the quality and quantity of food supplies and how these are likely to fluctuate, determining the extent of the period(s) of shortages of moisture, and provide general estimates of prevailing humidity conditions. Knowledge of the rainfall in a given area, including its intensity and duration, together with the temperature of the air and its movement will permit predictions of the mitigating effect that the rain would have on different animal coats.

4. Some measure of mean monthly values for humidity, wet bulb temperature, dew point, vapor pressure, or even relative humidity will aid in decisions on studies of the type and location of housing as well as in estimating the prevalence of disease and parasitism.

5. The mean monthly rate of air movement and prevailing direction—particularly with respect to duration of extremes ($<5$ km/hr or $>25$ km/hr)—will be helpful in planning research on the value of mechanical means of increasing air flow or on decisions regarding the need for protection of animals from prevailing high velocities.

6. Mean monthly radiation estimates, obtained either from direct measurements, hours of sunshine, or cloud cover, will also be of value

in planning research to test the effects of shades for animals. Estimates of radiation will be useful in assessing the extent of the total thermal heat load on the animal.

7. Some knowledge of the soil characteristics will prove helpful in estimating feed supplies and in determining if mineral deficiencies are likely to be confounding factors in characterizing animal responses.

8. Information on the incidences of prevailing diseases and parasites will be essential in determining the control measures required for preventing these factors from becoming the overriding element in the animal's response to the environmental variables under test.

9. Information on the prevailing photoperiod during various seasons will be particularly desirable in conjunction with experiments on poultry and turkeys and, to some extent, for sheep and buffalo. It is desirable to include length of photoperiod as part of the general description of the climatic environment.

10. Calculation of THI by months will be a means of describing with precision seasonal trends that are likely to be important in scheduling of experiments.

11. To best characterize the climatic environment, it would be valuable to derive estimates of the percent of time (day or longer periods) that the climatic conditions are expected to fall outside the comfort or neutral zone for the animals. For thermal stress, this is usually characterized as a proportion of time the dry bulb temperature would be expected to exceed 27°C; dew point temperature would exceed 21°C; solar radiation (langley per day) would equal or exceed 700 langley per day; the time rate of air movement would be $<5$ km/hr or $>25$ km/hr; and the number of days per month when rainfall might be expected. Evidence accrued to date indicates that the length of time the elements of climate exceed the limits enumerated will largely determine whether these variables are likely to have a detectable influence on the performance of animals (Johnston *et al.*, 1959). For example, at two locations or for a period at the same location, the mean daily dry bulb temperature may be as high as 28°C, but the number of hours when it exceeded 27°C may vary from 6 to nearly 24. For the short time span of about 6 hr, the maximum temperature might be high, perhaps 35°C or even higher, but the minimum temperature would most probably be relatively low. Under such conditions, significant changes in animal responses could not be detected if measured once in 24 hr. In contrast, where the ambient temperature may exceed 27°C for most of the 24-hr period, significant changes in animal performance might occur as compared to some cooler period; therefore, the frequency of measurement required to characterize the animal response for a day would be less.

One of the primary elements of the animals' environment not included in the listing is the food supply. Knowledge of currently available food supplies, both in terms of quantity and quality, are primary requisites for planning satisfactory experimentation. Often, in warm climates or during summer periods in the higher latitudes, animal responses attributable to changes in climate have been almost entirely confounded with fluctuations in the quality and quantity of food the animal ingests. When the ambient temperature is high, plants mature quite rapidly. With advancing maturity the energy that the animal derives per unit of plant ingested may decrease markedly; thus, the major response of the animal to the environment results from a decrease in energy intake. When the food comes from mature plants, usually the heat production resulting from the processes of digestion increases. This further confuses interpretation of the animal's true response to the prevailing climatic regime.

The initial reaction to the development of an extensive profile of the likely physical environment is a pessimistic one; e.g., is it worth the effort? The counter is that defining the expectancy of the environment determines the feasibility of the proposed research and brings about recognition of the elements, i.e., food supplies that must be standardized during the experimental period in order to assess the influence of the element(s) under consideration.

Certain complementary data that should be included will be helpful in interpretating research results and in drawing inferences for application to similar situations. These include (1) housing arrangements; (2) space allocated per animal—particularly where the animals are confined; (3) the height of the shelter along with its orientation in relation to the prevailing wind direction and the type of roof covering; (4) the nutritional level prevailing before and during the experiment; (5) the frequency of feeding; (6) the health status of the animal, if it is likely to be in a situation where parasitism may be a problem; (7) the age of the animal; and (8) animal relationships, e.g., identical twins, paternal-sibs, or random relationships. Additional factors that should be considered will become evident as other topics are discussed.

### Meteorological Data Requirements

The length of time for accumulation of meteorological data and the frequency of recording within period to establish a satisfactory profile largely depends on the precision desired and the location. Monthly mean dry bulb temperatures, derived by averaging daily means from one year's data, will give a predictability for most locations that is about 30 percent repeatable. Monthly means based on 5 years of data will usually have a standard deviation that will encompass the average conditions 66 percent of the time. The standard errors of the means from 10-year results will be repeatable with about 95 percent confidence, but the variance will be reduced very little by adding 20–40 more years of data (American Institute of Architects, 1949; U.S. Weather Bureau, 1955, 1956).

Average temperatures for a 24-hr period, based on either one half the sum of the maximum and minimum temperatures or a mean from more frequent observations, are much less repeatable, of course, than monthly means. In fact, it is virtually impossible to establish mean daily averages that would serve for a day-by-day prediction or even for weeks, beyond about 50 percent reliability. Nevertheless, with 5 years of data such values generally fall within a workable range (±3°C). It is often desirable to know the daily diurnal variation or the day–night differential. This can never be established with a high degree of precision, but predictions within a reasonable range can usually be determined from about 5 years of data.

Precise predictions of patterns of humidity, wind velocity, solar radiation, cloud cover or precipitation are more difficult than for dry bulb temperature; nevertheless, reasonable estimates of expectancy, at least by months, can be made from several years of data.

Near the equator and the poles the repeatability of weather conditions are higher than for the midlatitudes. Conditions in the midcontinent regions are the least predictable (Critchfield, 1966). However, continental areas sheltered or influenced by high mountains may have conditions as repeatable as for coastal areas. Except for storms or extreme disturbances, the climatic conditions for coastal areas are repeatable within fairly narrow ranges. In spite of all the factors that influence the climatic environment, estimates of the extent of the extremes that might be expected to influence animal behavior and performance can be established in a satisfactory fashion with several years of meteorological data. It must be kept in mind that these estimates will serve for a general description of the climatic environment, but they will not be precise enough to expect significant correlations with animal responses. Therefore, meteorological data collection must at least parallel animal observations. Preferably, the meteorological data should be more extensive. For instance, the level of ambient temperature for a previous period (1–2 days) will usually have a higher correlation with variations in milk production than the temperature conditions on the day production is measured. The reason is largely that environmental conditions have had an earlier influence on feed intake.

In Chapter 2 are described equipment of various degrees of sophistication for collecting meteorological data. Selection of equipment depends on the type and duration of the planned measurements of animal responses (Lee, 1953). Thus, in the planning and conduct of research experiments in the field, careful consideration should be given to describing the environment; but the extent of the description of the influences needed will depend on the objectives of the experiments.

It is evident that not only should the level of intensity of the various components of climate be measured but also their duration if assessments are to be made of the effects of that climate on the animal. Many of the effects will be of short duration, e.g., the effect of solar radiation on skin temperature. Others will develop over a relatively long term, e.g., the influence on the estrous cycle in females or decreased spermatogenesis in males. Also in assaying the impact of climatic stress on most animals, it must be borne in mind they are homeotherms. If, for example, the prevailing environment does not cause a change in body temperature for a considerable time, it is doubtful that the animals will be distressed to a significant degree. To the physiologist the displacement of body temperature represents a displacement in homeothermy, but to a livestock producer the animal under stress is one whose performance is depressed below the expected, no matter what its other reactions. Thus, in field research the measure(s) used to assess stress become exceedingly important.

## PHYSIOLOGICAL FUNCTIONS AND THEIR MEASUREMENT

In earlier chapters, techniques for identifying responses by animals to various stresses that influence homeothermy are described. Because of the usual wide fluctuation in field environments, many of these techniques cannot always be utilized satisfactorily. For instance, blood eosinophils of cattle have a wide variation among individuals and a low within-animal repeatability even under closely controlled conditions; therefore, significant correlations between eosinophil counts are likely to have a near-zero correlation with changes in air temperature in the field environment (Ellmore, 1954; Bianca, 1965).

Although stresses may be imposed on the organism in the field environment by any or all the elements of the climate acting directly or indirectly, the only way in which the stressed can be judged on warm-blooded animals is to measure the parameters of dysfunction of homeothermy. The purpose of this section is to discuss further the precision of measurement, the magnitude of error variance that may be expected, and the necessity for inclusion of allied observations to fully substantiate the observed responses.

### Productive Functions

To relate production functions to environmental factors, there must be an awareness that the accuracy of measurement is important because these are expressions of the net interaction effects of many physiological functions. Measures of production are also influenced by inherent differences among animals, and there are certain degrees of procedural variability. However, not every environmental abnormality results in a noticeable change, for example, in the daily milk yield of cattle. Further, it is unlikely that some environmental change with no clearly detectable immediate effect would seriously influence the total yield for a lactation period. In seeking to define a point of disturbance based partially on a change in daily milk yield, it is necessary to seek a limit far enough above or below the usual variations to distinguish differences due to changes in the environment. For Holsteins, as well as perhaps all other groups, daily yields vary about ±7 percent even in standardized environments; therefore the expected change must exceed these limits to be detectably associated with recorded changes in the environment. This percentage is 2.326 times the standard deviation between 2-day averages of milk yields (McDaniel *et al.*, 1959). Thus, to obtain a "treatment" effect, some reasonable prediction of the variation in performance must be considered.

Another factor that is frequently slighted in planning is the number of observations required to establish differences with a reasonable degree of confidence. To have some appreciation of the size of the expected or predicted error variance would be helpful. Also, if repeated observations are made on the same animal, some estimate of the likely repeatability is highly desirable in the planning process. If the commonly accepted standard deviations for milk yield, the estimated heritability of milk yield, and the calculated repeatability of lactation milk yield are used, about 21,490 cows, each with a single production record, would be required for 95 percent confidence of detecting a 45-kg difference in yield at $<0.05$ percent probability, and 36,470 records for $<0.01$ percent probability with a confidence level of 99 percent. The numbers required, of course, vary with the magnitude of the difference. Even with differences of 454 kg per lactation, 215 records would be required for 95 percent confidence and 365 records for

99 percent confidence that the differences were really attributable to the "treatment effects."

Similarly, body weights have a certain degree of "normal" variability depending on when the weight was taken in relation to the time of ingestion of feed and water. The weight of a ruminant may vary up to 5 percent within a 24-hr period, due to fill. Even when weights are repeated on subsequent days at the same hour—with all other variables nearly constant—variability will be on the order of ±1–2 percent of total body weight. In this case it becomes a matter of using predicted performance of animals to determine whether or not the expected treatment will bring about differences that will exceed reasonably well-established variations. Therefore, when there is an interest in the relationship of performance traits to climatic variables, it is clear that in most situations undertaking field research is not feasible unless the climatic variables are expected to vary enough to bring about stresses upon the animals that will be reflected in physiological functions over and above the range of the usual variations. On the other hand, some animal behavioral modifications, such as changes in feed intake or seeking shelter from extremes of wind or high temperature, are closely related to climatic variables; thus, they need not be considered as having as much variability.

Even a measure such as body temperature is influenced to a significant degree by a diurnal rhythm. Hence, to compare the body temperature of one animal in the early morning to that of another during mid-afternoon would not be valid. The same is true in relation to intervals following feeding. If two or more animals are to be compared, certainly the variables, whether they be rectal temperature, respiration rate, pulse rate, or some other change in body function, the measurements must be made in a short-time span.

Still another aspect important in field research and bioclimatology is the "lag time," defined as a difference between the occurrence of the most extreme in the climatic variable(s) and the maximum change or effect that can be detected in the animal. For instance, it has been found that the relationship of maximum ambient temperature or even the average for several hours during the day, and milk yield, is rather low. The ambient conditions of the previous day usually have a higher correlation, and maximum relationships are attained in most instances with about a 2-day differential. Even body temperature, as measured rectally, often does not reach its peak until 2–4 hr after maximum changes in ambient conditions have occurred. Other examples could be cited, but the central point is to emphasize the need for characterizing the environment over a long enough period to ensure assessment of the relationships intended.

Importance of lag time depends on the animal observations employed. Feed intake and respiration rate, for example, may be classed largely as the results of "moment function," meaning that the changes correspond closely with the fluctuations in the environment. These measures should, therefore, be determined with as close relation to the prevailing ambient conditions as possible.

## Body Temperature

Measures of body temperature should be recorded in relation to the intent of the experiment. Most frequently the means of describing body temperature is rectal temperature measured either by thermocouple or rectal thermometer. Temperature taken in the rumen, blood, by telemetry through implants in organs, or at the tympanic membrane of the inner ear may be used just as effectively (Guidry and McDowell, 1966).

If the animal is confined during the period of observation, rectal temperature is the easiest and most rapid means. Generally, rectal temperatures recorded over several hours or once per day will have a high correlation with the averages for the other measures of body temperature enumerated. But if the objective of the experiment is to evaluate speed of response or minute fluctuations, it would be highly desirable to put forth the extra effort required to employ more sophisticated equipment for obtaining temperatures from a more sensitive location, such as the tympanic membrane. If a cow ingested 100 liters of water that was within ±3°C of the "body" temperature, the impact of the differences in temperature would very likely not be detected in rectal temperature, whereas a change in tympanic temperature would be detectable within 2 min after the ingestion of the water (Guidry and McDowell, 1966).

## Heat Production

The heat production of the animal's body is another function that is closely related to level of nutrition, the type of foods, time after eating, the climatic conditions, and numerous other factors. Certainly efforts should be made to have observations on experimental animals and control groups coincide as nearly as possible, particularly with respect to time after feeding. The schedule of availability of water and the level of roughage feeding are all factors of importance in making comparisons of heat production.

### Body Fluid Composition and Endocrine Functions

Thus far, studies of body fluids such as corticoid secretions have not shown much promise of having significant associations with variations in climatic variables that normally occur under field conditions. The repeatability of observations on the same animal for such items as blood ketones, glucose, red cells, white cells, eosinophils, erythrocytes, creatinine, cholesterol, and rumen volatile fatty acid levels are rather low even under carefully controlled regimes (McDowell, 1958). These components are also markedly influenced by food intake; therefore, studies of their changes should be largely restricted to closer controlled conditions than are usually feasible under field situations. This does not entirely discount attention to these variables; laboratory results would be useful in interpretations of findings in the field relative to other measurements on the same animal. With advancements in technology, the picture may change, however.

The within-animal repeatability of estimates of blood hematocrit and hemoglobin components, such as glutathione, ascorbic acid, and mineral elements, seems considerably higher than for the other blood constituents enumerated. Therefore, changes in these may be noteworthy in field investigations. Careful recording of feed consumption and water intake would be part of the required correlated information.

Changes in endocrine function associated with elements of the natural environment are currently among the least defined. It has been demonstrated that high and low temperatures influence thyroid secretion rate. The pituitary and adrenals are no doubt also involved functions in heat balance of the body, but the extent to which the other glands are influenced is not clear at present. There is limited evidence that extremes of ambient temperature have some effect on prolactin and 17-ketosteroid secretions. Probably the most significant action of the endocrine glands is in the maintenance of balance of heat by shifting of electrolytes with vasoconstriction and vasodilation of blood vessels. The probable endocrine responses to changes in temperature for most animals are summarized in Table 8.2.

Two major problems are apparent in properly determining the impact of environment on the endocrine glands: (1) it is very difficult to develop methods of direct measure, and to obtain any measure is usually a tedious task requiring more elaborate equipment than can be

**TABLE 8.2 Summary of Probable Endocrine Changes under Low and High Ambient Temperatures[a]**

| Gland | Temperature[b] | |
|---|---|---|
| | Low | High |
| Pituitary | >thyrotropic hormone | <thyrotropic hormone |
| | ? | <prolactin |
| | >electrolytes in interstitial fluids | >shifts in electrolytes |
| Thyroid | >thyroxine | <thyroxine |
| Adrenal medulla | >adrenalin | no change? |
| Adrenal cortex | >17-ketosteroids | ? |

[a]From McDowell (1970).
[b]>, increase; <, decrease; ?, no clear evidence for change.

handled satisfactorily in the field; (2) although estimates of change may be made from the identified normal state of functioning, one is never sure whether the extent of change is due to the direct action of the heat-regulating centers on the glands or to decreased food intake, which, in turn, may be a normal function related to energy balance rather than an adjustment in body heat regulation. Much better techniques of assay are required before very good relationships with expected changes in natural environments can be delineated.

## Behavior

In the broadest sense all modifications of physiological processes in heat exchange could be classed as changes in behavior. In this discussion it refers to shifts in the usual patterns of body posture, movement, and food intake that might take place under extremes of climate. Generally, alterations in behavior are made by the animal either to reduce the level of heat production, to increase heat loss from the body, or to avoid adding heat from the environment. The avoidance of mounting during estrus and a decrease in the consumption of roughage in proportion to concentrate are two examples of changes when animals are under thermal stress. The tendency for animals to seek shade represents a measure to prevent the heat load from being increased.

Some very striking illustrations of changes in behavior under hot conditions are described by Schmidt-Nielsen (1962). He indicates that many animals, such as the kangaroo rat, manage to survive under extreme climatic conditions largely by their patterns of behavior. This suggests that animals living under extreme conditions become "escapists." In hot areas they avoid exposure to high temperature by burrowing into the ground, hiding in depressions or seeking shade by any and all means to prevent the body from taking on heat or preventing excessive loss of body water.

*The Behavior of Domestic Animals* (Hafez, 1962) describes some seasonal changes in patterns of behavior that are often overlooked as being commonly associated with domestic animals.

Although there is a paucity of data regarding the extent of changes in behavior on the part of most animals under conditions of extremes, it is clear that alterations such as in posture, activity, and eating habits are important adaptates employed to reduce the effect of environmental stresses. Thus, it seems that a fruitful area of field research would be systematic recording of changes in behavior. If behavior patterns are not the direct objectives of the experiment, they certainly should be included as correlated information for use in the interpretation of other observations.

### Field Measurements of Heat Exchange between the Body and Its Environment

This section deals with the problems of measuring exchanges by radiation, conduction, and evaporation.

So far as the animal is concerned, the climatic elements may operate conjointly or interact; i.e., the effect of one can be initiated by another, or a change in one may be offset by a corresponding change in another from the opposite direction. A decrease in air movement, for instance, may have approximately the same effect as a certain rise in dry bulb temperature. Unfortunately, the extent of the interactions have not been determined; thus, proportionate weightings, for example, of an increase of 5 km/hr in air movement in relation to 3°C change in air temperature is not clear. Figure 8.3 shows a system for determining the effect of changing air movement on the responses of animals.

The rate of heat exchange by conduction centers on two avenues: the interchange with surrounding ambient air and with objects in contact with the animal body. If the animal is lying or reclining, this may become a significant avenue of heat exchange. If the animal's body is in direct contact with a surface, the rate at which heat is transferred is

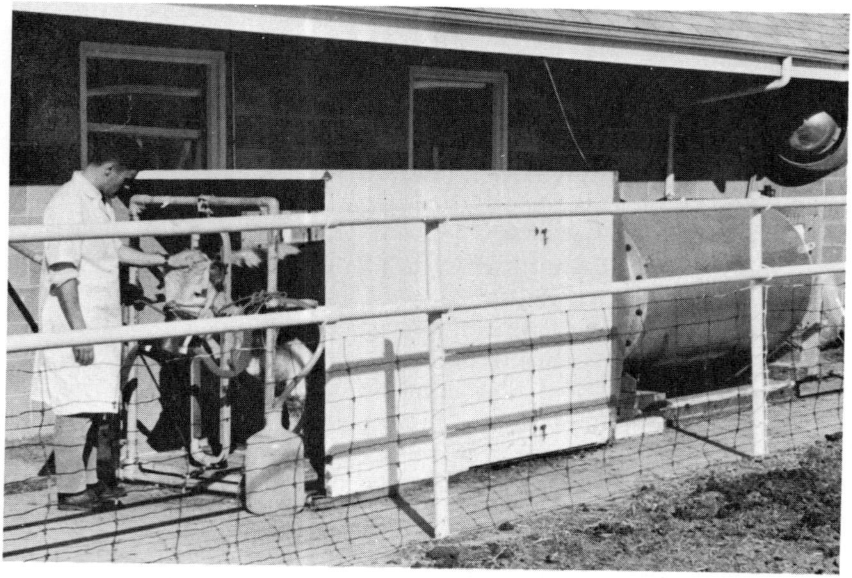

FIGURE 8.3. Variable speed wind tunnel at the Environmental Physiology Center, University of Alberta, Edmonton, Canada. Maximum air velocity attainable is 15 mph. Mean January air temperature at the outdoor recording site is about −15°C. Photo courtesy of A. J. F. Webster (1968).

a fairly simple process immediately after contact is made, but the rate at which heat will go on being transferred depends on a number of factors, resulting in a rather complex situation. The rate at which heat will continue to be transferred depends on the rate more heat can be passed from the deeper parts of the object or the animal body to the contact surfaces. This in turn is influenced by the temperature differential at the time of initial contact and the means for the transfer of heat from deeper surfaces to the external surface. This being the case, proper assessment of the rate of exchange by conduction through direct contacts can be very problematic. This problem is generally avoided by keeping the animal in the upright or standing position while the experimental observations are made.

The usual process for estimating the rate of heat flow from the body by conduction is the skin or surface temperature measured by thermocouple or some other suitable equipment. If the skin temperature and the ambient air conditions are known, the calories of heat lost per unit of area of surface in relation to a time span can be estimated. Such estimates are somewhat gross at best because of the rapid changes in air movement, particularly where there is turbulent flow, and other elements that create a very unstable state in the microenvironment surrounding the animal's body. The thin layer of air in immediate contact with the skin rapidly comes to skin temperature. When this condition exists, a further exchange of heat between the contact layer and more distant air will be slow, unless the air in proximity to the surface is changed. Since the air movement about the body is often in a continuous state of flux in the field, estimates of the rate of heat loss by conduction cannot be determined with certainty.

In addition to the heat exchange by conduction from the external surface, there are internal surfaces—the respiratory and alimentary tracts—that also play a role. The ambient air entering the respiratory tract comes into equilibrium, or near equilibrium, with the temperature of the body, depending on the temperature of the ambient air entering the tract and the time in the tract. The latter depends upon respiration rate and tidal volume (liters per breath). When the air temperature is 20°–32° C, the exchange by conduction, that is, the heating or cooling in the respiratory tract, is of little consequence. Under cold conditions or extreme high temperature conditions, the rate of exchange via this avenue—loss or gain—may be significant. The heat exchange through the alimentary canal may be quite variable, depending on the temperature of the food and water ingested. If, for example, a 454 kg cow were to consume 18 kg of 21° C water, she would lose about 18 Mcal of heat in warming the water to body temperature. This would constitute a significant amount of heat; on the other hand, if the same water were evaporated from the skin, the heat lost would be approximately 16 times greater.

In summary, it is clear that estimates of heat exchange between the animal and its environment by conduction is very difficult to evaluate. If attempts of estimates are undertaken in conjunction with other observations, then it is evident the animal should be kept standing to avoid confounding the picture by heat interchange through direct contact. If skin temperature is to be used to assess the relative rate of exchange with the environment, correlated air temperature must be recorded (Wiersma and Nelson, 1967). Rate of heat loss by conduction through respiration can be obtained by difference in the temperature of expired air and the air entering the tract along with respiration rate (Barrada, 1957). Heat loss through the alimentary tract can be estimated by temperature differentials between intake and urine and feces.

Heat loss by evaporation takes place from the respiratory tract and skin surface. Water on the skin surface comes there by one of three means: simple transudation through the superficial layers of underlying tissues to the surface; via the sweat glands; and by external application. The process of transudation or simple diffusion is generally referred to as insensible perspiration and occurs relatively independently of external environmental conditions. The action of the sweat glands seems to be largely under the control of the heat regulation centers. The external application of water can be by artificial means, rain, or in some cases by the animal submerging itself in water such as is commonly done by swine and water buffalo. Like heat loss by conduction, continued loss of heat into the air by evaporation is markedly dependent upon the extent and type of air flow about the body—directional or turbulent. The thin layer of air adjacent to the skin rapidly becomes equal in humidity with the skin. Thus, the further exchange of water vapor by diffusion between the air in contact with the skin will be slow when air movement is low (Joyce and Blaxter, 1964; Morrison *et al.*, 1968a).

In recent years a number of techniques have been proposed for esti-

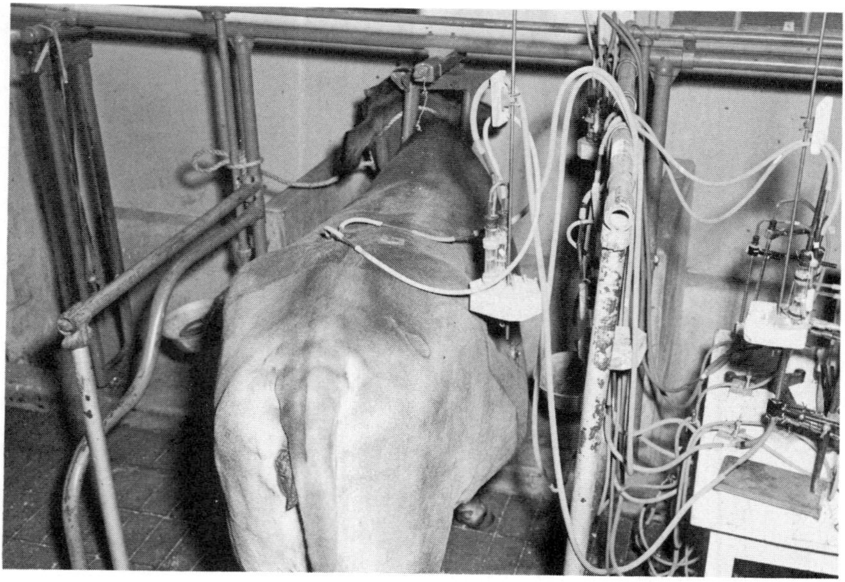

**FIGURE 8.4. Illustration of capsule technique for determining rate of evaporation from the body surface using a closed system of passing air through a saturated salt solution. (After, McDowell *et al.*, 1954). Photo courtesy USDA.**

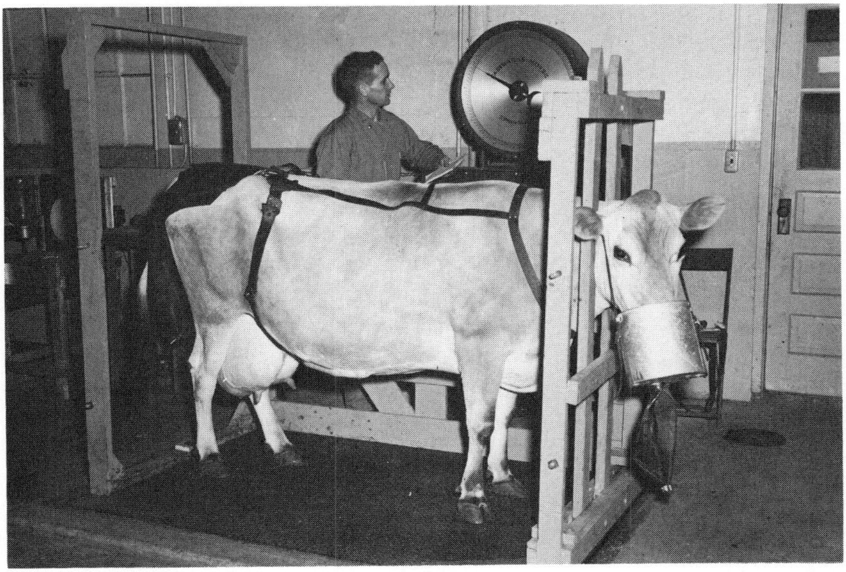

**FIGURE 8.5. Illustration of the use of balance for obtaining estimates of rate of evaporative losses.**

mating rates of surface evaporation. These have been primarily oriented toward estimating losses by sweating. The more popular of these are the use of calcium chloride disks and the capsule technique (McDowell *et al.*, 1954; McLean, 1963). The use of the capsule technique is demonstrated in Figure 8.4. These techniques have the limitation of being confined to spot checks on certain body areas. They therefore do not give a very good estimate of the evaporation rate from the entire body surface. Measurements of evaporation from selected skin areas, coupled with measures of body surface area, have been used on numerous occasions to estimate overall body surface losses. The capsules have limited field application due to the requirements for animal confinement and mass of equipment. The calcium chloride disks seem most satisfactory for field use.

Another system that may be used in certain situations for estimating evaporative losses is shown in Figure 8.5. In this case the animal is weighed at identified time intervals. The feces, urine, and saliva are collected. The differences in weight provide measure of evaporation from the skin and respiratory tract. If respiratory losses are determined, then surface losses can be obtained by difference. The degree of sensitivity of the scales is important in the precision of the estimates.

Currently, use of a hygrometric tent is the most satisfactory technique for assessing total surface evaporation in a system initially developed by Yeck (1960) and later modified by Guidry and Hofmeyr (1968); see Figure 8.6. Although this represents a fairly sophisticated assemblage of equipment, it can be useful in some field situations.

A major problem in properly determining the rate of heat loss from the animal body by evaporation is the variation in type and extent of body covering–hair or wool. Hair density (number of follicles per area), the depth of the hair coat (lay of the hair, erect or flat), and oiliness of the hair all influence the rate of evaporation. Therefore, measurements of these, along with the precise recording of climatic variables, is essential for anywhere near satisfactory estimates of surface evaporation under field conditions.

The temperature of surfaces evaporating from the respiratory tract into the air stream stays fairly constant. The surfaces are also sufficiently supplied with water to ensure that the inspired air becomes saturated in spite of its temperature. If the inspired air contains only a small concentration of water vapor–characteristic of hot–dry climates–

**FIGURE 8.6. A hygrometric tent as a system for measuring surface and respiratory evaporative losses. Photo courtesy of USDA.**

the opportunities for evaporation via respiration may be substantial. Conversely, if the air has a high water vapor content, evaporation rate is much lower. Evaporation from the respiratory tract increases with an increase in rate and volume of air expired. Thus, high rates of respiration observed in many livestock under thermal stress will result in a substantial rate of heat loss into dry air but less under humid conditions. If the temperature of the ground or floor is below the skin temperature of the animal and if the air is also below this temperature, then humidity will be of reduced consequence because the animal will lose heat largely by nonevaporative means (McDowell, 1967).

The amount of heat loss through evaporation from the respiratory tract may be estimated by using respiratory volume, the corresponding body temperature, and certain corollary information about ambient conditions. Techniques for determining respiratory volume may be used to a reasonable degree of satisfaction in the field (McDowell *et al.*, 1953) or some of the other techniques described in the previous chapters. A formula for estimating respiratory evaporation in grams per square meter in 24 hr is

$$R = k[\nu(P_b - P_a)]/m^2$$

where $R$ = respiratory evaporation g/m$^2$/24 hr
$k$ = 0.000941 = grams of water vapor in 1 liter of air/mm Hg water vapor pressure
$\nu$ = calculated respiratory volume (liter/24 hr)
$P_b$ = saturation vapor pressure at the temperature of the body taken as 55 mm Hg for a maximum body temperature of 40°C
$P_a$ = vapor pressure of the inspired air in mm Hg
m$^2$ = square meters of body surface area

As indicated above, a satisfactory assessment of the total radiant heat load on the animal in the field is difficult to derive. Although measurements of sources of radiation are complex, this is by no means the whole picture in determining the heat load imposed on the animal. Its posture, surface area profile (body configuration), size, frequency of movement, length of hair coat, angle of the sun, and other factors make satisfactory measurement of the incidence on the whole body little more than a gross estimate. When the sun is overhead on a clear day, the amount of radiation may be high, but the proportion of the surface presented to it by an animal may be small as compared to the total body area. At noon, the area that a man would have exposed to direct radiation would be small, whereas a sheep would have the full length of the top surface of its body exposed. Thus, man would have the greatest surface exposed in mid-morning or afternoon. For the sheep, the greatest exposure would be between 10 a.m. and 2 p.m. solar time. Usually the total heat load is greatest in the afternoon because ground radiation and the air temperature are higher. The maximum incidence received by most animals is when the sun's angle is about 50° above the horizon. Hence, the animal never receives the full impact of the direct radiation over more than about 30 percent of its surface. On this basis, the calculated heat load of 662 kcal/m²/hr for an animal in the sun in California (Figure 8.7) does not mean, as suggested, that the animal's entire body is subject to this amount of heat. The actual net heat load is considerably less because of the configuration of surface, reflection, and reradiation.

Although the amount of radiation impinging on an animal that will be effective in increasing the heat load depends on the emissivity of the coat and the type of coat, solar radiation may be looked upon

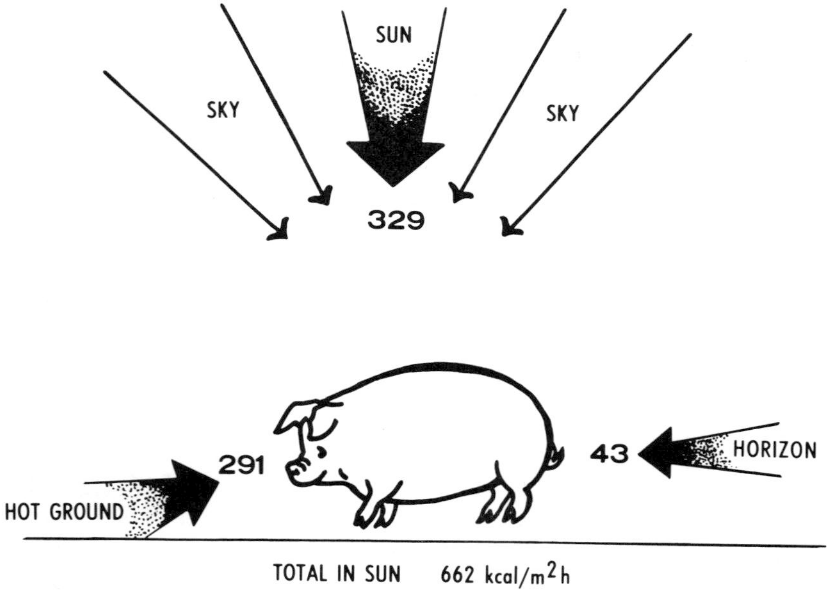


FIGURE 8.7. Radiation received by an animal in the sun on a typical August day at El Centro, California. (Adapted from Bond, 1967.)

largely as an additional heat load in terms of temperature. Ultraviolet radiation penetrates very little beyond the epidermis; however, the visible segment of the spectrum penetrates into the dermis and infrared radiation beyond the dermis. In the humid tropical areas the concern will be with the part of the radiation spectrum that contains a moderately high heat component, namely the visible and the infrared. Contrarily, in the dry tropics, where ground cover is sparse, the color of the soil is light, and the soil dry, the total radiant heat load is much higher due to the longwave thermal radiation (Bond, 1967; Johnston *et al.*, 1966). In the humid tropics, differences in coat color are of some importance; they are of less significance in the dry areas. Stewart and Brody (1954) give examples of reflectances from hair coats of various animals at different wavelengths. Hence, in assessing the effects of solar radiation on animals, it is desirable to include assessment of the color of the coat (Lee, 1953) and also a description of the surface area configuration.

What the true rate of exchange may be is not known, and there has been reluctance about spending a great deal of time attempting such assessments because researchers realize that estimates of the radiant heat load derived would apply to only micro conditions in relation to time and the environment. Even if an animal could be represented by a horizontal cylinder, the radiation exchange would still be three-dimensional, consisting of a horizontal component, a vertical component, and the incoming radiation from the sun and sky. If the other possible sources of radiation, such as reflection from the terrain and the losses by reradiation from the body are considered, the calculations become extremely complex. Thus, it does not seem that, for the present state of technology, collection of solar radiation data to more than assess the general environment is warranted.

One means that may be used to acquire approximate estimates of the total radiation exchange—other than the direct solar—on an animal is using globe thermometer temperatures (Bond and Kelly, 1955; Pereira *et al.*, 1967). From simultaneous measurements of the black globe temperature, the air temperature and the rate of air movement, the amount of radiation gained or lost to the atmosphere, can be determined from monographs (Bond and Kelly, 1955; Lee, 1953). Due to the difficulty in obtaining all the needed correlated information on atmospheric conditions to estimate rate of radiation exchange with these scales, most researchers who have used the black globe thermometer in animal experiments have not followed the procedures outlined but instead have used the globe temperatures to further describe the thermal environment. The black globe temperature is usually about 6°C above the air temperature in the shade. This has been used in lieu of regular shade temperatures as a means of incorporating some estimate of the rate of heat gain by the animal.

It seems, therefore, that the great effort to obtain estimates of the rate of heat exchange by radiation in the field is scarcely worth while. In most instances it would be satisfactory to utilize only general rules. For instance, from 6 to 12 months of the year in the lower elevations of the north–south 30° latitudes, the intensity of direct and reflected radiation will be such that for 5–10 hr per day the animals may receive significant amounts of heat. It should also be kept in mind that the duration of the radiant-energy-stress period may extend beyond the daylight hours if the animals are housed under a shelter that has a high emissivity.

To summarize: Any study of physiological functions under field conditions must deal with the homeothermy of the animal. Although precise measurement of the rate of heat exchange is difficult, the roles of conduction, evaporation, and radiation in heat loss must be accounted for. (Convection has not been discussed because information in relation to animals in various environments is inadequate.) The rapid fluctuations that occur in the field environment make the task even more discouraging. Body temperature, then, becomes the best index of the animal's ability to keep in balance heat gains and losses from the body. The value of all this information is to use it in planning and designing practical experimental procedures. Since body heat is very closely associated with production functions, every effort should be made to assist the animal in maintaining its normal balance or to minimize the inhibition of the stress on its heat balance mechanisms.

## ANIMAL RESPONSES TO FIELD TESTING

### Selection of Animals

The importance of using healthy animals cannot be overemphasized. The extent of the problem depends upon location—dry or wet—and the husbandry practices employed. Numerous other factors may, under certain circumstances, prove important, but it should be clear that careful selection of the animals is essential, particularly if responses for a short-term period are under evaluation, If the duration of the experimental period is a matter of a few hours or a few days, then the uniformity among the animals groups is critical. This problem is explored in Chapters 5 and 6.

### Adaptation and Compensation

Frequently the field researcher describes one or more environmental variables as the factor(s) imposing "stress" on the animal and what he measures as the animal response is the resulting "strain." But the stress–strain relationship is an oversimplification of the association of external disturbances and their influences on animal functioning. Body functioning and the sequence of events that an animal goes through in response to changes in environment are in themselves very closely interrelated. Adolph's hypothesis (1964) on the course of adjustment made by the individual seems to depict more clearly certain sequential events and their interplay in response to changes in the environment than the stress–strain phenomena. Adolph uses the term "adaptagent" to identify the force applied, and the resulting strain is measured as an "adaptate" (the measurable changes). This is expressed as ΔA (the force) →ΔE (the resulting strain). If ΔE fails to compensate for ΔA, then there is a shift in the homeothermy. Since many processes are involved in maintaining homeothermy, ΔA infrequently goes directly to ΔE. What more logically transpires are progressive stages; e.g., ΔA produces an initial adaptate that may be measured as ΔB. This in turn produces a chain of events before ΔE compensates for ΔA (McDowell, 1971). Thus, the chain of events is more appropriately written:

$$\Delta A \longrightarrow \Delta B \longrightarrow \Delta C \longrightarrow \Delta D \dashrightarrow \Delta_n \longrightarrow \Delta E$$

The sequential development of the compensations or attempted compensations by the animals to an increase in ambient temperature would likely take place in the following order: (1) increased blood flow to the skin; (2) initiation of sweating; (3) increased respiration rate; (4) changes in hormone or endocrine activity; (5) changes in patterns of behavior; (6) increased intake of water; (7) elevated body temperature; (8) changes in the use of body water; and (9) partial dehydration. If these steps failed to renew the animal's thermal equilibrium or shift it to a new plateau, then progressive stages of failure of the heat regulation mechanisms would be in evidence: diarrhea followed by a general weakness, staggers, convulsions, and death.

Probably the main reason that the correlations obtained in field experiments for some of the physiological processes previously discussed (such as rate of evaporation from the skin) in relation to changes in climatic variables have been low is that the environmental conditions are likely to change before new plateaus in animal response are in evidence. Also there is an action–counteraction that tends to modulate the magnitude of the change in the various processes. For instance, respiration rate may reach its maximum and then decline before the ambient temperature reaches its highest level. Sometimes this has been associated with fatigue of the muscles involved or possibly to changes in the $CO_2$ combining power of the blood, resulting in respiratory alkalosis (Barrada, 1957). On the other hand, it may be that the initiation of sweating decreased the need for the increased respiration rate. It has also been assumed that under thermal stress a shift in body temperature was highly undesirable. It has served as an estimate of the breakdown of the functions of other mechanisms. An elevated body temperature could also be the resulting interaction response to enhance heat loss by conduction and radiation in a hot environment. Therefore, changes in several processes should be observed simultaneously and over sufficient time for the experimenter to appreciate the action–counteraction employed by the animal. Seldom do the extremes of the day in the field environment prevail long enough to identify a modification in the use of body water and state of hydration, but variations in the other functions can usually be detected in a matter of hours.

When field research is undertaken, a knowledge of the possible interplay of the various compensatory mechanisms is highly desirable, particularly where a limited number of processes are to be measured. It is also evident that a clear picture of how the animal responds to changes in environment would of necessity require simultaneous measurements over a period of a number of the variables indicated.

### State of Acclimatization

Most dictionaries define acclimatization as a process of adapting to a new temperature, altitude, or climatic environment. Physiologists have used the term to refer to the state occurring when an animal in response to repeated or continuous exposure to an environment develops functional or structural changes that enhance its ability to get along with a minimum of distress. These definitions imply changes that result in new physiological states for the animals and are thus far too rigid to be applied to the functioning of heat regulation and other centers when the animal's environment fluctuates. Immediate or short-time fluctuations in homeothermy and acclimatization merge imperceptibly because at any one moment the temperature of the body depends not only on the heat load but also to some extent on the preceding heat load. Therefore, it is obvious that regulation of homeothermy or heat balance and acclimatization are continuous and interdependent no matter what the external environment may be.

A distinction should also be made between the state of acclimatization and the ability of the animal to acclimatize. These might be described as differentiating the conditions under which the initial responses were measured vs. those measured at some later time when there was a distinct change in the environment. In this concept it seems possible that animals that remain for a long period at a fairly constant environment, as, for example, in the lower altitudes near the equator, are less able to acclimatize to a new environment than animals accustomed to changes of climate. In other words, the state of acclimatization to hot conditions may be good among animals indigenous to the area, but their ability to acclimatize may be poor. There is some experimental evidence to suggest that fluctuations brought about by seasons improve the ability to acclimatize (McDowell, 1968). In any case, it is clear that acclimatization is a fluctuating condition taking place within the body; therefore, attempts should not be made to rigidly describe the state of acclimatization of an animal or its capability to acclimatize to changes in the environment.

All animals have certain capabilities of adjusting to hazardous conditions. This varies with species, and also with age at the time the animal is initially subjected to the environment (McDowell, 1971). The methods of measuring the adjustments or lack of adjustments are also important in assaying how the animal functions in a new environment (McDowell *et al.*, 1969). For instance, short-term exposure of cattle to high temperatures in controlled laboratories has resulted in substantial rises in body temperature in some animals and relatively small ones in others (McDowell, 1966b). The differences in magnitude of response between animals or groups in these experiments have been used to predict suitability to a hot climate, but it is evident from the hypothesis of adaptagents–adaptates that decisions based on one measure may not reflect the real suitability of an animal to a given environment. It is conceivable that an animal with the more elevated body temperature was lower in suitability, but, as often happens with lactating cows, those with the highest yields have the higher body temperatures under thermal stress. Under such circumstances, functional adaption would be a more satisfactory measure of acclimatization from the livestock producer's standpoint.

Another factor that often confounds the distinction between acclimatization and response is that, except for conditions very close to the equator, the climatic environment changes by seasons as a progressive rise or decline. If the animal observations are made in the early stage of the onset of warmer or cooler weather, responses may be quite different than when measured after the animals have become accustomed to the extreme warm or cool conditions (McDowell, 1968). There is also the other extreme, when an animal may have been protected from the gradual trends in the season and then is suddenly introduced to the near maximum of the season. Responses measured at that time indicate lack of acclimatization but provide little information on the ability of the animal to prepare itself to meet the extremes.

Studies conducted under both field and laboratory conditions have shown that the environmental situations to which the animals were accustomed prior to exposure to the test conditions markedly influenced the measured responses of the animal to the experimental procedures (McDowell, 1966b). This is the reason for the earlier emphasis in this chapter on the need for developing a profile of the physical environment and also the strong emphasis on the necessity for characterizing the environmental fluctuations as a requisite to the execution of satisfactory field experiments.

## Nutritional State

The nutritional state of the animal is also very important in the assessment of response in field tests. Most of the time animals under field conditions are in one of three states of nutrition: a more or less stable body weight; gaining weight; or losing weight. All three states have an influence on the animal's general metabolism and hence on its response to other environmental variables. The level of nutrition, the quality of feeds available, and the health status of the animal markedly influence its being placed in one of these three categories. A general characterization of the nutritive state of an animal can be derived by recording body weights over a period of time—at least two intervals 10–30 days apart. The general characteristic of the pelage is another way of grossly estimating whether the animal is gaining or losing in body weight. If the hair coat has a glossy appearance and when brushed in the opposite direction to its "normal lay" responds immediately by returning to the original position, the animal is considered to be in a good state of nutrition and no doubt will be gaining weight. Conversely, if the appearance of the coat is dull and the ends of the hairs dry, the animal is probably losing weight. The response to a counterstroking of the hair will also be slow. In choosing animals for experiments, it is important that their nutrition history prior to the experiment be similar. It is also highly desirable that the level of nutrition during the experimental period be such that the animals are either maintaining a near stable state in body weight or gaining. If the animal is losing weight, this may confound the desired test of the influence of the environmental variables under evaluation.

## Age

The confounding effects of age in measuring responses to controlled environmental regimes was indicated in Chapter 4. Under field conditions, age as an interaction with nutrition could be quite an important variable, particularly where animals may have gone through one or more cycles of severe fluctuations of body weight. The difference between chronological age and physiological age may be wide in cases where animals have experienced one or more seasons during which they lost significant amounts of weight and were forced to regain or restore their status during the experimental period. This often happens with grazing animals, such as sheep and cattle running on extensive grazing lands in parts of the world including the western portion of the United States.

## Size

Size *per se* as well as size for age is an additional variable that creeps into selection of animals and the interpretation of experimental results. Many reports have appeared in the technical literature of the past two decades about thermal tolerances of zebu vs. European breeds of cattle. In most of the reports the observed differences in tolerances of thermal stress were largely attributed to differences in certain physiological and anatomical traits of the two groups. Few, if any, of the experiments employed animals of similar body size from the two groups. In a number of instances a closer evaluation of the data indicates that differences attributed directly to breed groups may be more associated with size than with physiological or anatomical traits peculiar to the animals. (McDowell, 1971). Therefore, it becomes highly desirable for the purpose of direct comparisons among animals or among breed groups that the animals be paired according to size, or preferably that the entire experimental and control groups be within a relatively small range in size. If variability in size is necessary, it should be distributed as nearly equally as possible across groups.

### Permitted Activity

The extent of activity permitted the animal while undergoing experimental treatment can influence the results particularly if the experimental procedure makes a marked change in the customary freedom of movement (Bond, 1969; Morrison *et al.*, 1968b). Thus, it is highly desirable in transpositions from one location to another that the animals have an opportunity to become adjusted to the new level of activity because of its influence on later reactions. Any changes in activity that influence the temperament of the animal will have, for example, measurable influences on pulse rates and other functions that may be directly related to the experiment.

### Reproduction and Lactation

Additional factors needing consideration for field research include reproductive and lactation status as well as biological rhythms, which exist in the animals selected (Hafez, 1962). These characteristics are associated in one way or another with a number of the physiological functions of animals. The importance of reproduction and lactation are mentioned in earlier chapters as forces contributing to heat production. The reproductive state as well as stage of lactation, frequently are principal factors that determine the efficiency of productive functions; therefore, great care should be taken in evaluating their influence upon the response of the animal to an environment. For example, if an animal is maintained primarily for milk production, this factor becomes of paramount importance in experiments that may require a reduction in milk yield. Drastic changes in milk yield should be avoided when other traits are of primary interest during the experimental period.

### Biological Rhythms

Biological rhythms are important in some instances. For instance, when animals are transferred from one photoperiodic environment to another, there may be a carryover of previously ingrained biological rhythms that may take one or more seasons to alter to a state that is more compatible to the new environment.

### Parasites

Where either external or internal parasites are likely to be problems, effective control of parasites becomes important both in the preexperimental and experimental periods. The importance of parasite control is exemplified by experiments in Brazil. To test the impact of parasites on growth rate and general efficiency, a group of heifers was treated at 20-day intervals for one year to control ticks, grubs, and internal parasites. A comparable group was used as a control. The feed consumed per kilogram of gain in body weight was 28 percent higher for the untreated group. The untreated group developed long shaggy hair coats, which were considered a contributing factor to the 50 percent lower breeding efficiency in this group the following year. Most of the untreated animals also showed positive response to a complement-fixation test for anaplasmosis. They had high fecal egg counts, suggesting problems of internal parasites. The conclusion was that effective parasite control is important in the efficiency of feed utilization and the ability of growing animals to become adjusted to subtropical conditions.

### Indices for Assessing Adaptability

In the past, physiologists were interested in the influences of various environmental changes on animals particularly where extremes of ambient temperature were above the comfort zone. In these tests attention was given to either (1) identification of breeds or strains that exhibited the least shift in bodily heat balance if the animal was exposed to thermal stress, or (2) attempts to identify physiological and anatomical traits associated with the promotion of heat loss. The hypothesis was that a reasonably high correlation existed between minimum shifts in homeothermy and performance under high-temperature conditions. Persons associated with animal production are always interested in hastening the selection processes by identification of individuals or groups of animals that would suffer least from a hot environment. To this end, a number of indices have been proposed for animal selection from both field and psychrometric laboratory experiments (Bianca, 1961; McDowell, 1966b).

The most prominent of the indices developed from field data was the "Iberia heat tolerance test" (Rhoad, 1944). The use of this index served to identify breed group differences among cattle in several locations. In general, the results showed that zebu-type cattle had the best tolerance, with pure European breeds ranking much lower. Based on these findings, a number of livestockmen and researchers have assumed that the zebu were most likely to withstand the high temperature–humidity conditions prevailing in many areas of the tropical and subtropical areas. However, analysis of a large volume of data from a station in Louisiana (Vernon *et al.*, 1959) showed the correlation between the heat-tolerance coefficient derived by the Iberia heat tolerance test and the animal's birth weight was very small (–0.05). When the animal's 6-month weight was correlated with its coefficient of heat tolerance, the relationship was also poor. Correlations between the heat-tolerance coefficient and the average birth weight of the calves of the cows was –0.11. When the average 6-month weight of the calves was used, the correlation was still very low (0.09). From this study, it was concluded that the suitability of cattle to the conditions in southern Louisiana could not be judged byh the level of body temperature and respiration rates collected on several days during the summer season. A further conclusion was that direct selection for performance should result in selection pressure not only for tolerance to thermal stress imposed by the locality but also for other characteristics necessary to achieve desired level of performance.

Several other indices (McDowell, 1966b) based on field experiments have been proposed, but none of these has been used extensively enough so far to determine their relationships to animal performance. Field tests based on hair coat assessment in Australia (Turner and Schleger, 1960) and South Africa (Bonsma, 1949) have shown reasonably good correlation between coat score and general performance. Hair coat quality, however, is not a measure of direct response to thermal stress, strictly speaking, but does provide valuable information on the animal's general capacity to thrive in a given environment.

A more recent proposal for establishing a climatic stress index for domestic animals (Lee, 1965) is based on the ratio of the rate at which heat will accumulate in the absence of compensatory evaporation to the maximum rate at which heat can be removed by evaporative devices available to the animal. The equation is as follows:

$$\text{Required compensation} = M + \frac{5.55(T_a - 35) + RI_a}{I_a + I_{cw}}$$

where $M$ = metabolic rate (kcal/$m^2$ hr)
$T_a$ = air temperature °C
$R$ = net radiant heat load (kcal/$m^2$ hr)
$I_a$ = insulation of air in clo* units
$I_{cw}$ = insulation of coat in clo units at maximum wetness

Possible application of the equation was illustrated (Lee, 1965) with data for a Jersey cow. In theory, the index would be useful to demonstrate the effectiveness of various types of ameliorations of the animal's immediate environment in relation to climatic stress, such as changes in rate of air movement or protection against radiant heat. It cannot, however, be used effectively to predict physiological responses to various environmental conditions. The difficulties in current application of this index to field use are numerous, mainly due to the paucity of acceptable values for metabolic rates in the field, a scale of clo units for various pelages, respiratory volumes, and evaporation rates. The use of values solely from laboratory experiments would have the same shortcomings as for many other measures when transposed to the field. Development of values to utilize such an index would be useful for later application to the field, however.

The indices proposed to date for use in the field may suffice for some degree of animal selection if only homeothermy is considered. They seem largely outmoded, however, for accomplishing the desired results. One major shortcoming has been that little consideration was given to males, who make a much greater contribution to future generations than individual females. Also, the proposed indices do not include measurement of performance in usable product. Furthermore, there is the high cost of obtaining data. Since the emphasis in this chapter is on experiments related to performance, the foregoing is presented to illustrate how more emphasis in future research should be given to the relation of physiological variables and performance traits. It also serves to illustrate that, in the planning stages, the objective of the research should be clearly defined as to intent, that is, whether changes in certain physiological traits are to be measured *per se* or whether responses are to be correlated between some physiological variables and animal performance. The inadequacy of the proposed selection indices does not preclude the desirability of evaluating changes in certain "physiological traits" to fluctuations in environ-

*One clo unit is the thermal resistance that will maintain a difference of 0.18°C for a heat flow of 1 kcal/$m^2$ hr (Lee, 1965).

ments. Such information is essential when man looks upon his animals as "biological machines" that have optimum operating efficiencies.

If man expects to obtain satisfactory economic returns from animal industries, he must be aware that productivity is reduced when animals are placed in environments not conducive to optimum efficiency (Bianca, 1965; McDowell, 1966a). If adaptates that alter the utilization of energy available for productive processes are necessary, the "physical cost" is a reduction in the ratio of energy input to output (McDowell *et al.*, 1969). It behooves those planning research to place strong emphasis on the predicted physiological limitations and the economics of losses in efficiency with changes in the environment.

## THE USE OF RADIO TELEMETRY IN FIELD RESEARCH

In field research it is highly desirable that the responses of the animals be evaluated as nearly as possible in its "normal state," meaning without changes in its usual habitat. The rapid developments in the capabilities of radio telemetry* indicates that this may be an excellent approach to at least some aspects of field study. Energy balance has been studied by this means. To date, however, the small number of complete energy-balance studies that have been reported reflects that existing methods are expensive, time consuming, tedious, and restrictive in terms of the number of subjects. In addition, the methods for energy balance study can be applied only with great difficulty to unconfined or field situations. At present the only techniques that can be applied without confinement (e.g., to grazing animals) are the body-balance methods, which involve slaughter and analysis of the whole body, and the respiratory exchange methods involving the use of cumbersome equipment mounted on a tracheotomized animal. Nevertheless, it is recognized that knowledge of the influence of environmental changes on energy utilization is highly desirable.

Radio telemetric methods already in use can effectively measure body temperature, pulse rate, and (to a limited extent) characteristics of cardiac function (Franklin and VanCitters, 1967; Franklin *et al.*, 1966; VanCitters and Franklin, 1969; VanCitters *et al.*, 1968). Two limiting factors are the distance that can be tolerated between the animal's transmitter and the receiver, and the restriction on radio frequencies by the U.S. Federal Communications Commission. When the

*See additional discussion in Chapter 2.

distance exceeds a few hundred feet or there is interference by hill masses and heavy foliage, the system becomes of little value. If the power pack for the transmitter and receiver is enlarged to increase the range, the power requirements increase geometrically to the point where the unit no longer meets the requirements for portability. Developments in the quality and compactness of equipment seem to be overcoming these restrictions.

Information on body temperatures and pulse rates would be useful in the field, particularly on pulse rates that have a close linear relationship to oxygen consumption, except under extreme stress. A recent and successful method measured the blood-flow rate in a number of major vessels, including the aorta, of such animals as the baboon, birds, turtles, running horses and dogs, and swimming sharks and alligators, all unrestrained (Franklin and VanCitters, 1967; VanCitters and Franklin, 1969). The measurement of blood-flow rate through either the pulmonary artery or the aorta would represent cardiac output, which in turn has a close relation to oxygen consumption. If to this could be added a suitable means of measuring total cardiac output and arteriovenous oxygen difference, the possibilities of studying bioenergetic problems concerned with the unrestrained animals in relation to changes in environment would be vastly expanded. Remote sensing is the most promising field for this type of study.

## HOUSING AND MANAGEMENT

Some modification of the environment is usually necessary for the most efficient animal production. Farm livestock structures frequently have been designed more for the comfort of the worker than for that of animals. A good example is the traditional, poorly ventilated dairy barn found in temperate areas. More recently, animal rather than worker comfort has been emphasized, although the latter cannot be overlooked.

The present concept of environmental modification centers on the creation of an environment that (a) minimizes all stresses on the animal that might reduce or limit productivity; (b) maximizes the efficiency of labor; and (c) minimizes cost per unit of animal product. This requires not only careful design of farm structures but also the development of general management systems, the protection of animals against diseases and parasites, and the production and/or purchase of balanced rations. While it is seldom possible to include all of these factors in a single field study, they must all be considered in its design. If the study is to be justified on the basis of animal performance, it is essential that attention be given to the productive functions of the animal and the effect that the experimental procedures will have upon them.

While some sort of environmental modification is almost always necessary, the complexity of the modulations required varies markedly from one location to another. In the higher latitudes or the deep tropics, these modifications are essential. For some of the subtropic areas, where climatic conditions rarely move out of the comfort zone of most animals, modification of environment would have minimum importance. On the other hand, it would be impossible to produce any animal product without shelter both for the animal and its feed supply in high latitudes. In the tropics this may not be quite so obvious. Many feel that the problems to be solved in improving animal production in the tropical zones are actually less than those that have been faced and solved in the temperate zones. In areas such as the northern United States or southern Canada, cattle production, either milk or meat, would be almost impossible without storage of feed to be used during nearly half of the year. But in the tropics, production of feed is frequently such that the animal can at least survive during the entire year on the available forages. These forages may, however, be of sufficient quality to maintain high levels of productivity only for periods approximately equal to those observed in the high temperate zones (McDowell, 1966a). Experience has shown that efficient production in either location can be maintained if sufficient environmental modification is employed (Johnson *et al.*, 1962; McDowell, 1968). The recent establishment of efficient poultry and pig industries in the humid tropics is a good example of what can be done through adequate environmental modification in order that highly productive animals may be utilized in such environments.

Virtually all of the environmental variables may be modified in some fashion or another. The determining factor on which variables to modify are costs and the improvement in productive efficiency that may be expected. Solar radiation, for example, may be limited through the use of rather simple shelters, but if temperature and humidity are to be varied, much more complex equipment is often required. The simplest case would be the utilization of the evaporation of water through some type of desert cooler. The other extreme would be the use of mechanical air conditioning in some type of structure. Rates of air movement may be modified in some instances by location or orientation of structures as well as by mechanical systems. Photoperiod can be simply modified, where electricity is available, by the use of lighting. *Air pollution*, such as dust, is difficult to control in areas where extremely dry conditions are existent, although dust filters can be employed when mechanical ventilation is practical. In the case of contaminates, such as ammonia, which occur in closed housing systems, the cooling or frequent removal of waste may assist in reducing the problem. Other variables such as nutritional status may be changed through alterations in frequency of feeding, and the time of day grazing is allowed. The type of feed also is subject to change but sometimes with difficulty depending on the ability to produce and/or store the types and qualities of feed required.

Requirements for heat loss may be varied through management, particularly where animal movement can be restricted or confined to periods when solar radiation loads are minimal and air temperatures are low. Lowering requirements for heat loss should, in most cases, be restricted to factors that do not reduce productivity. Reduction of muscular activity, such as decreasing the distance to walk in feeding, is an example as it does not require a reduction of feed intake.

### Use of Shelters

*Shelters* for livestock, irrespective of location, are justified on one or more of the following needs:

1. Protection of animals and their handlers from the extremes of climate without introducing additional elements of stress.
2. Improve the efficiency of handling animals for feeding, milking, breeding, or treatment.
3. To ensure sanitation of animal and product and minimize the hazards of health problems (Johnston, 1968).

Sheltering is usually justified in terms of economic efficiency, therefore, emphasis would be toward simplicity and cost. Esthetic values cannot be ignored, however, since they may have economic value, i.e., appearance to customers. The type of shelter that best meets the needs is determined by climatic conditions, quality and cost of available labor, as well as cost of materials and equipment. Since climate is of major importance, the initial step in planning research should include characterization of the local climate to include wind direction. Coupled with the evaluation of the influence of shelters on animal performance should be the development of systems of management that will maximize their value. Research on shelters, along with methods of use, are needed because what often appears suitable does not provide the desirable end-product.

For example, studies in Canada on the value of solid-wall, three-side enclosed shelters for beef cattle proved them to be rather ineffective because of the discomfort to the animals by the turbulence of the air within the shelter. When the solid wall of the long axis of the shelter, which was exposed to the wind, was replaced by louvered siding the air turbulence was reduced and the animals made much more effective use of the protection (Williams, 1967).

In hot, dry areas, shelters have proven highly beneficial in the performance of beef cattle on feedlots, lactating dairy cows, and for swine. Hereford steers in California with shade consumed more feed, gained 0.5 kg more per day, and the feed required per unit of gain was about 25 percent less than for animals without shade. Similar observations have been made with swine in this area (Bond, 1967, 1969). In Arizona cows having access to shades with evaporative coolers produced more milk (+ 1.8 kg per day) and had higher breeding efficiency than cows with only conventional sheltering (Wiersma and Stott, 1965).

Contrarily, summer shelters did not prove effective in the more humid southeastern part of the United States when the responses of lactating Jersey cows were measured by feed intake, water consumption, body temperature, and yields of milk, milk fat, and solids-not-fat. In Georgia, there were no significant difference between sheltered and nonsheltered cows. The addition of fans and/or sprinklers did not significantly improve the effectiveness of sheltering (Johnson and Givens, 1962). Studies with Holsteins in Louisiana also showed no significant benefit from protection during the summer months. In the Louisiana investigations, the level of fiber in the ration was more important than shelter effects (Guthrie *et al.*, 1968). Sheltering did not prove effective for beef cattle on pasture or drylot in Georgia.

Moveable shades 3.7 × 4.9 × 1.8 meters high proved sufficient to support satisfactory gains and feed conversion for swine in Georgia. Although the addition of sprinklers increased average daily feed intake and rate of gain, the treatment did not decrease the amount of feed needed to produce a unit of gain.

In the humid areas there are no doubt several reasons for the less spectacular benefits from sheltering. The rate of air movement is usually lower than for dry climates. Air turbulence is further reduced when animals are in close proximity under the shade, therefore, rate of convective cooling is decreased. The shelter may offer protection from the solar heat load, but this is seemingly offset by the increasing of the time it takes for the animals to restore their heat balance after sunset. Another reason is that when animals are on relatively good pasture and do not have to cover large areas to secure food needs, they tend to fill rapidly and spend most of the day under the shelter. This reduces feed intake as compared to animals without protection. Therefore, in humid areas access to shelter during the hot part of the day may help but this should be followed by forcing the animals into the open near sunset.

It appears that shelters 2–3 meters in height are more suitable for warm, humid climates than the 4–5 meters recommended for hot, dry climates.

### Confinement Rearing

The outlook is for continuing increase in confinement rearing of livestock in all countries. Confinement rearing often constitutes a major change in the animal environment, therefore, investigations on space per animal, ventilation needs, type of flooring and the influence of waste products are needed (USDA, 1964).

For example, in one experiment cattle limited to 3 $m^2$ of slatted floor space required 20 percent more feed per unit of gain than those allowed 60 percent more *space* or cattle on dirt floors with 9 $m^2$ per animal (Maddex, 1967, Mahoney *et al.*, 1967). Two other tests in California showed that space allotment below 3.7 $m^2$ per head reduced body weight gain and feed efficiency (Morrison *et al.*, 1968b). In some studies type of flooring influenced animal performance, while in others the accumulation of waste and buildup of ammonia decreased production.

Under high humidity conditions and warmer temperatures, spacing, waste accumulation, and ventiliation would be more critical for efficiency of animal production.

### Systems of Feeding

It is clear that the feed supplies, including quality and quantity are important environmental modifications, especially in warm climates. Permanent pastures as the source of roughage gave rather low milk yields in southern Louisiana. The average yield per cow increased over permanent pasture by 30, 40, 49, 74, and 105 percent, respectively, when the forage sources consisted of: (1) permanent pasture with supplementary grazing in the summer months; (2) permanent pasture with supplementary grazing in the summer and winter; (3) permanent pasture and supplementary grazing plus silage; (4) supplementary grazing and silage without pasture; and (5) ad libitum feeding of silage plus 2 kg of hay per cow per day. In point 5, the cows were kept in drylot. For all treatments, the cows received concentrates at the ratio of 1 kg concentrate per 3 kg of milk (McDowell, 1971).

In Georgia grazing 1 hr in the morning and late afternoon for about 100 days per year and supplementing with silage made from local forages increased milk yields 20 percent per cow per year over permanent pasture and supplementary summer grazing, which was available about 9 months per year.

Young dairy and beef stock in southern Louisiana had 27 percent increased breeding efficiency on pasture supplemented with silage vs. pasture alone. On a similar feeding regime beef cows had 30 percent higher calving percentage and the length of breeding season was reduced approximately 35 percent.

Much of the drier grazing areas are too low in protein to supply the needs of cattle. Some form of supplement is needed but the major deterrent to efficiency is the labor required. In New Mexico it was demonstrated that feedings spaced at weekly intervals are satisfactory and and even 2- and 3-week intervals were sufficient to provide the necessary protein for range livestock.

These are only a few examples of manipulations of feed resources as a means of modification of the animal environment, but they serve to demonstrate the importance of research in this segment of the environmental regime.

## Reproduction

Reproductive inefficiency constitutes the greatest problem in man obtaining the highest economic returns from his animals. This is due to the stresses of the direct and indirect effects of the climate and the indirect influences through quantity and quality of feed. The term inefficiency has been used rather than sterility since most losses for reproduction failure are "economic sterility." There is evidence that season of breeding plays a role in conception, embryo survival, vigor of the young, and mortality. In regions of long, hot summers, the stress imposed by high ambient temperatures is a limiting factor in the time of estrus, services for conception, and length of breeding period as well as regularity and intensity of estrus and embryonic mortality. In the tropics, rainfall distribution, as it influences feed supplies, is probably more important for conception rate than the direct effects of climate; nevertheless, the high temperatures influence the frequency of estrus, its intensity and duration, time of ovulation, and fertilization.

Considerable progress has been made but it is evident that a somewhat different type technology about handling routine breeding is needed for areas where high temperatures impose stress on livestock.

Research on methods of modification of the environmental influences to reduce animal losses through morbidity resulting from disease and parasites appears a fruitful area for research. Furthermore, means of reducing the inefficiencies of animal production due to health problems of a noninfectious nature, such as nutritional disorders, is important.

In short, the technology on the most suitable and practical environmental regimes for various areas is far from complete. But it is evident that changes in traditional systems of housing, feeding, and other measures are needed to enhance the efficiency of animal performance.

Field investigations of methods of modifying the animal environment ought always to include data on cost of the modification(s) and their effect on production level and efficiency. They should be carried out under the environmental conditions and using animals of the same type and level of productivity as those to which the data will be applied. In other words, these types of studies should be the final link in the research chain that begins with laboratory experiments on mechanisms of animal response and ends with improvements in the efficiency of animal production in a specific environmental situation. All too often the animal physiologist has been satisfied to make projections of expected economic benefit based on physiological change alone. Such extrapolations have been inadequate for justifying expenditures by commercial operators. The final study, frequently the most difficult but certainly one of the most rewarding investigations, is the one that proves the effectiveness of a soundly based theory by increasing the economic efficiency of production.

## REFERENCES

Adolph, E. F. 1964. Perspectives of adaptions: Some general properties. Sect. 4. *In* Handbook of physiology. Amer. Physiol. Soc., Washington, D.C.

American Institute of Architects. 1949. Regional climate analyses and design data Bull. A.I.A., Washington, D.C.

Barrada, M. S. 1957. Responses of dairy cattle to hot environments with special emphasis on respiratory reactions. Ph.D. thesis, The Johns Hopkins University, Baltimore, Maryland.

Bianca, W. 1961. Heat tolerance in cattle—Its concept, measurement and dependence on modifying factors. Intern. J. Biometerol. 1:5.

Bianca, W. 1962. Relative importance of dry and wet bulb temperatures in causing heat stress in cattle. Nature 195:251.

Bond, T. E. 1967. Microclimate and livestock performance in hot climates. *In* Ground level climatology. AAAS Pub. No. 86, Washington, D.C.

Bond, T.E. 1969. Livestock environment research review. Trans. Am. Soc. Agr. Eng. Phoenix, Arizona.

Bond, T. E., and C. F. Kelly. 1955. The globe thermometer in agricultural research. Agr. Eng. 36:251.

Bond, T. E., C. F. Kelly, S. R. Morrison, and N. Pereira. 1967. Solar atmospheric and terrestrial radiation received by shaded and unshaded animals. Trans. Am. Soc. Agr. Eng. 10:662.

Bonsma. J. C. 1949. Breeding cattle for increased adaptability of cattle to tropical conditions. J. Agr. Sci. 39:204.

Critchfield, H. J. 1966. General climatology. 2nd ed. Prentice-Hall, Englewood Cliffs, New Jersey.

Ellmore, M. F. 1954. Blood eosinophils of dairy cows during the parturition period. Ph.D. thesis, University of Maryland, College Park, Maryland.

Findlay, J. D. 1968. Climatologic data needed to specify climatic stress. Report of 2nd meeting of FAO Expert Panel on Animal Breeding and Climatology. 25–29 Nov., Rome.

Franklin, D. L., and R. L. VanCitters. 2967. Blood flow in meseneric artery of chimpanzee monitored via radio telemetry. Aerosp. Med. 38:926.

Franklin, D. L., N. W. Watson, and R. L. VanCitters. 1964. Blood velocity telemetered from untethered animals. Nature 203:4944.

Franklin, D. L., N. W. Watson, K. E. Pierson, and R. L. VanCitters. 1966. Technique for radio telemetry of blood flow velocity from unrestrained animals. Am. J. Med. Elec. 5:24.

Guidry, A. J., and H. S. Hofmeyr. 1968. A hygrometric tent system for measuring total respiratory and surface evaporation from sheep. Proc. S. Afr. Soc. Anim. Prod. 7:195.

Guidry, A. J. and R. E. McDowell. 1966. Tympanic membrane temperature for indicating rapid changes in body temperature. J. Dairy Sci. 49:74.

Guthrie, L. D., J. B. Frye, Jr., and J. A. Lee. Effects of exposure to shade or sun and level of fiber on the efficiency of energy utilization by lactating Holstein cows. J. Dairy Sci. 51:969.

Hafez, E. S. E. 1962. The behavior of domestic animals. Bailliere, Tindall and Cox, London.

Hahn, G. L., and J. D. McQuigg. 1967. Expected production losses for lactating Holstein dairy cows as a basis of rational planning of shelters. Am. Soc. Agr. Eng. Paper No. MC-67-107.

Hahn, G. L., and J. D. McQuigg. 1970. Evaluation of climatological records for rational planning of livestock shelters. Agr. Meterol. 7:131.

Johnson, J. C., Jr., and R. L. Givens. 1962. Single versus frequent observations for estimating some summer climatic conditions in South Georgia. J. Dairy Sci. 45:695.

Johnson, H. D., A. C. Ragsdale, I. L. Berry, and M. D. Shanklin. 1963. Temperature–humidity effects including influence of acclimation in feed and water consumption of Holstein cattle. University of Missouri Res. Bull. 846.

Johnson, J. C., Jr., B. L. Southwell, R. L. Givens, and R. E. McDowell. 1962. Interrelationships of certain climatic conditions and productive responses of lacating dairy cows. J. Dairy Sci. 45:695.

Johnston, J. E. 1968. Housing and management problems in cattle production in humid tropics. Report of 2nd meeting of FAO Expert Panel on Animal Breeding and Climatology. FAO Meeting Rpt AN:ABC/68/2, Rome.

Johnston, J. E., R. E. McDowell, R. R. Shrode, and J. E. Legates. 1959. Summer climate and its effects on dairy cattle in the Southern Region. Sou. Coop. Ser. Bull. 63. Louisiana Agr. Exp. Sta., Baton Rouge.

Johnston, J. E., E. J. Stone, and J. B. Frye, Jr. 1966. Effects of hot weather on the productive function of dairy cows. 1. Temperature control during hot weather. Louisiana Agr. Exp. Sta. Bull. 608.

Joyce, J. P., and K. L. Blaxter. 1964. The effect of air movement, air temperature and infrared radiation on the energy requirements of sheep. Brit. J. Nutr. 18:5.

Kibler, H. H. 1964. Thermal effects of various temperature–humidity combinations on Holstein cattle as measured by eight physiological responses. Res. Bull. Mo. Agr. Exp. Sta. No. 862.

Köppen, W., and R. Geiger. 1936. Handbuch der Klimatologie. Vol. 1, Part C. Burntraeger, Berlin.

Lee, D. H. K. 1953. Manual of field studies on the heat tolerance of domestic animals. FAO Dev. Paper No. 38. FAO, Rome.

Lee, D. H. K. 1965. Climatic stress indices for domestic animals. Intern. J. Biometerol. 9:29.

Maddex, R. L. 1967. Evaluation of research on confinement beef feeding systems. Trans. Am. Soc. Agr. Eng. No. 67–910.

Mahoney, G. W. A., G. L. Nelson, and S. A. Ewing. 1967. Performance of experimental close-confinement (caged) cattle feeding systems. Trans. Am. Soc. Agr. Eng. No. 67–405.

McDowell, R. E. 1958. Physiological approaches to animal climatology. J. Heredity 49:52.

McDowell, R. E. 1966a. Problems of cattle production in tropical countries. Cornel Intern. Agr. Dev. Mimeo. No. 17. 24 pp.

McDowell, R. E. 1966b. The role of physiology in animal production for tropical and sub-tropical areas. World Rev. Anim. Prod. 1:39.

McDowell, R. E. 1967. Factors in reducing the adverse effects of climate on animal performance. *In* Ground level climatology. AAAS Pub. No. 86, Washington, D.C.

McDowell, R. E. 1968. Climate versus man and his animals. Nature 218:641.

McDowell, R. E. 1971. Improvement of livestock production in warm climates. W. H. Freeman and Co., San Francisco. In press.

McDowell, R. E., and L. V. Cundiff. 1957. Genetics and environmental interactions. Proc. 6th FAO Inter-Am. Conf. Anim. Prod. and Health, Gainesville, Florida. September 10–20. 16 pp.

McDowell, R. E., D. H. K. Lee, M. H. Fohrman, and R. A. Anderson. 1953. Respiratory activity as an index of heat tolerance in Jersey and Sindhi × Jersey ($F_1$) crossbred cows. J. Anim. Sci. 12:573.

McDowell, R. E., D. H. K. Lee, and M. H. Fohrman. 1954. The measurmeent of water evaporation from limited areas of a normal body surface. J. Anim. Sci. 13:405.

McDowell, R. E., E. G. Moody, P. J. VanSoest, R. P. Lehmann, and G. L. Ford. 1969. Effect of heat stress on energy and water utilization of lactating cows. J. Dairy Sci. 52:188.

McDaniel, B. T., C. A. Matthews, and N. D. Bayley. 1959. Screening records for inheritance studies. U.S. Dept. Agr. Mimeo. April 6. Beltsville, Maryland. 7 pp.

McLean, J. A. 1963. Measurement of cutaneous moisture vaporization from cattle by ventilated capsules. J. Physiol. 167:417.

Morrison, S. R., T. E. Bond, and H. Heitman. 1968a. The physiological response of swine to wetting. Trop. Agr. 45:279.

Morrison, S. R., V. E. Mendel, and T. E. Bond. 1968b. Influence of space on performance of feedlot cattle. Trans. Am. Soc. Agr. Eng. No. 68–919.

Pereira, N., T. E. Bond, and S. R. Morrison. 1967. Thermal characteristics of a table-tennis ball used as a black-globe thermometer. Trans. Am. Soc. Agr. Eng. No. 66–328.

Rhoad, A. O. 1944. The Iberia heat tolerance test for cattle. Trop. Agr. Trin. 21: 162.

Schmidt-Nielsen, K. 1962. Desert animals. Oxford Univ. Press, New York.

Stewart, R. E., and S. Brody. 1954. Effect of radiation intensity on hair and skin temperatures and on respiration rates of Holstein, Jersey and Brahman cattle at air temperatures 45°, 70° and80°F. Res. Bull. Mo. Agr. Exp. Sta. No. 561.

Thornthwaite, C. W. 1948. An approach toward a rational classification of climate. Georgia Rev. 38:55–94.

Turner, H. G., and H. V. Schleger. 1960. The significance of coat type in cattle. Aust. J. Agr. Res. 11:645.

USDA. 1964. Confinement rearing of cattle. U.S. Dept. Agr. ARS 22–89, Washington, D.C.

U.S. Weather Bureau. 1955. Manual of surface observations. 7th ed. U.S. Dept. Commerce Circular N, Washington, D.C.

U.S. Weather Bureau. 1956. Monthly normal temperatures, precipitation and degree days. U.S. Dept. Commerce Tech. Paper No. 31, Washington, D.C.

Van Citters, R. L., and D. L. Franklin. 1969. Cardiovascular performance of Alaska sled dogs during exercise. Circ. Res. 24:33.

VanCitters, R. L., W. S. Kemper, and D. L. Franklin. 1968. Blood flow pressure in the giraffe carotid artery. Comp. Biochem. Physiol. 24:1035.

Vernon, E. H., R. A. Damon, Jr., W. R. Harvey, E. J. Warwick, and C. M. Kincaid. 1959. Relation of heat tolerance determinations to productivity in beef cattle. J. Anim. Sci. 18:91

Webster, A. J. F. 1968. Heat losses from cattle in cold and wind. pp. 31–32. Ann. Feeders' Day Proc., Univ. Alberta, Edmonton.

Wiersma, F., and G. L. Nelson. 1967. Nonevaporative convective heat transfer from the surface of a bovine. Trans. Am. Soc. Agr. Eng. 10:733.
Wiersma, F., and G. H. Stott. 1965. Micro-climate modification for hot weather stress relief in dairy cattle. Trans. Am. Soc. Agr. Eng. No. 65–404.
Williams, C. M. 1967. Livestock production in cold climates. *In* Ground level climatology. AAAS Pub. No. 86, Washington, D.C.
Yeck, R. G. 1960. Evaporative heat loss of dairy cattle at high environmental temperatures. Ph.D. thesis, Univ. of Missouri, Columbia.

# INDEX